Advance Praise for
When the Dragon Wore the Crown

Mythology has been confusing to most people because the keys to unlocking the symbols have been obscured. It is rare to find a book that deciphers large chunks of the mystery, tying elements together into a cohesive vision.

When the Dragon Wore the Crown is such a book. It introduces an astrological orientation in the context of large cycles of time. Much of what we learned in school and what is presented to us in mainstream media leaves pivotal secrets concealed as though keeping us in the dark has hidden value. Nevertheless, threads embedded in mythology have endured hidden in plain sight. *When the Dragon Wore the Crown* reveals much, pointing to connections that become obvious once one journeys into the book.

As a renowned astrologer, Don Cerow has penetrated deep into the collective unconscious. Linking symbolism from ancient cultures around the world, Don discovered patterns pointing to a science in the sky. Curiously, it seems the ancients were fully cognizant of these crucial cosmic relationships, well understood in times past. However, a darkness overcame the world and secrets of magic, oracles, cosmology and knowledge were driven out of our consciousness.

Compiling the research and the multiple images for this book has been a true labor of love for Don. Finally, an integrated and well-illustrated revelation is available for us all to appreciate. Bravo to Don for his persistence and for his vision which is of enormous import to humanity in these changing times.

—Carmen Boulter, PhD
 www.interactive-u.com
 www.pyramidcode.com

"Our Dragon is a child of Time, and each culture that invokes the Dragon is wrestling with a deeper understanding of 'Time' and how it worked" says Cerow —and takes us on a most exhilarating journey through the millennia in pursuit of this truth. Richly and lucidly illustrated, this book is a delight and an inspiration. I love it!
 —REV PAM CRANE (author of *The Draconic Chart*)

"Don's investigation of precessional motion provides a fascinating look at the correlation between the stars, myth and history. Wonderful work!
 —WALTER CRUTTENDEN
 BinaryResearchInstitute.org

In *When the Dragon Wore the Crown*, cosmo-linguist Don Cerow draws us into the mythic realm as it has been dreamed into existence by sky-sages from around the world. He carries us to the dawn of consciousness, when the skies were known to be an integral part of our reality, both pragmatic—pinned to seasonal variations and agricultural cycles—and symbolic—showing the evolution of humanity. Don's unique deciphering skills lead us on a journey into that consciousness. We get the chance to peek over his shoulder as he decodes the myth, story, symbol and structure that reveal this cosmology, reawakening our awareness from some ancestral space and allowing its relevance to unfold before our eyes, alive and enlivening. It is a profound and magnificent experience that should not be missed.
 —Terry Lamb, cosmo-linguist and author of
 Born to Be Together: Astrology, Relationships, and the Soul

When the Dragon
Wore the Crown-
Center and Circle

Fig. 0

Photo credit Tony Jones of Horsham, UK

When the Dragon Wore the Crown—

Center and Circle

*Putting Starlight
back into
Myth*

*

Don Cerow

IBIS PRESS
Lake Worth, FL

Published in 2013 by Ibis Press
A division of Nicolas-Hays, Inc.
P. O. Box 540206
Lake Worth, FL 33454-0206
www.ibispress.net

Distributed to the trade by
Red Wheel/Weiser, LLC
65 Parker St. • Ste. 7
Newburyport, MA 01950
www.redwheelweiser.com

Copyright © 2013 by Don Cerow

All rights reserved. No part of this publication may be reproduced or transmitted in any form or by any means, electronic or mechanical, including photocopying, recording, or by any information storage and retrieval system, without permission in writing from Nicolas-Hays, Inc. Reviewers may quote brief passages.

ISBN 978-0-89254-205-5
Ebook ISBN 978-0-89254-594-0

Library of Congress Cataloging-in-Publication Data

Cerow, Don.
 When the dragon wore the crown-center and circle : putting starlight back into myth / Don Cerow.
   p. cm.
 Summary: "When the Dragon Wore the Crown-Center and Circle covers a period of approximately six thousand years, focusing primarily on what astrologers would call the Ages of Gemini, Taurus and Aries, and the classical astronomy of the Greeks and Romans. It opens and closes with the Chinese tradition, and touches on Sumerian, Babylonian, Hindu, Norse and Mayan cultures and their mythology and astronomy weaving together many of their celestial serpentine similarities. The book primarily focuses on the role of the Dragon, the astronomical marker of the North Celestial Pole for literally thousands of years. The earthshaking importance of this simple astronomical fact helped open the door to navigation, farming, and social organization. The cycle of the seasons was marked by this cadence, and the moving center protected the secret of the circle, the Divine Circle, for literally thousands of years. Without the correct center, you did not have the correct circle. Don Cerow shows how the Dragon was the solitary guardian of the secret of the center, of the magic forged by the circle, a gem of incalculable worth"--Provided by publisher.
 Includes bibliographical references.
 ISBN 978-0-89254-205-5 (alk. paper)
 1. Astronomy, Ancient. 2. Dragons--History. 3. Mythology--History. 4. Constellations. 5. Spherical astronomy. I. Title.
 QB16.C47 2013
 520.93--dc23
                                                                2013008812

Front cover: Fiske Planetarium Art Department *M-51 Spiral Galaxy*
Back cover: *St. George* photo by David Cooley. Author photo by Lisa Michel

Book and cover design by Don Cerow
Book production by Studio 31

Printed in China

And the light dwelt in the darkness

and the darkness knew it not.

—John 1:5

Christian

Coupe Nocteum

Thank you for coming

Acknowledgements

This has been a wide-ranging path which began years ago. Many have helped along the way. First, to Katie and Andy, for being so patient, understanding and supportive of their Dad.

I love you both, very much.

To Mom and Dad for everything (Here's the book, Dad)!

More recently, thank you Scott Silverman for your enthusiasm in helping to open the publishing door. To Yvonne Paglia, Jim Wasserman and Scott, my editors, and to the many friends who have helped with the book's evolution: Linea Van Horn, Mark Taylor, Cynthia Cwynar, Martha MacBurnie, Marie Massud, Suzanne Keating, Rich Ware, Larry Shea, Michael Ford, Bob Spohn and Willow LaMonte, to name a few. To the Longmont, Colorado Astronomical Society, for sharing their passion. To Walter Cruttenden and Dr. Carmen Boulter for their trust and creative encouragement, and to Nancy and Patrick Hiester for providing shelter from the storm.

Naturally, to the staff and volunteers at Fiske Planetarium, with special thanks to Tito Salas, Julie Carmen, Will Fleming, Zane Lyon and Dr. Doug Duncan.

To John Thomas, for your warm and enduring friendship.

And finally, I would like to thank my wife, lover and best friend, Lisa Michel, for her wisdom and understanding over the years. Her encouragement and support have helped make the telling of this tale possible.

*

The disclaimer* at the bottom of this page represents the specific wording requested by Heritage Malta, but is true for all of those who contributed materials and images, including, but not limited to the countries, institutions of learning, or individuals who have so graciously agreed to allow their work be shared within these pages as part of a larger story-line.

A heart felt thank you, One and all.

"This document* includes materials/images made available by Heritage Malta. The views expressed herein are those of the author and can therefore in no way be taken to reflect the official opinion of Heritage Malta."

Dedication

Athena

Katie & Andy

**Lisa,
Gail,
&
Karen**

Invocation

The Grey-Eyed Goddess

I begin to sing of Pallas Athena,
the glorious goddess, bright-eyed, inventive,
unbending of heart, pure virgin,
saviour of cities, courageous Tritogeneia.
From his awful head wise Zeus himself bare her
arrayed in warlike arms of flashing gold,
and awe seized all the gods as they gazed.
But Athena sprang quickly from the immortal head
and stood before Zeus who holds the aegis,
shaking a sharp spear:
great Olympus began to reel horribly
at the might of the bright-eyed goddess,
and earth round about cried fearfully,
and the sea was moved and tossed with dark waves,
while foam burst forth suddenly:
the bright Son of Hyperion stopped
his swift-footed horses a long while,
until the maiden Pallas Athena
had stripped the heavenly armor
from her immortal shoulders.
And wise Zeus was glad.

Hesiod's Hymn to Athena [00]

γνωθι σαυτον

Contents

page	Title	Culture
viii	Acknowledgements	
ix	Dedication	
x	Invocation	
xi	Inscription at Delphi	
xv	Preface	
xvi	Turning on the Tides of Time	
1	**The Answer**	Back to School

The Astronomy

page	Title	Culture
3	**Dragon's Gold**	Heaven and Earth
	Axis Mundi	As the World Turns
5	Center and Circle	Equator & North Celestial Pole
6	**Spring Stars**	8,000 years of precessional motion
	Precession of the Equinoxes	The Parade of Heaven
7	**The Range of Draco's Realm**	Rise and Fall of the Dragon
8	**Draco's Key**	Dragons and New Year's Day

Opening the Circle

page	Title	Culture
11	**The Imperial Dragon**	Chinese
	Xia Dynasty	Chinese

Gemini

page	Title	Culture
15	**Time of the Twins**	Central European
17	Snake Goddess and Bird Goddess	Central European
19	Egg of the World	African
20	Odin and the 'Shell of Existence'	Norse
	Vishnu and the 'Shell of the Cosmic Egg'	Hindu
	Dan Ayido Hwedo, the Rainbow Serpent	African
21	Nabta Playa	Egyptian
22	Cosmogony, Philo	Phoenician
	Aer & Aura, Eudemus	Phoenician
	Aether & Air, Mochus	Phoenician
	Pelasgian Creation Myth	Pre-Greek
23	Ophion and Eurynome, the Universal Egg	Pre-Greek
	Kusor & Hasisu	Phoenician
24	Tale of the Two Brothers	Egyptian
25	Nut	Egyptian
27	Helical Rising of Sirius	Egyptian
28	Shu, God of the AIR	Egyptian
	P'an-Ku, born of the Egg	Chinese
30	**Megalithic Island**	Megalithic
	Newgrange, Knowth and Dowth	Megalithic
31	Senchus Mor and the Ring of Power	Megalithic
	Sunlight and Shadow	Megalithic
32	Gnomon	Megalithic
36	Mnajdra	Malta
37	**Dracospito**	Pre-Greek
39	*Works and Days*, Hesiod, Pleiades Rising	Greek
	Red Paint people	Maritime Archaic

page	Title	Culture

Taurus

(page 41)

page	Title	Culture
42	**Toro Time**	Egyptian
43	Apis	Egyptian
47	Palette of Narmer	Egyptian
49	**The Sumerians**	Sumerian
50	Gilgamesh	Sumerian
	Hanging Gardens, Seven Wonders	Babylonian
50	Fertile Crescent and Abraham	Hebrew
52	Autumnal Equinox	Sumerian
	Hyades, Pleiades	Sumerian
53	Neck of the Bull	Sumerian
54	Marduk and Ti'amat	Sumerian
	Marduk spreads his net	Sumerian
55	Thuban, 2788 BC	Sumerian
56	The Uraeus and the Eastern Fort	Egyptian
58	Thor and Jormungand- the Midgard Serpent	Norse
59	Vasuki and Surabhi	Indian
60	Takshaka	Indian
62	Utanka, Takshaka and the Nagas	Indian
63	Hammurabi	Babylonian
64	Inanna	Sumerian

The World Tree

(page 65)

page	Title	Culture
65	Yggdrasil, the World Ash	Germanic/Norse
66	Nidhogg	Germanic/Norse
67	Ragnarok and the Rainbow Bridge	Norse
69	Yule Log	Germanic/Norse/Celtic
71	**The Mayan World Tree**	Mayan
	Popol Vuh	Aztec
74	Huracan	Mayan
77	Story of the Maiden and the Tree	Aztec
79	Serpent in the Tree	Hebrew/Christian
81	**A Moment in Time**	Egyptian
84	Celestial Meridian	Egyptian
86	**More Mayan World Tree**	Mayan
91	Venus	Mayan
	Great Year	Mayan
92	Julian Days	Astronomy
93	Ages of Man	Greek
94	Islands of the Blessed	Greek
95	Golden Age	Roman
96	Anatomy of the Bull	Celestial cuts
98	Minoa	Minoan
99	Minoan Snake Goddess	Minoan
100	Wadjet	Egyptian
102	Cat	Sumerian/Egypt/Minoan/Norse
104	Pepy II	Egyptian
105	The Curse of Akkad	Akkadian
106	**A Pastoral Post Card**	Fertile Crescent Cultures

Center and Circle

page	Title	Culture

Aries
107

page	Title	Culture
108	**Mythological Ramifications**	Sumerian/Greek/Hebrew
111	Apollonius of Rhodes	Greek
113	Athena and the Dragon	Greek
114	Stellar Wisdom from the East	Phoenician/Greek
116	Fruit from the Tree	Hebrew/Christian
119	Moses on the Mountain	Hebrew
123	**Hooves Lock Horns**	Canaanite/Hebrew
	Ba'al and Bel	Canaanite
126	No other Gods	Hebrew
129	Phoenician Roots	Phoenician
130	Baal Worship	Pre-Celtic
135	Rites of Bit Akitu	Babylonian
136	Nebuchadnezzar and Daniel	Babylonian/Hebrew
139	The Pits	Mayan
140	Pleiades	Mayan/Greek
144	Hyksos chariots	Egyptian
148	The Covenant with Abraham	Hebrew
150	Agni, Alexander and Cernunnos	Persian/Hindu/Greek/Celtic
152	**Time of the End**	Astronomy
153	Harappan Civilization	Harappan
154	Book of Daniel	Hebrew
158	Antiochus IV Epiphanes	Greek/Hebrew
159	Roman Republic	Roman
162	**Rainbow Dreams**	Australian/African/Native American
164	Wing of the Dragon	Sumerian/Egyptian
165	Scorpion King	Egyptian
166	Heracles	Greek
168	Hera	Greek
169	Achelous	Greek

Pisces
170

page	Title	Culture
171	**Reflections**	One evening's sky
176	**The Imperial Pearl**	Chinese
179	**Twilight**	Pan

Closing the Circle
180

page	Title
182	Star Crossings
187	Star Positions of Draco, Greek key
188	Tale of Two Brothers
194	Quiz on the Tale of Two Brothers
196	Time Lines
198	Figures and Illustrations
206	Roots of the Tree
208	Foot*notes*
215	Manilius

Preface

One night, long, long ago, I had a dream.

It was a sunny day beneath a bright blue canopy, without a cloud in the sky.
There were streams and trees spotting the landscape here and there.
All this stretched across a great, green vista of rolling hills,
fading into the distance as far as the eye could see.

Passing by in front of me single-file was a long line of dragons,
each with their left arm placed on the shoulder of the dragon in front of them.
This serpentine column fanned out ahead, over the hill and into the valley beyond,
only to reemerge in the distance and coil over the next hill
and the next, until they too merged into the horizon.

As they moved, the long line pulsed, a visual reptilian ripple along its length.
Together they chanted in unison a forgotten, hypnotic rhythm,
repeated over and over again. Two steps forward, one back,
roll out the hips down, around and back with a gutteral "hun-unh".

Each happy to be there, undulating to the beat
of an ancient serpentine secret.

Day after day, world without end.

Amen.

Blessings,

Athena's Web

Center and Circle

Fig. 1
Fig. 2
Fig. 3
Photo credit Luis Benkard

Turning on the Tides of Time...

When Theseus journeys across the Mediterranean as one of seven youths, he is part of an annual debt to Crete. His life is on the line. According to the terms of an old war treaty, Athens had to pay a yearly tribute to the Minoans. None of the other youths had ever returned to tell of their trials. In volunteering for this deed, Theseus does so over the earnest protests of his father, the Athenian king.

There are two dangers our brave lad must face in this far-off foreign land. First he must find, confront and defeat a fearsome creature legend has dubbed the *Minotaur* (Bull of Minos). Should he prove victorious, he must then work his way out of a mysterious labyrinth. Success in this quest would bring fame, while defeat means ignominy, dishonor, and possibly even death.

How to know which way to go?
Our youthful hero is helped on his quest by a beautiful, Minoan princess. She is immediately smitten by this daring foreign prince, falling in love with him at first sight, and decides to help him however she can.

Together, between excited whispers of love and affection, they devise a secret plan that just might help accomplish his aim.

xvi

Athena's Web

Center and Circle

One cool, clear evening just before the New Year ceremonies, the prince and princess quietly make their way to the walls outside the labyrinth. Here they fix a string she had hidden precisely for this purpose, securely attaching it so it can not be unintentionally undone. Taking firm hold of the other end of the life-line lest he become lost in the twists and turns of its dark, unfamiliar passages, Theseus then makes his way into the maze.

The rest, as they say, is history...

Fig. 4

Photo credit petrus.agricola

Minotaur and more

...or at least legend.

Just as Theseus did, we too will make use of a golden thread to weave our way through the heart of a maze containing both Minotaur and more. If you know where to look, the way can still be seen even after all these centuries. Adventurous souls willing to undertake this journey will travel a path once attempted only by the heroes of yesteryear. Ahead of us lie great opportunities, but together with them come great challenges.

Although these ties to the past are fascinating, what's more important is where they lead us. As you are about to discover, this thread is knotted in the not-so-distant future.

The trail we choose is not some forgotten backwoods footpath overgrown in underbrush and thorns, but rather the principal mythic highway of time. Yet while the maze may be complex, its solution is simple. Locked inside the enigma of the centuries is the key to both past and future.

Truth smiles with a simplicity that gently illuminates from within. Quiet reflection, repeated logic and a steady light will help to make our path clear.

As we prepare for our journey, know that we are getting ready to enter another realm; one in which the creatures of legend reclaim their place in the Underworld, on Earth and in Heaven above. We were once familiar with the mythical beings that eternally encircle us; but their watery reflections were perturbed, their outline confused long, long ago. We endeavor to walk a highway of diamonds with nobody on it, though these paths once teemed with traffic.

Yet even after centuries of neglect, these hearts beat with the stoic determination of those who cannot die.

They are immortal.

We are about to ride across the centuries on the back of a dragon. Athena will illuminate the way, as she has so often before.

For those who continue to feel that myths are merely the imaginative playthings of the fanciful, long ago uprooted from reality and cast into the compost of ignorance to be forever forgotten- think again. The lid of Pandora's box is being reopened.

Do you dare gaze upon what lies inside?

The time spoken of by the prophets is close. For those who have ears to hear, the call now goes out. Heed the Earth's drumbeat about you.

Witness Gaia's theater unfolding both near and far.

What was prophesied shall be revealed; the sands are running out.

We approach the gates.

As we look down on yesterday's pathways from our Slytherin seat, we will orient ourselves using recognizable landmarks. Viewed from above we will see familiar vistas from fresh perspectives. Most of the maps that once showed the way over these trails crumbled to dust long ago, lost to time. But we haven't been left entirely without guiding light. There have always been a few who have been able to unravel these knots and make clear the path. But there have also been those who have watched to make sure this information remained hidden, secreted not only *in* the shadows, but *by* the shadows.

What has changed is that events spoken of long ago are now at hand. Like a reservoir filled too full, the dam can no longer hold back the coming floods. The sands of our hourglass quicken their pace.

The hour draws near.

Fig. 5

DaVinci's Dragon

Our story describes the journey along this path, both its remote past and more immediate future.

The time grows short.
We must begin...

Athena's Web

Artwork credit Craig B. Musselman

Fig. 6

"It does not do to leave a live dragon
out of your calculations, if you live near him."

—*J.R.R. Tolkein*

Parselzunge hier gesprochen

xviii

When the Dragon Wore the Crown-

Center and Circle

*Putting Starlight
back into
Myth*

*

Dragon Eggs hatch as storms before ascending to heaven

Center and Circle

Fig. 8
Photo credit Claire Whittenbury

In our school texts while growing up, it was often the case that the answers were in the back of the book near the index. You read the chapter, did the problems, and then checked the correct answer. If the student copy didn't have this answer-key, then the teacher's edition certainly would.

In this work, however, the answer to the riddle is right on page one. In the many years it has taken to pull these tales together, I have thought about this matter often and see no way around it. It would seem that the answer should come only after a great deal of preparation; after painstakingly making your case, after sitting through the story line. Tensions build, and in theater, books, on TV, in movies and even jokes, the answer comes at the end. It just doesn't seem right otherwise.

Having said all that, the answer is this:

The Serpent is in the Center.

The *Center* of what, you might ask? The *Serpent* is in the *Center of a Circle* and it is in the *Center of the Sky*, for all the people who, a long, long time ago, looked up and used the *Heavens* to tell time.

The *scientific method* states that we should establish a *hypothesis*, and then check it *empirically*. So, let's transform the answer into a hypothesis in order to see how well it flies.

Is the Serpent in the Center?

From about 7,000 BC down to the collapse of the classical tradition during the Roman Empire, this is what people saw in the nocturnal skies. For them, there was a giant Snake going round and round, eternally it seemed, guarding the *Central Core of Heaven*.

Naturally, this simple equation has many variations; variations within cultures, mythological traditions and in the powers the *Serpent* was said to possess- but when we patiently observe its motion across the centuries, the answer becomes apparent. This is simply what was seen night after night, throughout the course of civilizations and across the millennia.

We will look at the astronomy and hear stories from various traditions around the world. We will see variations and we will see similarities, but in the end, it will all boil down to just one thing.

Can you guess?

The Serpent is in the Center.

Athena's Web

Center and Circle

The Astronomy

The astronomers

FIG. 9

Center and Circle

Dragon's Gold

Fig. 10

Now that we have the answer, it's time to work on the question.

The *Serpent in the Center* equates to the *Dragon in the Circle*. Dragon is an ancient Greek word (Δρακων) which translates into English as "*serpent.*" From the Greek comes the Latin term from which we derive our name for the constellation, *Draco*. It's simply another word for "*snake*".

Approximately 5,000 years ago, the star pictured as the 'heart' (*Thuban*) of the Dragon stood at the *North Celestial Pole* of heaven. Currently another star occupies this location. Its name derives from its location of central importance. It is the one visible star around which all others turn. It is called either *alpha Ursa Minoris* or *Polaris*, our pole star. It's also called the *North Star*, because it illuminates that cardinal direction.

Polaris is the last star in the tail of *Little Bear* (*Ursa Minor*). Five thousand years ago there was no *Polaris*. *Alpha Ursa Minor* was in the sky with all the others, but it didn't occupy the pole position.

There's a recorded memory of when we see *Ursa Minor* growing in importance as a 'polar' indicator. *Ursa Minor* is not mentioned by either *Homer* or *Hesiod*, our oldest Greek authors, but he is mentioned by *Strabo* (63/64 BC – ca. AD 24). According to *Strabo*, it was not admitted to the pantheon of Greek constellations until about 600 BC

Fig. 11

All Creation turned on the 'heart' of the Dragon

when *Thales*, inspired by its use in Phoenicia, suggested it to Greek mariners as a way to determine *North* in place of *Ursa Major*. *Thales* saw the seven stars of the *Little Bear* as part of the ancient *wings of the Dragon*.

In the *3rd millennium BC* the pole was marked by the star known as the '*heart*' of the Dragon. For more than a thousand years, *Thuban* had no rival. It was sole contender as *North Star*. The Arabs name the entire constellation *Thuban* after its *alpha* star.

The Earth's Tilt

Most people know what the *North* and *South Poles* are. They're locations of extreme cold in the Arctic and Antarctic, respectively. The '*pole*' that runs through these two points is *the axis around which the world turns,* the **Axis Mundi** (Fig. 12). It is often depicted as an *Earth* with a column, arrow, stake or spear running through its middle, like an giant *shish-ka-bob*. The angle of this axis is not ninety degrees to the *ecliptic* (the path of the Sun), but ninety degrees to the *Earth's equator*. Because of the *Earth's* rotational speed, there's a slight bulge at the equator as it spins, as if the Earth had consumed a few too many carbs. Because the *Polar Axis* is inclined at an angle of 23 1/2 degrees, it looks as though our planet is slightly tipped when viewed against the plane of our *Solar System*.

Because of this 'tip', there are seasons when the northern hemisphere is more inclined towards the *Sun*, and other times when the reverse is true. It is this angle that engenders our seasons and not our

Athena's Web

Center and Circle

distance from the Sun. When the northern hemisphere is tipped towards the Sun, it experiences summer; away from the Sun, winter. Our atmosphere is partly to blame for this phenomena. It acts as a blanket, wrapping and insulating us like the oceans. The more directly the Sun's rays hit the atmosphere 'straight on,' the greater the amount of Solar radiation that penetrates this blanket. The more oblique the angle at which the Solar rays hit the atmosphere, the more of its effects are thinned, with additional Sunlight being deflected and returned to space. Although it may seem counter intuitive, the northern hemisphere is actually farther from the Sun in summer, and closer in winter.

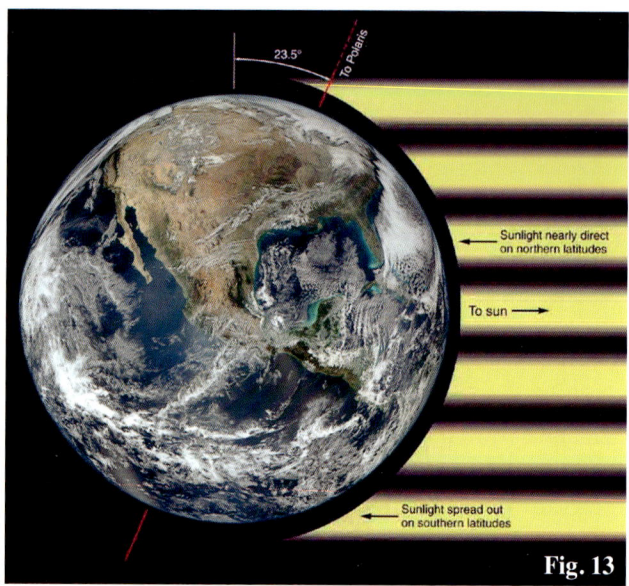

Summertime in North America

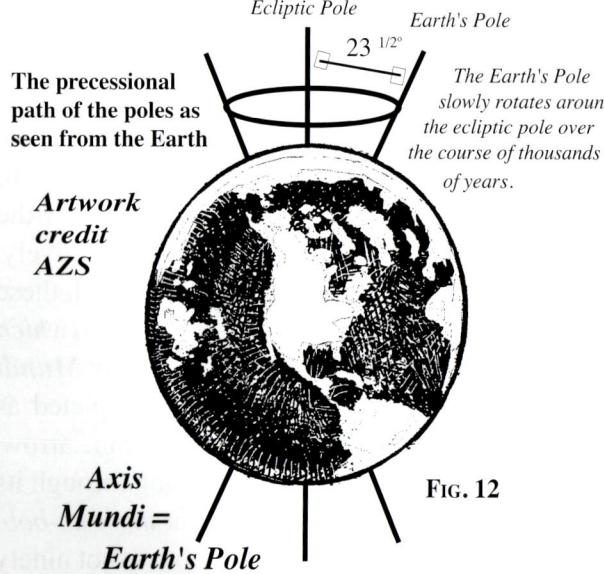

In the southern hemisphere, the trend is reversed. There the Sun is closer in summer, further away in the winter. One would suspect the southern hemishpere's seasons are more extreme because they are turned toward the Sun *and* closer. But, since there's less land mass in the southern hemisphere, the cooling provided by the collective southern oceans counterbalances these temperature extremes.

There's more than one way to cool a planet.

This imaginary axis which runs through the poles is often extended into space, marking the North and South "*Celestial*" Poles. If you were to go to the North Pole and look straight overhead, this would be the point around which all the stars rotate. Right now, we happen to be in a place in time when there is a star marker which helps us to identify this point in the night sky, and that marker is of course *Polaris, the Pole Star*. But such was not always the case.

Because of precessional motion, our pole, this huge imaginary shaft that runs through the planet, traces out a circle in the northern and southern skies. We speak of the *Southern Celestial Pole* less often, simply because it can't be seen unless one travels south of the *Equator*. Currently there is no pole star marking the South Pole. The *Southern Cross* is used to help determine its visual location.

Many myths around the globe tell stories of the *great winged Serpent* which flies overhead, sometimes seen hiding in the clouds, sometimes making himself invisible, but one for whom great fame and glory from time immemorial has been promised.

The notion of heaven turning on its axis was well known in the ancient world

Athena's Web

Center and Circle

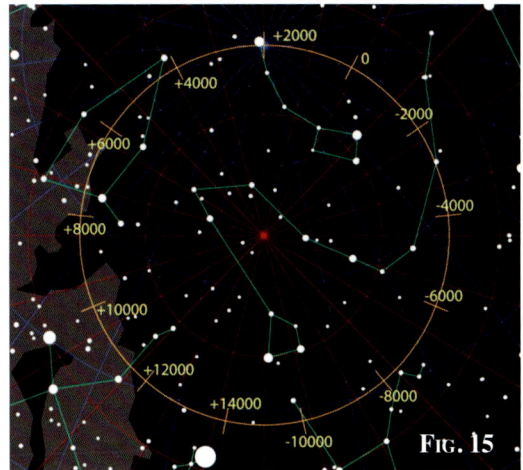

Photo credit Tau'olunga

FIG. 15

The path of the North Celestial Pole through the northern constellations (see Fig. 278)

If we look to the northern circumpolar stars, we see the circle being traced out by the polar path of precession (Fig. 15). It takes 25,765 years for this circle to be traced out once. *Alpha Ursa Minor* (*Polaris*) is located close to the top of the circle and where the precessional pole is now.

In 2788 BC, *alpha Draconis* (*Thuban*) was close to the precessional pole. Neither *Thuban* or *Polaris* are ever *exactly* on the center, but they come closer than any other bright stars. *Polaris* is brighter than *Thuban*, and thus would have attracted the interest of navigators, but *Thuban* had centuries of tradition to help secure its reputation as pole star. *Thales* gives us a clue as to when interest in *Polaris* began to gain ground around 600 BC as precession carried our pole closer towards this star.

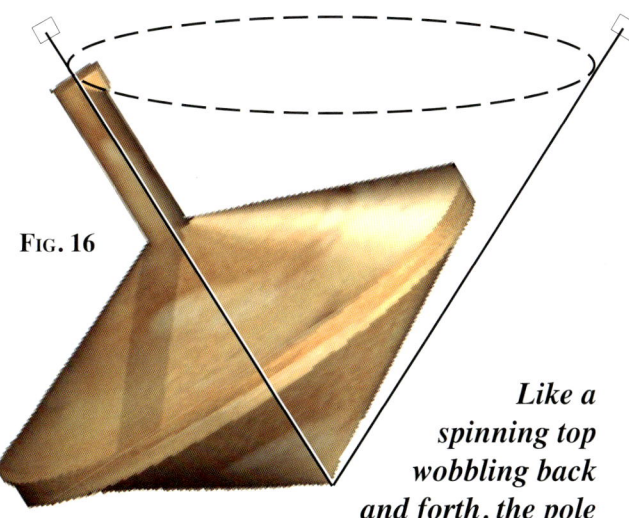

FIG. 16

Like a spinning top wobbling back and forth, the pole of the Earth traces out a circle in heaven

The *North Pole* tracing out this circle among the stars is due to the gravitational attraction of both the Sun and Moon on the Earth. The Earth spins on its axis in a daily rotation; but like a top, as it spins on its axis, there is a slow back and forth rocking motion (Fig. 16).

As we can see from Fig. 17, the *Polar Axis*, indicated by the letters '*N*' and '*S*' are *perpendicular* to the *Earth's Equator*. Because the pole is slowly rockin' (and a rollin') to point to the stars indicated in Fig. 15, means that the *Earth's Equator* must be moving along with it. The *North Pole* is the *Center of the Circle* traced out by the *Equator*. The pole cannot move without affecting the position of the circle. The pole defines the "*Circle*", commanding its "*Center*",

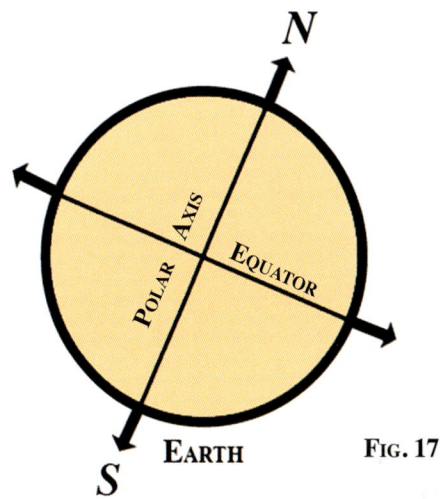

FIG. 17

The pole and equator are often thought of in terms of the Earth, but an extension of these positions into the skies is what gives us our celestial pole and celestial equator

but it is the *Equatorial Circle* which we most often hear about when discussing precessional motion. It's known as the *Precession of the Equinoxes*. **Center and Circle** are locked together in this great dance like a mantra:

Center and Circle; Center and Circle.

The *Equinoxes* are highlighted on the days the *Sun crosses* the *Celestial Equator*. In practical terms this means the hours of daylight and darkness stand in balance twice a year. On these two days, there are twelve hours of daylight and twelve hours of darkness. The *Equinoxes* are further defined as the intersection between the *Circle of the Equator* and the *Circle of the Ecliptic*, the apparent path of the Sun.

Athena's Web

Center and Circle

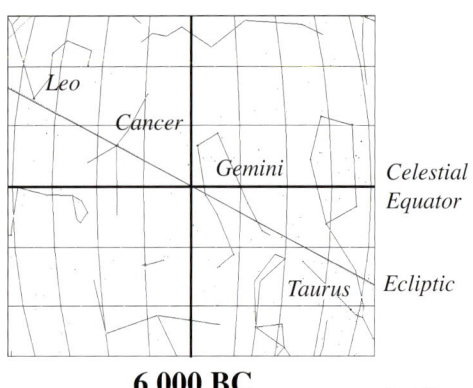

6,000 BC

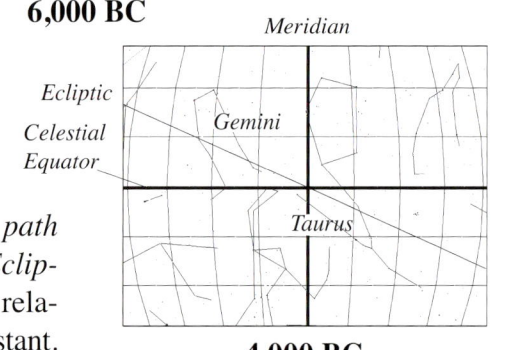

4,000 BC

Spring Stars
Across the Millennia

The path of the Ecliptic is a relative constant. Over the centuries, there is little deviation in the plane of the Sun and Earth, but because of the Earth's wobble as it spins, *the Celestial Equator* (horizontal line in diagrams) 'slides' down the path of the Sun, visiting different stars along the zodiac as time marches on. Because the Earth is round, we superimpose circles on the Sky so they reflect our reality- it's how we see things from the Earth. The *"Equator"* is a Circle. The *"Ecliptic"* is a Circle. The *Celestial Meridian* (the vertical line ascending through the diagram leading to the *North Celestial Pole*) is a Circle. Each one of these three lines are segments of different Circles.

All three intersect at the *Vernal Equinox*.

Although we could call this motion the *Precession of the Poles* or *Solstices*, we don't. We term it the **Precession of the Equinoxes**. Strictly speaking, either is correct. But it is the *Vernal Equinox* (VE) which we use as our astronomical indicator, our point of origin. *Zero hours of Right Ascension* is marked by this intersection, just as zero degrees of terrestrial longitude runs through Greenwich, England. The *Vernal Equinox* marks the beginning for many who chart the courses of heaven, marking celestial coordinates, the agricultural year, and the calendar; the Heavens gave each their birth.

Because the Sun-Earth relationship is a fairly stable one, ancient temples that reflect solar observation have not been adversely influenced by precessional motion. The *heel stone at Stonehenge* still successfully marks the *Winter Solstice*, as does the shaft of sunlight coming down the darkened chamber at Newgrange, or in Karnak, Egypt. Since a full cycle of precession clocks in at a little short of 26,000 years, and since we've divided the heavens into twelve zodiacal constellations, it takes a little over 2,000 years for precessional motion to pass through a single constellation of the zodiac. The figures at left illustrate this passage of time over a period of 8,000 years. Notice how over two-thousand-year intervals, one constellation enters, and one leaves.

(*0 BC/AD is relative, and not really a true date in time.)

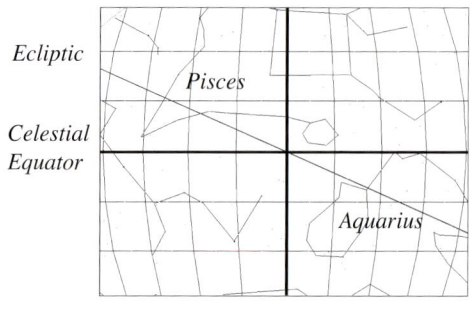

2,000 AD   FIG. 18

Athena's Web

Center and Circle

The range of Draco's Realm

Now, rather than watching the motion of the *Vernal Equinox*, we turn our gaze north to observe the flight of the Dragon across time. Unfortunately, none of what we are examining here can be proven. You cannot radiocarbon date myth. The barometer by which contemporary investigation gauges its results is the written record; the ability to read the minds of individuals and hear what they thought. Having said that, even the written record has been subjected to exhaustive scrutiny and criticism, with much of what ancient authors claimed being summarily rejected.

Which is precisely why mythology becomes so important. Myth *is* the media through which ancient peoples recorded developments in their civilization. We shouldn't reject myths simply because they weren't scientific, or because they don't meet our standards; yet that is just what many scholars have done. Rather, we should endeavor to understand myths in their own 'write'. It is *their* way of life and of looking at the world we are attempting to uncover, not the other way around.

Let's now test our hypothesis and see if it holds up to scrutiny. We'll begin by examining the astronomical record. Just how long has the constellation we now think of as *Draco* held the preeminent pole position, 'territorially' commanding the 'high ground'? When did it first begin to move into the highest reaches of Heaven, and when did precessional motion start to carry it out of these realms?

Voyager v. 1

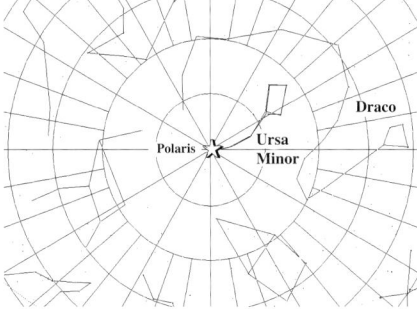

North Celestial Pole — present

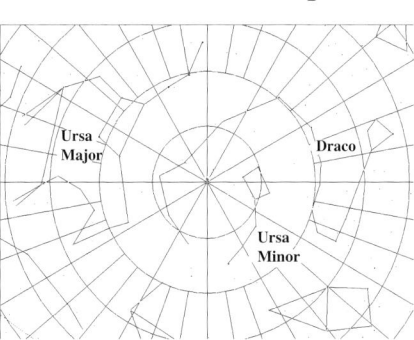

North Celestial Pole — 600 BC

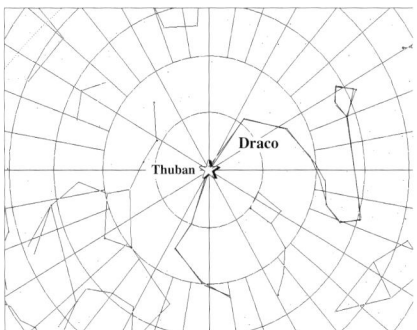

North Celestial Pole — 2788 BC

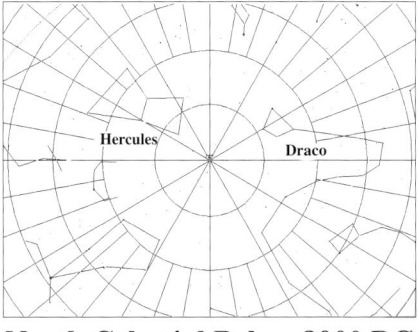

North Celestial Pole — 8000 BC

FIG. 19

Fig. 19 shows the progression of precession. The top map is of the *present* and shows "*Polaris*" occupying the pole position. The next diagram shows the pole around 600 BC, when *Thales* first suggested that Greek sailors follow the example of their Phoenician neighbors and begin to use *Ursa Minor*, rather than *Ursa Major*, as a polar indicator. The next map, 2788 BC, shows the circumpolar constellations when *Thuban* was at its closest approach to the pole. From there, we jump back to 8,000 BC, when the pole position was moving between the constellation we now call *Hercules* and the *Dragon's head*. Even here, the two stars in the *head of Draco* are brighter than any of those in the lower body of *Hercules* (closer to the pole). This approximates some of the earliest periods when *Draco* was first commencing its reign of 'constellational command' over the *Northern Celestial Pole*. Were people watching that far back?

We don't know. If they were, herein may lie some of the earliest origins of the legend.

Center and Circle

Photo credit Georgelazenby

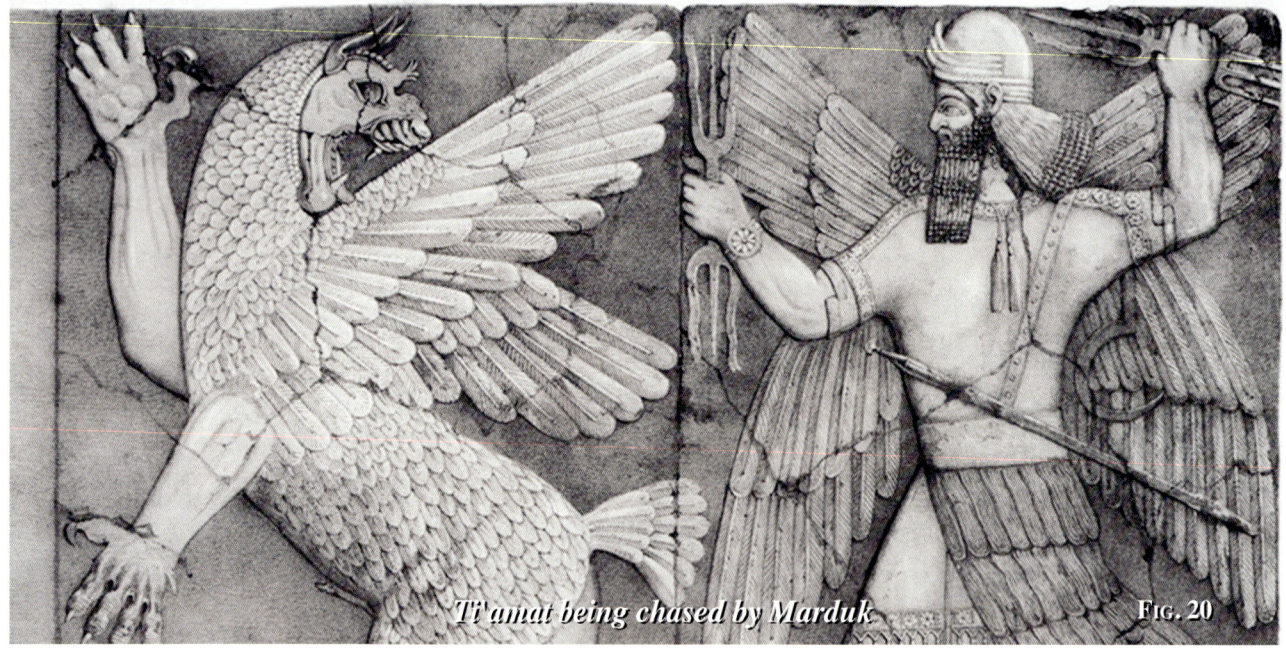

Ti'amat being chased by Marduk   FIG. 20

Draco's Key

The Dragon is an ancient keystone, unlocking the vault of heaven to mortal contemplation. In a language known to both mariners and the initiates of temple mysteries, mythological choreographers were creating a record of their society's greatest 'mathematical and scientific' (to use modern terms) achievements for posterity. They observed and recorded the motion of the *Precession of the Equinoxes*; describing the correct *Celestial Pole* and *Vernal Equinox* among the stars in pictorial form. Over time, sky positions shifted, and the mythic imagination kept pace right along with them. The 'evolution' of the dragon myths describe the changes of these markers through the years.

During the early 3rd millennium, the *North Celestial Pole* passed by *Thuban*, listed as *Draco's* primary star. This precise pinpointing pierced the venomous viper, and he squirmed on the observational and mythological lance of these civilizations for centuries.

FIG. 21

Victorian print of a May Day festival with the Dragon

"It is possible that the Sumerian dragon myth reached Egypt towards the end of the third millennium BC and inspired the legend of the huge sea-serpent Apophis (or Apep or Apop), enemy of the sun god, Ra. Another text containing a curse against the enemies of the Pharaoh says, 'They' (i.e. the king's enemies) 'shall be like the snake Apophis on New Year's morning.' Here the snake symbolizes darkness which the sun defeats every morning as he begins his journey through the heavens, and especially on New Year's morning. We have here an interesting parallel with victory of Marduk over the dragon Ti'amat at the Babylonian New Year Festival." [0]

This interpretation, while catching the drift, misses the mark. Light defeats darkness and the Dragon hugging the *North Celestial Pole*

Center and Circle

seems to be the champion of the night, and therefore the opposite of everything the Sun represents. What is being missed is an image; simple, effective, and to the point. *Shaft them through the middle until dead*, just as the sighting on the Dragon was astronomically, geometrically and symbolically pierced by various priests every New Year's.

As our quote illustrates, this motion was also known to the Sumerians, and may have been 'passed on' to the Egyptians, or it could have been the other way around. The Egyptians celebrated their New Year about the same time we do, shortly after the *Winter Solstice* with the 'birth' of the Sun.

There are English woodcuts from the 19th century showing Dragons as part of a *May Day festival* (Fig. 21), the ancient Druidic New Year.

The Chinese continue to associate the Dragon with the start of their calendar to this day. Their New Year follows the New Moon of February.

Unlike the Egyptians, the Sumerians celebrated their New Year on the *Vernal Equinox*. Each of these is 'cutting' the Circle at a different point for different reasons we will explore later; but they *all* invoked the Dragon.

Around the globe Dragons and New Year's go back a long way together.

Because of this precessional shift, all of the stars of heaven are slowly 'drifting' across the sky. During the course of a lifetime, we do not see much change in this picture. Of the 360° in a Circle, precessional motion shifts a degree every 72 years, which is to say that in the course of a human lifetime it only moves a single degree- not something that most folks would notice. By observationally re-calibrating where the exact pole was each year, ancient priests may have kept pace with this shift.

To pinpoint the *Center* is to command the *Circle*; to know how, when and where the stars fused to that *Circle* will rise and fall.

FIG. 22

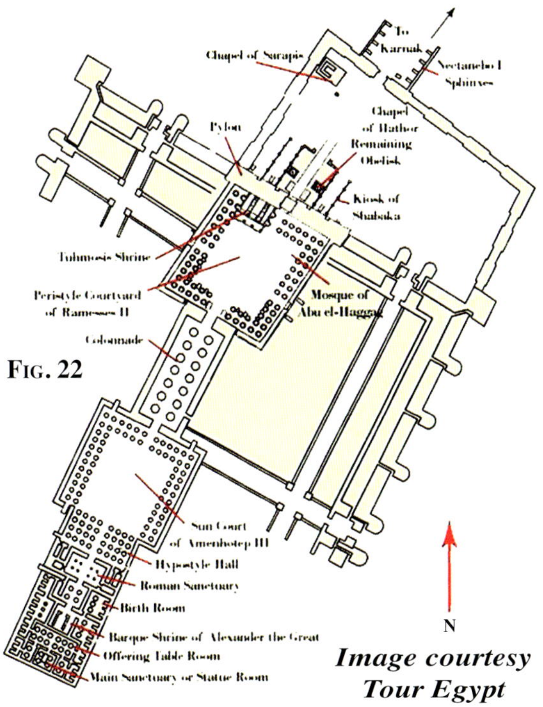

Image courtesy Tour Egypt

Three successive changes in alignment at Luxor in Egypt

To lose touch with precession is to forsake accurate calculations. The longer this motion is neglected, the greater the compound discrepancy. We know that during the early 2nd millennia, star records begin to fall centuries out of alignment for this reason.

Surprisingly, it is not so much in Heaven that we learn this, as on Earth. We can see in certain Egyptian temples realignments in their axes. Like Newgrange in Ireland where long, darkened tunnels were constructed to 'catch' sunlight on a single day of the year (see Fig. 65), some Egyptian temples were aligned to observe specific stars rising on the horizon.

Fig. 22 illustrates three successive changes in the alignment of a temple in Luxor over the centuries. As precessional drift would carry a star out of its optimal position, coming down through a darkened entry way, a new 'hall' would need to be built to follow this shift. As precession continued, the process was repeated. Naturally, this was a slow development, and the re-alignments would only occur after decades of use.

This temple realignment is not for solar temples, but only for those that sought to monitor the stars.

Having fortified our hypothesis, we're now going to expand it. For cultures around the globe, the serpent was the *Center* for centuries. We will discover that the image of the Dragon was universal, but has faded with time. It once held the supreme position in the sky, the heavenly high ground of Creation, above whom no planet or God could fly. But for now, let's examine the path of the Dragon and test his course for ourselves.

Athena's Web

Center and Circle

Opening the Circle

Fig. 23

Stained Glass Compass Rose by TR Corwin

Center and Circle

It would seem natural that the constellational importance of the Dragon was highest through the 2nd millennia when *Thuban* was closest to the *Celestial Pole* in 2788 BC (Fig. 11). We will come to examine Mediterranean cultures in depth, but let's begin by looking at an example of observational mythology and how it works.

The following is what the *Chinese Classics* have to say.

Fig. 25

Fig. 24

The Imperial Dragon

Images in this section courtesy of Starry Night Education

Fig. 26

First, a little background on the Dragon in *China*. The **oldest known Chinese dynasty**, the **Xia**, were a tribal people that dominated much of the area from 2205 BC to about 1500 BC. Their totem was the *Snake*, and it appears in many of their myths. Later, this totem evolved into the Dragon, in the same fashion that his distant counterpart evolved in the West.

Since the people of China are considered to be *Children of the Dragon*, the Chinese have long held a special place in their cultural heart and soul for this creature.

Athena's Web

Center and Circle

The Dragon is associated with floods in many cultural legends. Here's one example:

The *Emperor Shun* appointed *Yu* the task of finding a means of controlling the great flood. *Yu* is depicted as a Dragon or man, with older myths inclined to the former. *Yu* worked so hard on the problem which the Emperor had given him that his hands grew worn and his feet thickly calloused. His skin was black, he was thin as a rake, and could barely hobble, but in the end his labors were rewarded. By making artificial canals he was finally able to drain the water from the floods into the sea.

In royal recognition of his efforts, *Shun* abdicated and *Yu* became the First Emperor of the Xia Dynasty, ruling from 2205 to 2197 BC.

This tale is one thin slice from the mythological background, but there are also astronomical themes concealed in the myth. The *Chinese Classics* taught that the Dragon is *thunder* (a common attribute of the Dragon worldwide), a creature of the waters who rested in pools in the winter, rising in the form of rain clouds in the *Spring*. In the *Autumn*, the dry season, he sank back into the pools where he slept as he waited for *Spring*.

There are two times of day that changing stellar activity is generally noticed; *Sunrise* and *Sunset*. At *Sunrise* new stars first appear in the east in the morning twilight. This is known as their *heliacal rising*. But all stars appear at twilight. If one faces *South* they climb through the nocturnal sky in a *clockwise rotation* as either evening or season progresses; appearing higher and higher overhead as time marches on. But, if we turn around and face *North* to examine the circumpolar constellations, they appear to rotate in a *counterclockwise* direction around the pole.

One spiral does not fit all.

The Seasons here are generic. Spring (Fig. 25) *is calculated for May 1st, 7:30 PM.* **Fall** (Fig. 26) *is Oct. 1st, 7:30 PM; and* **Winter** (Fig. 27) *is Dec. 31st, 6:30 PM; all calculated from Beijing in 1200 AD.*

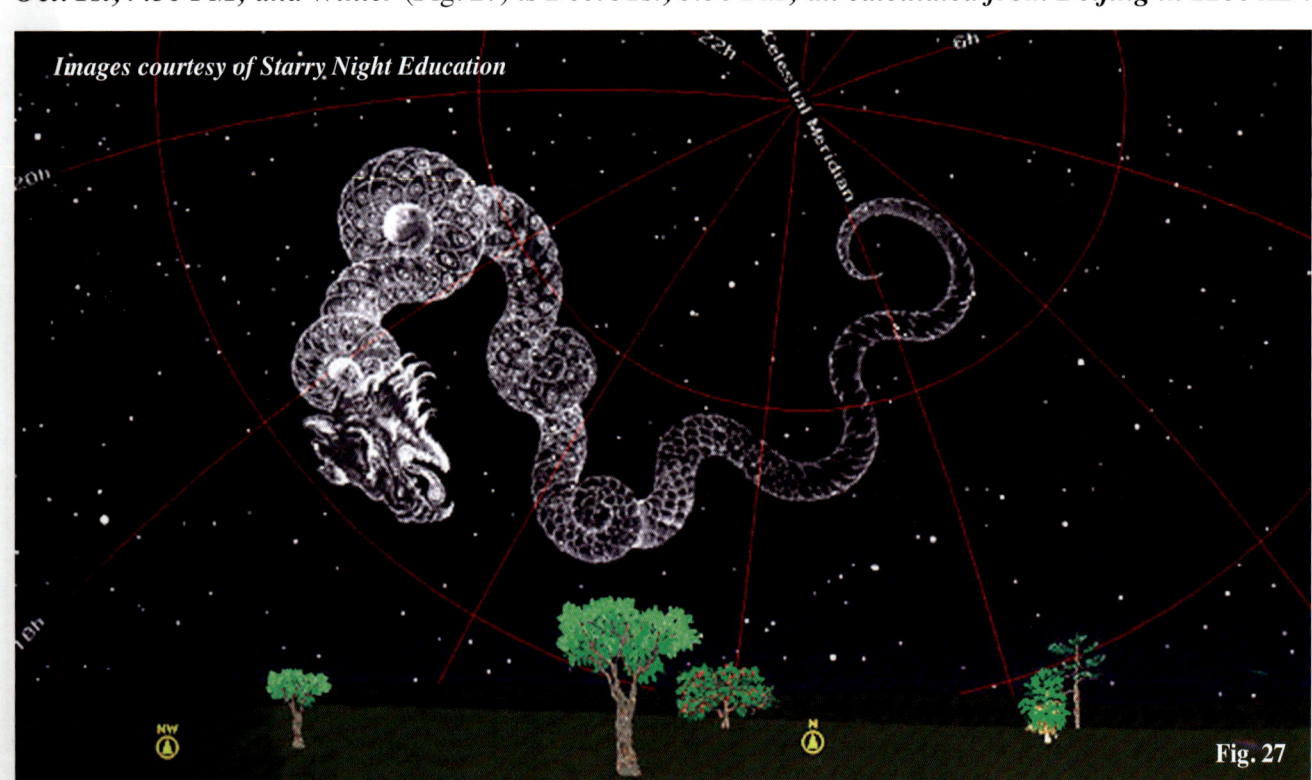

Fig. 27

The Chinese classics teach that the Dragon is...
... a creature of the waters that rested in the pools in the Winter...
... where he slept as he waited for the Spring

Center and Circle

Fig. 28

If we examine the early evening hours of *Spring* at twilight in China circa 1200 AD from the latitude of Beijing (Fig. 25), we would see the constellation *Draco* just beginning to appear and ascend in the darkening sky in the north-northeast, starting its great climb high over the *Pole of Heaven*. In the *Fall* (Fig. 26), the Dragon's first nocturnal twilight appearance would be as his head begins to descend toward the northern horizon, this time in the north-northwestern sky. Once again, through the course of the evening (or season) the head and body of the dragon will rotate in this counterclockwise direction. Finally (Figs. 28-29), the constellation is seen with its head resting upside down upon the 'pillow' of the Earth through the winter,

"*...where he sleeps as he waits for the spring.*" [1]

In this case, it was the annual 'hour hand' of this circumpolar constellation which was woven into myth to chart the seasonal weather changes. It's a handy time-tool.

But if this image makes sense astronomically, it also binds this motif to its western counterpart; linking it to a universal image. While we know the star position of the plebeian (common) Dragon, we do not know the location of the *Imperial Dragon*.

One simple image, twisting for all to see. Naturally, seasonal weather patterns change, depending on your location.

The Dragon was the keystone[2] to the arch for cultures whose roots extend far enough back in time to have witnessed the *Great Serpent* spinning and rotating, twisting and squirming upon the *Axis of Heaven*. They did not record these highlights in doctoral theses or in encyclopedic collections, nor measured by statistical analysis nor democratic vote. They painted them into their myths and passed them on as a practical tool for telling time, whether day or season. The Chinese Classics illustrate one example of what we will see repeated in various cultures around the world, across the centuries; over and over again.

Center and Circle;
Center and Circle.

It's been said that East is East, and West is West and ne'er the twain shall meet; but that's not true. A long, long time ago, in lands far, far away, they did indeed meet in a common cause and purpose; located right at the summit of heaven where all who had eyes to see, could.

Fig. 29

Athena's Web

Center and Circle

Gemini

Fig. 30

From the "*Atlas Coelestis*" by Sir James Thornhill (1729), Plate 13

Center and Circle

Voyager v. 1

Time of the Twins

The Serpent and his relationship to the pole in 5000 BC

We enter our time machine and take a giant leap back into the archaeological record; before what we think of as writing and history began. Our markers of precessional motion are *Center and Circle*; the *North Celestial Pole*, the *Axis of Heaven* around which the stars rotate, and the *Vernal Equinox*, the start of *Spring*. It was important for evolving farm communities to know when to sow one's crops; not so much for the individual (although that was important too) as for the community. It was therefore in the interest of the community to ascertain precisely when the seasons would occur.

Vernal Equinox through Gemini

The epoch we are examining is the 7th through 5th millennia BC. The region under scrutiny is central Europe. The quotes in this section are from Dr. Marija Gimbutas' work, *The Gods and Goddesses of Old Europe, 7000-3500 BC*. If we look closely at just a few of the artifacts from these times, we will see some interesting patterns beginning to emerge.

By 6000 and 5000 BC, the *Celestial Serpent* had clearly staked out his claim to the polar regions astronomically (Fig. 32). Although the stars of the *Serpent* are not particularly bright, their proximity to the pole at this time far outshines their lack of individual brilliance. Between approximately 6300 and 4800 BC (Fig. 31), the *Vernal Equinox* was 'sliding' along the *ecliptic* as it passed through the constellation *Gemini, the Twins*.

Since *Spring* is the season of the Earth's renewal, it might seem obvious to agrarian societies that this was the source of not only

Center and Circle

Nature's creation, but all Creation, with a capital 'C'. In later centuries, we know that many civilizations would begin the cycle of their year at the *Vernal Equinox*, in the *Spring*. Indeed, this equation can be boiled down to a fairly simple formula:

Vernal Equinox + stars = Creation + myth

As *Spring* slowly slid through the stars with the changing constellations, myth reflected what was happening in Heaven, mirrored by the cultures of their day. Myth, tradition and ceremony reflect Heaven's will. This was a huge component of pagan religion. The earliest evidence is in the archaeological record, which can be dated to specific periods, providing a framework in time from which we can draw some chronological order.

Gemini's association with duality is legendary. We know that the *Vernal Equinox* was passing through the constellation of the *Twins* between roughly 6,300 and 4,800 BC. Among the predominate images of this period are figurines of *Twins*, both in human form (with breasts, Fig. 33), and as a double-headed bird deity (Fig. 34). The common thread running through this imagery is clear. The artifacts on this page all date to the 5th millennium BC.

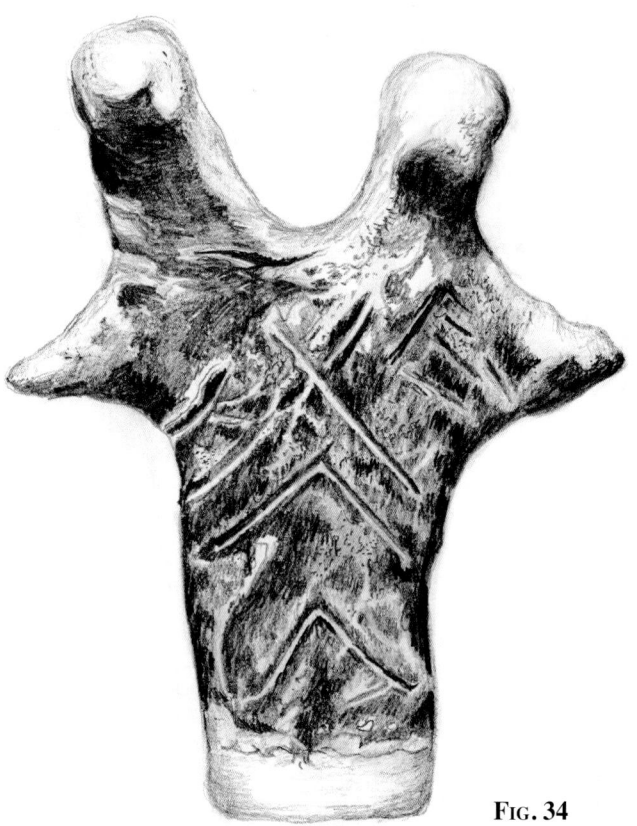

Fig. 34

Double-headed bird deity

But we also stated that it was the *North Celestial Pole* and the *Vernal Equinox* which were important; *Center and Circle*. Astronomically speaking, the *Serpent* (Draco) was wrapped around the pole in 5,000 BC. Fig. 35 is an artist's rendering of the spiraling serpent from this period. If this indeed is a stylized representation of what they were seeing nightly in the Sky, it is an apt one. It's not a literal image—it's metaphorical, derived from a 'living' source. We may have here an artist's image of an astronomical weave, depicting mythological events and forming the basis of a calendar. Note the 'dots' (stars?) stamped into the body of the *Spiraling Serpent*.

Artwork Jeanne Fallot

Fig. 33

Two-headed Goddess

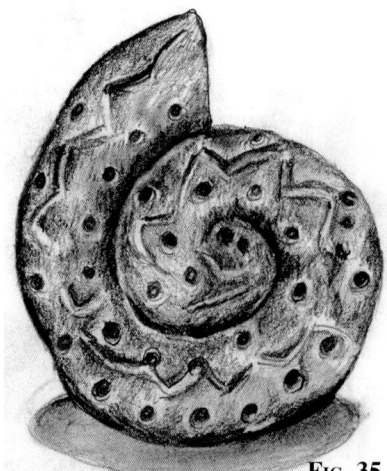

Fig. 35

Spiral Serpent

* Early Bird Goddess	* Peak of culture	*Early solar calendar	Egyptian Unification *
6000 BC	5000 BC	4000 BC	3000 BC

Center and Circle

The earliest astronomical writers of Greece and Rome speak of the element 'AIR' as being associated with this constellational image. It was common knowledge among these writers that many of these themes were incredibly old. Not only the concepts, but even the materials themselves were dated. For instance Ptolemy comments on the extreme age of the following manuscript. He writes,

"Recently, however, we have come upon an ancient manuscript, much damaged, which contains a natural and consistent explanation... The book was very lengthy in expression and excessive in demonstration, and its damaged state made it hard to read, so that I could barely gain an idea of its general purport; that, too, in spite of the help offered by the tabulations of the terms, better preserved because they were placed at the end of the book." [3]

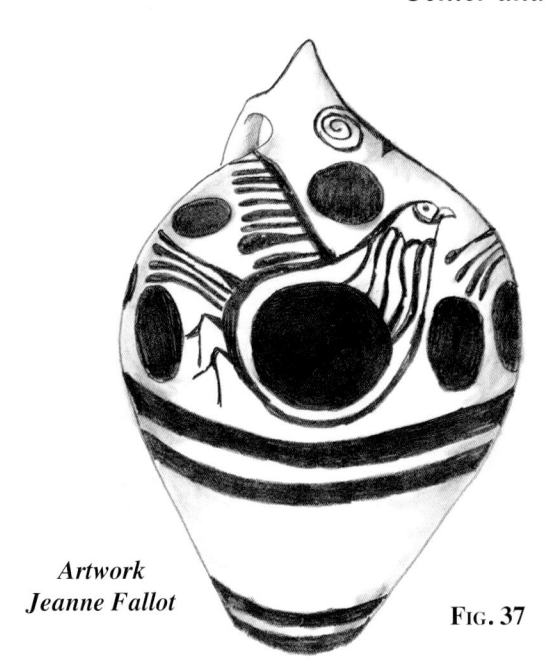

Artwork Jeanne Fallot

FIG. 37

Bird Goddess of 6000 BC

FIG. 36

Birds represent creation. Eggs are creation of the bird. This single image depicts dual creation.

In his footnote the translator assumes that the manuscript Ptolemy is reviewing is a scroll, which may explain why the final pages, encased by the outer wrappings, were better protected, while a book might have been damaged at each end.

The astronomical and mythological notions were old when writers whose works are still extant began to record their views. The question is, how far back do ancient beliefs, from a time we consider ancient, go? If 'AIR' themes mean '*creatures of the air*', it would help explain the preponderance of bird drawings throughout this period. *Spring* is the time of origin, of new beginnings. The stars said new *Twins*, new AIR, and/or new Birds. Birds and their origins, the *Egg*, are a common theme from this period.

If this is an ancient association, then later literature certainly illustrates them together. We *see* the imagery in the art and *hear* about it in the myths. In Fig. 34, this theme of *Birds* and *Twins* blends into one. We will watch as it takes root, and then grows to entwine the serpent.

Each of these images are important snapshots of Heaven as depicted by their moment in time. Each is distinctive, and yet they are often artistically rendered as a single concept. As a circle is 'fused' to its center, so one cannot move without impacting the other, *Center and Circle*. From Dr. Gimbutas's work:

*"The **Snake Goddess** and the **Bird Goddess** appear as separate figures and as a single divinity. Their functions are so intimately related that their separate treatment is impossible. She is one and she is two, sometimes snake, sometimes bird. She is the goddess of waters and air, assuming the shape of a snake, a crane, a goose, a duck, a diving bird."* [p. 112]

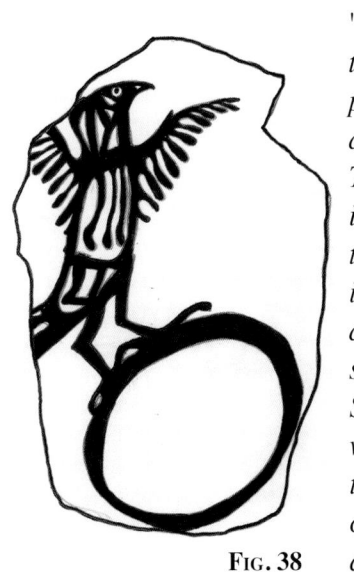

FIG. 38

* Pyramids	* Abraham	Troy *	* Homer	Julius Caesar *
3000 BC	2000 BC	1000 BC		0 BC

Center and Circle

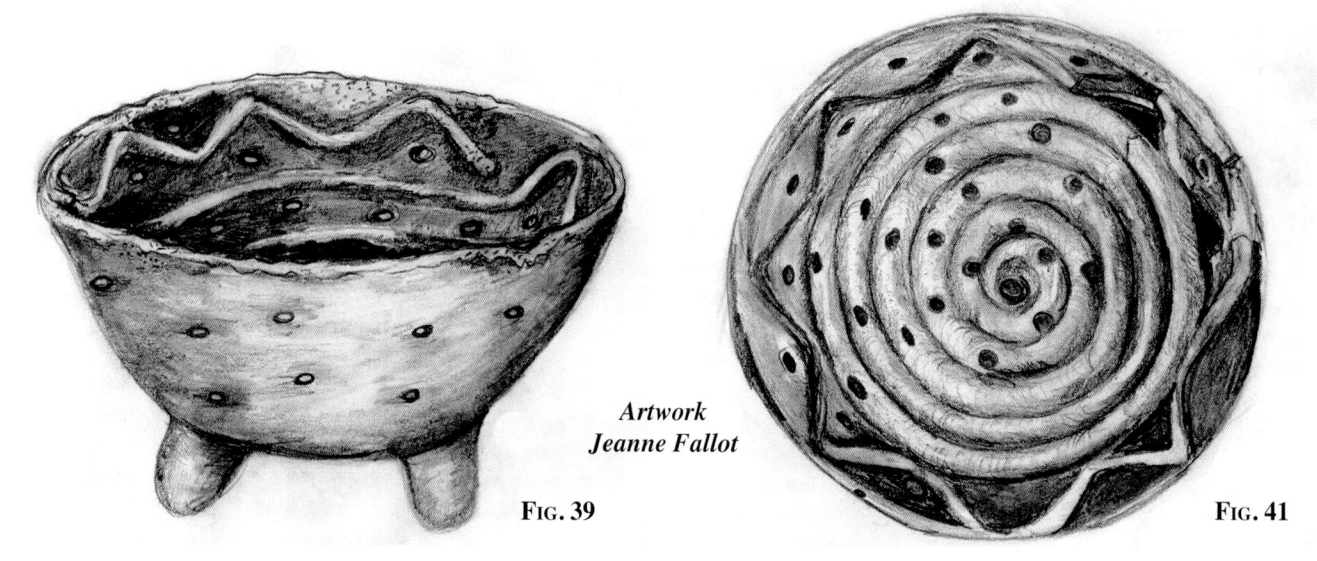

*Artwork
Jeanne Fallot*

FIG. 39

FIG. 41

*The Spiraling Serpent, as seen from two different perspectives of the same bowl.
Once again, notice the dots or 'comb-like stamps' running throughout the body of the serpent.*

"The snake and its abstracted derivative, the spiral, are the dominant motifs of the art of Old Europe, and their imaginative use in spiraliform design throughout the Neolithic and Chalcolithic periods remained unsurpassed by any subsequent decorative style until the Minoan civilization, the sole inheritor of Old European lavishness [...] "This art reached its peak of unified symbolic and aesthetic expression c. 5000 BC." [p. 93]

"'Symphonies of snakes' appear in colours and in graphite or white-encrusted incisions on cult vases, lamps, altar tables, hearth panels and house walls [...] Snakes, their bodies marked by dots or comb-like stamps have been found incised on a number of vases from the Vinca mound." [p. 94]

FIG. 40

"The mysterious dynamism of the snake, its extraordinary vitality and periodic rejuvenation, must have provoked a powerful emotional response in the Neolithic argriculturalists, and the snake was consequently mythologized, attributed with a power that can move the entire cosmos. Compositions of the shoulders of cult vases reveal pairs of snakes with opposed heads, 'making the world roll' with the energy of their spiralling bodies [...]

"The organization of the motifs demonstrates that the imagery is genuinely cosmogonic... the snake compositions appear in bands occupying the middle of the vases, associated with the belts of the upper skies containing rain clouds... On some vases snake coils in the upper bands have diagonal stripes probably to indicate torrents of rain. In some cases the snake is portrayed winding across the double-egg." [p. 94]

"The snake was the vehicle of immortality. Some vases flaunt a gigantic snake winding or stretching over the 'whole universe,' over the sun or moon, stars and rain torrents..." [p. 95]

Dr. Gimbutas sees these images as being fused as one; so much so that their *'separate treatment is impossible'*. These are the *'dominant motifs of the art of Old Europe'*. Understandably so, if these were representatives of a period of time, and what they looked up to see in the skies. *'This art reached its peak of unified symbolic and aesthetic expression c. 5000 BC.'* The timing is perfect, gaining momentum as it approached its cultural culmination c. 4800 BC.

She continues, the snakes are *'associated with the belts of the upper skies'*. Indeed.

Nowhere in her work does Dr. Gimbutas link these images to specific stellar groupings, yet her sentiments perfectly encapsulate the artwork being depicted here. These images dominate their period, and for good reason. The distant fires of heaven may have kindled the spark of spiritual reverence, ritual form and the practical, material, and agrarian concerns of their day, marking their 'moment in the Sun.'

Center and Circle

But, we're not limited by physical evidence in our hunt for stellar themes. Myths provide some interesting *dual images* as clues. What can we learn from legend about the *Twins*? Where do we find *Twins* as the *Author of Creation*? For this next tale, we turn to a legendary tribe, the **Dogon of Africa**.

According to the *Dogon*, in the beginning was the "**Egg of the World**." This *Egg* contained *two pairs* of mixed *Twins*. One of the *Twins* threw himself into the *Void*, attempting to take his *Twin* with him. His quest was in vain and he returned *'to the sky'*. But *Amma* (God) had already given his place to the *couple* remaining in the other half of the *Egg*...

Amma wished to start his creation over by creating *new pairs of Twins* "*in the sky*." [4]

Are these references to the sky generic; or do they have specific stars in mind? Do the *Dogon* give us any clues that might suggest, perhaps, they were indeed monitoring the motions of heaven?

It turns out, the *Dogon* have "*...elaborated astronomical systems of considerable complexity...*

"*At Sanga the apparent solstitial and equinoctial positions of the sun were formerly measured by sightings made by the use of three altars of 'the union'*...

"*The operator placed a small staff vertically at the top of the altars and used a known landmark on the horizon to observe the rising sun. This measurement was taken four times a year by one or another of the chiefs of the joint family; it was called the 'measure of the direction of the moment...'*" [5]

Fig. 42

Nommo, the Celestial Twins of Heaven
Artwork Jeanne Fallot

Twins are a dominant theme throughout many West African mythologies. For the populations that consider themselves part of the *Mande* (*Malinke, Bambara, Dogon, Bozo*, etc.), *Twins* recall the mythic times when the first living creatures were *pairs of Twins of the opposite sex*. According to the indigenous tribes, *Amma* (God) created the universe from the infinitely small, a sort of initial atom, represented by the smallest of all seeds. "*This 'Seed of the World' contained the strength of the four elements (Fire, Earth, Water and Air).*" The seed then exploded to become the '*Egg of the World*',[6] which contained the *two sets* of *mixed* (male and female) *Twins*.

The *Dogon* use *two* calendars, *one marked by the Sun, the other by the Moon*. The Lunar calendar commenced with the start of harvest, but the Solar calendar began with the *Winter Solstice*. It was an Egyptian (African) mathematician who advised Caesar to use a calendar based on the *Winter Solstice*. Is this part of an ancient African tradition? Thousands of years later, *Twins* would be remembered for founding Rome in an apparently already well-established theme.

During this time of the *Twins*, we find a '**celestial shorthand**' developing, weaving *dual images* as *Center and Circle*; *Serpent and Twins*. We see *Eggs, Birds, Snakes*, and *Twins*. Dr. Gimbutas spoke of the *Serpent* wrapping the '*double-egg*'. Although from different traditions, it's not a question of whether central Europe or western Africa were 'first' with these concepts.

Each shared the same sky.
If you're living, you belonged.

Center and Circle

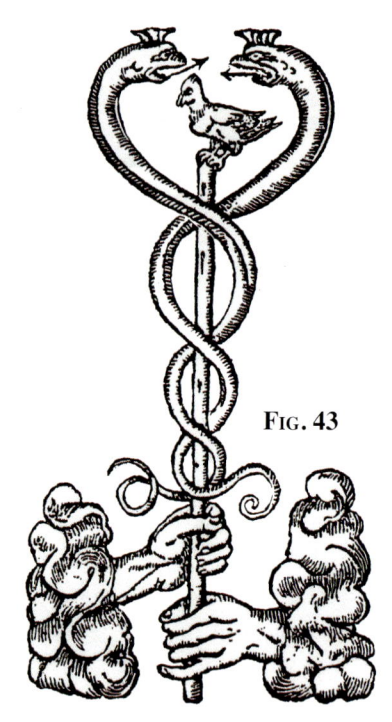

FIG. 43

Some of our earliest representations of *Two Serpents* entwined around a single shaft (*the Axis of Earth, the Axis Mundi*, Fig. 43) stem from this millennia in a figure now familiar to us as the **Caduceus**.

If we explore *Creation Myths* using the *Egg* our search widens considerably. There are **Norse myths** of **Odin** and the *Egg*, with its "**Shell of Existence**" being split into *Two*, one half becoming the *Heavens*, the other the *Earth*.

In **Hindu mythology**, **Vishnu** is the preserver of cosmic order (dharma). He might be found sleeping on the back of a coiled *Serpent*, within the '**Shell of the Cosmic Egg**' (Fig. 44).

In a *symbolic shorthand*, this was the mantra of the time; *Center and Circle*; North Celestial Pole & Vernal Equinox, *Two* fused as *One*. Vishnu on a *Serpent*, in an *Egg*; *Serpent* wrapped around the *Double-Egg*; *Two Snakes* wrapped around a *Shaft*.

It's all part of the same *Celestial Code*, a snapshot of their moment in time; a picture of their sky. Once the fusion of *Center and Circle* is grasped, this 'overall' design begins to become obvious.

Tales of the **Rainbow Serpent** come from the **Dahomey** tribe, where he's called *Dan Ayido-Hwedo* (Fig. 45). He's existed since the beginning of time. He served *Nana-Buluku*, the one god. *Ayido-Hwedo* carried *Nana-Buluku* in his mouth, and the curves of rivers, mountains and valleys were created because that's the way *Dan Ayido-Hwedo* moves.

After all was created by *Nana-Buluku*, the load was so heavy He asked *Ayido-Hwedo* to coil beneath the Earth to help cushion it. Because *Ayido-Hwedo* could not stand the heat, *Creator* made the *Ocean* for him to live in and there he has remained since the beginning of time, "*cushioning the Earth with his tail in his mouth.*" [7]

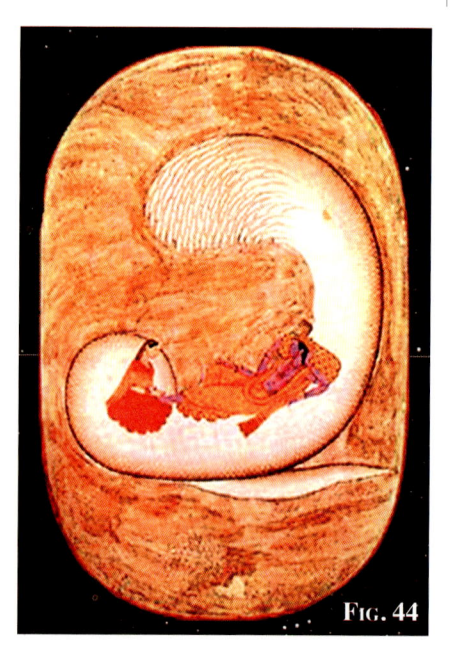

FIG. 44

Vishnu, sleeping on the Serpent (Center) within the Egg (Circle)

The *Dahomey* are from modern Benin in Africa. As seen from that latitude, a good portion of Draco lies below the Earth's horizon. This tribal myth 'remembers' this as the *Serpent* lying beneath the Earth.

He still does (Fig. 295).

Dan Ayido-Hwedo

FIG. 45

Stories of the Rainbow Serpent: The rivers curve because that's how Dan Ayido-Hwedo moves
Artwork Jeanne Fallot

We have begun to unravel a few early strands of the weave.

When the *Great Bird* laid the *Egg of Creation*, all Life was pictured as our familiar 'Oval', an apt metaphor for the *inner curved dome of the sky* revolving ceaselessly overhead throughout time. The sky measured the planting of crops and defined compass points for travel. For nomadic peoples it marked when to leave with the herd for summer grazing and when to return.

The *Equinoxes* split time in half in a dance of the seasons; Heaven and Earth, Spring and Fall, night and day, woman and man, peace and balance, *Yin* and *Yang*. Twice each year we are given the chance to begin again.

At the start of the year these peoples would seek to know,
What is Heaven's will?
Look to the stars.

Center and Circle

Photo credit Raymbetz — FIG. 46

Nabta Playa, Egypt

They left behind evidence of a higher degree of urban planning than did their northern-Nile neighbors. Buildings were aligned with one another, deep wells were dug to help maintain a more constant water supply, there was above and below ground stone construction, *and* they fashioned the oldest acknowledged astronomical megalith in the west, one considerably older than Stonehenge.

"An exodus from the Nubian desert at 5,000 years ago could have precipitated the development of social differentiation in pre dynastic cultures through the arrival in the Nile valley of nomadic groups who were better organized and possessed a more complex cosmology." [8]

We see archaeological evidence of astronomical interest between 5500 and 4800 BC in a remote location known as **Nabta Playa** in southwestern Egypt.

From approximately 10,000 BC rainfall patterns were such that a lake formed in the area. Nomads started to bring their cattle to *Nabta Playa* through the summers. Then around 5500 BC something changed.

The people starting sacrificing young cows and burying them in clay-lined roofed chambers covered by rough stone.

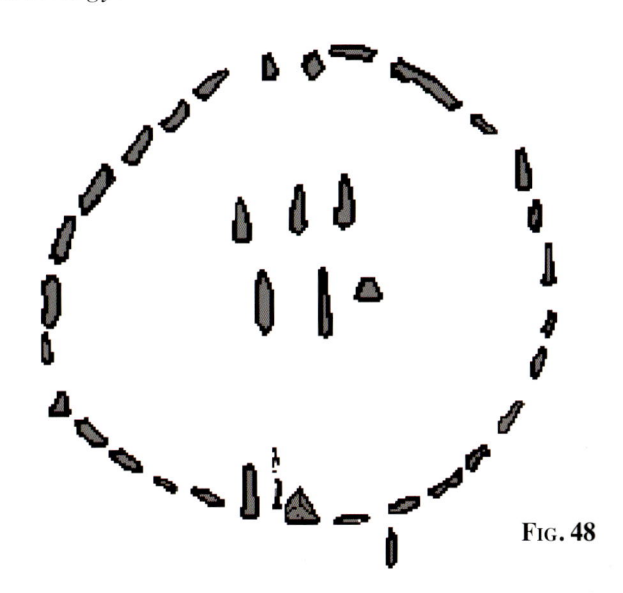

FIG. 48

Astronomical alignments verified by Prof. Kim Malville, University of Colorado, Boulder

FIG. 47 — *Photo credit The University of Texas at Austin*

This 'exodus' was part of a nomadic tradition summering at *Nabta Playa* as a 'regional ceremonial center', but then wintering, where? Deeper inside the heart of Africa, in what we now considered sub-Saharan Africa? Does our earliest evidence of star observation originate in darkest Africa?

In 1996, *Nabta Playa* was investigated by Professor Kim Malville of the University of Colorado and found to be the oldest currently known archaeological site used to determine the date of the summer solstice.

From 4800 BC onward, the *Vernal Equinox* was leaving the *Twins*, and beginning to **approach the Horns** of a *Great Celestial Bull*. But before we commence that cattle drive, there are other myths which also speak of *Twins, Birds, Eggs* and AIR.

Athena's Web

Center and Circle

We turn to **Philo of Byblos**, a *Phoenician* who undertook to show that *Greek myths* were based on *Phoenician sources*.[9] Although we have only fragments and titles of his works mentioned by numerous authors, the *Cosmogony* of Philo remembered several Creation Myths that seemed to evoke our airy origin themes.

It is a world in which "*troubled and windy air or a breath of wind and dark chaos*"[10] presides.

In those days, "*the breathing air became enamoured of its own principles.*"[11]

We should not be surprised to learn that it was accompanied together with the *Cosmic Egg*.

According to another Phoenician cosmology by **Eudemus**, "*From the union of these two first principles were born Aer (air) and Aura (breath).*"[12] This couple then produced the *Cosmic Egg*. Duality is applied to 'Air' itself; rather than *Twins*, *Twin Birds*, *Twin Eggs*, or *Twin Snakes*, now we have *Twin* AIRs, *Air* and *Breath*.

We are seeing variations on a theme born of the stars. In the 2nd century BC, **Mochus** stated that in the beginning was a double principle- **Aether** and **Air**. Then came the Wind and then the two winds "*Lips*, *Notos*... *and the Egg*."[13]

Another legend, which the Greeks themselves say is among their earliest myths, is the...

Pelasgian Creation Myth

*In the beginning, Eurynome,
the Goddess of All Things,
rose naked from Chaos.*

*She found Nothing upon
Which to rest her feet, and thus,
She divided the Sea from the Sky.*

*She danced lonely upon
The waves of the Sea.*

*She danced towards the South, and
the Wind set in motion behind Her
seemed something new and strange
with which to begin a work of creation.*

*Wheeling about, She caught hold of
this North wind, rubbed it between
Her hands, and behold!
The Great Serpent Ophion.*

*Eurynome danced to warm Herself,
wildly and more wildly,
until Ophion, enchanted,
coiled about Her divine limbs
becoming One with Her.*

*As she lay with the Ophion,
Eurynome was got with child.*

*Eurynome assumed the form of a Dove,
Brooding upon the Waves
and with time,*

She laid the Universal Egg.

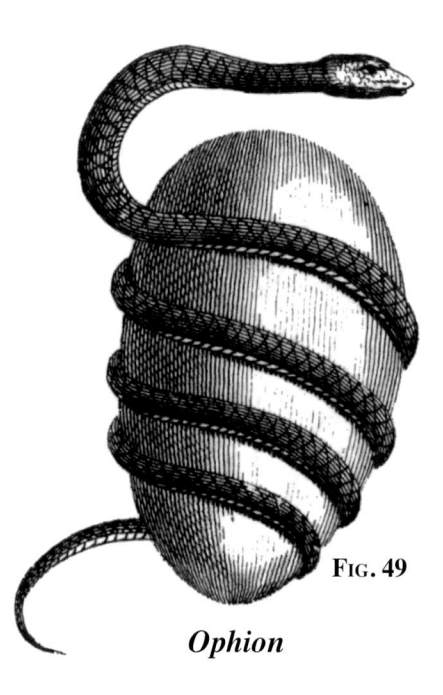

FIG. 49

Ophion

*At her bidding,
Ophion coiled seven times
About this egg, until it hatched and
split into Two.*

*Out tumbled All Things that exist,
Her Children: Sun, Moon, planets,
stars,
Earth with Her Mountains
Rivers, Trees, Herbs,
and All Living Creatures.*

*Eurynome and Ophion made their
home upon Mount Olympus
where He vexed Her by
claiming to be the
Author of the Universe.*

*Forthwith,
She bruised His Head with Her Heel,
kicked out His Teeth,
and banished him to the
dark caves below the Earth.*

*Eurynome opened Her gaze
and Her arms to Her children,
giving each its name
which She read off its own
singular power and being.*

*She named the Sun, Moon,
planets, stars and the Earth
with Her Mountains and Rivers, Trees,
Herbs and Living Creatures.*

*She took joy in Her Creation,
but soon found Herself alone
desiring the face, voice, ear
and warmth of another of her Own.*

*Eurynome stood up
and once again
Began to dance
alone
upon the waves.*[14]

-Robert Graves

Center and Circle

Out of Chaos, Creation was formed, dividing *Sea* from *Sky* and introducing our *Divine Duality*. Through dance *Eurynome* stirred first the South and then the North Wind into being (notice just two of the four winds), and in *Her Hands* she then discovers *Ophion*, guardian of the *North Celestial Pole* in his correct cardinal location. We have now mythically established our *North Celestial Pole* and *Vernal Equinox- Serpent and Wind, Center and Circle*.

Photo credit Daniel Fernandez

FIG. 50

A holy enclosure for Ba'al. Early observatories?

Ophion and **Eurynome** conceive (*Center and Circle*). As a *Dove* (creature of AIR) *Eurynome* gives birth to the **Universal Egg**. *Ophion* wraps himself around the *Egg* (*Center and Circle*) seven times.

We have all the elements of our astronomical and mythic 'world view' from 6300 to 4800 BC. Our *Creation Myth* begins with a stillness over the *Waters*. It returns us to that same moment one cycle later as *Creation* (the New Year) begins again.

We have seen this same motif in various locations around the Mediterranean; and also in Scandinavia and India. Learning to tell time is a way to harness civilization to pull together, to meet as *One*, to rejoice as *One*, to harvest as *One*. This ability to tell time is woven into the fabrics of each of these cultures, whether they are shaped by *Sea* or *Sand*; they still need to know how to mark chronological cadence. The *Sky* is the *Father* (or *Mother* in the Egyptian tradition) of Gods and men. From *Heaven*, the *Seasons* and the *Hours*; *All Things* are born. We live within it.

It is the living reflection of *Heaven on Earth*.

Athena's Web

Geography molds civilization's personalities. Barren land forces folk to turn to the *Sea* for sustenance, while rich soils and river valleys bind traditions to the Earth. But *being able to tell time* is central, no matter what the societal differences. It was in the interest of communities to be able to do this well. Recent research has suggested that the paintings of Lascaux Caves indicate the lunar phases of the month.[15] If correct, these ephemerides were recorded some 13,000 to 17,00 years ago. How many thousands of years does it take for people to go from lunar phases to the stars upon which they rest?

As the *Vernal Equinox* slipped from *Gemini* into *Taurus*, the following myth speaks of *Twins* and *Bull* appearing in the *Sky* next to each other. According to a **Phoenician myth**, **Kusor** and **Hasisu** are building a holy enclosure for *Ba'al*, a *Farm God* who made fertile family and fields, flocks and herds. Like many other deities of his day (*Hathor, Apis, Bel*), *Ba'al* is often depicted in the *Bovine*, as *Bull* or *Cow*.

"*A house shall be constructed for Ba'al, for he is a God; and a holy enclosure...*"

"*And Kusor-and-Hasisu said: 'Listen Aleyin, son of Ba'al, mark our words, O Rider of the Clouds. Behold, I shall put a sky-light in the sanctuaries, a window in the middle of the temples.'*"

"*Kusor, the mariner, Kusor, son of the law (?), he shall open the window in sanctuaries, the sky-light in the middle of the temples. And Ba'al shall open a fissure in the clouds, above the (face) of Kusor-and-Hasisu.*"[16]

FIG. 51

Chalk Mt. Observatory in Colorado

Center and Circle

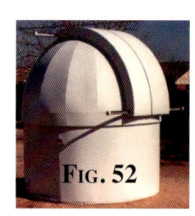

The myth refers to the segment of the sky we've been watching, the *Twins* and *Bull*. This is where the *Vernal Equinox* was from 4800 to 3000 BC, as the *Spring Sun* slid from the toes of the *Twins* to the *Horns of the Bull*. They are building a 'temple' to observe the stars, opening a sky-light to observe Heaven. The sky-light in the middle of the roof is the same way we monitor the heavens with modern observatories today. An observatory is essentially a window on the sky. Whether you use the naked eye or telescope, you're framing the segment of the heavens into which you wish to look.

There's another **myth**, found in both **Egyptian** and **Phoenician** versions, which describes this celestial passage and the shift from one constellation to the next.

FIG. 52 Doors closed

FIG. 53 Photo credit Leo Cundiff
Tulsa's Gladys-Leo Observatory- doors open

Obviously enough, it is called a **'Tale of Two Brothers'** (p. 188).

"Once upon a time there were Two Brothers, so the story goes, having the same mother and father." These *brothers* live and work together, but due to the *duplicity* of the older *brother*'s wife, the younger *brother* is killed, and then reincarnates as a beautiful *Bull*.

"Then Beta said to his elder brother: Look, I shall become a large Bull that has every beautiful color and whose sort is unparalleled...

I shall become a great marvel, and there shall be jubilation for me in the entire land..." [17]

This is a mythic memory of a stellar passage. *Spring* is moving from the *Twins* to the *Bull*. We have a *brother* dying and coming back as a *Bull*. Gemini is over, *Taurus* has begun. The Sky picture tells the time; not as time of day, month or year, but as millennia. It was their time. It was their 'world view'.

In various mythologies the sky is the source of the *Creator*. From the *Sky* comes the Seasons, which in turn produces the calendar, together with its divine directions.

Photo credit Isle of Man Astronomical Society **FIG. 55**

The newly capped Isle of Man Dome (1999)

Many before have claimed that what is being personified here is *Nature* in her *death and rebirth cycle*. *Osirus* dies, but is re-born through his son *Horus*. The younger *brother/Bull* is the agricultural year; planting, growing and harvesting, and then starting all over again. The story line speaks of the celestial shift.

What do we know about time and the calendar at this early date? We know that the *Moon* was the planet of preference for telling time. Many of the oldest African calendars are based upon the *Moon* and consisted of 360 days (12 months at 30 days per month). The month did not begin with the astronomical '*New Moon*' as we now know it, when the Sun and

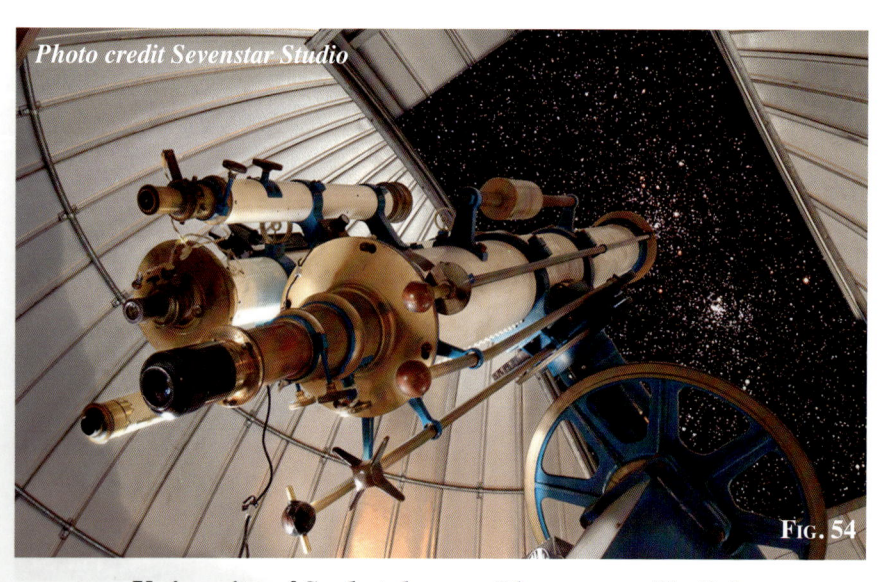

Photo credit Sevenstar Studio

FIG. 54

University of Saskatchewan Observatory Skylight

Athena's Web

Center and Circle

Photo credit Ahmed Rabea

Moon 'align' in the heavens during the 'dark' of the *Moon*. Their *New Moon* was watched for by the Babylonian and Hebrew priests in the Levant, by the Romans, Germans and Druids in Europe, and even across the ocean by the Maya in Central America. In a ritual that was common to them all, when the first thin sliver of a *crescent Moon* appeared in the western skies shortly after Sunset (which is generally a day or two after what we now consider to be the *'New Moon'*), they would call it out, often from a mound or hilltop for better Sky viewing. In Latin, this came to be known as the *Kalends* or *Calends*, from which we get 'calendar' (*calare;* to announce solemnly, to call out). This ritual is still practiced by those using the *Islamic calendar* to this day.

We also know that even though the Egyptians started with a *Lunar calendar*, they were the first to switch to a *Solar calendar*, using a 365-day year (based upon an approximation of the amount of time it takes for the Sun to go around the Earth) by simply inserting five extra days at the end of the 360-day year. These extra five days seemed to exist *'outside of time'*. It was akin to the week between Christmas and New Year's; a holiday of gaming tucked away at the end of their cycle. Curiously, it was considered to be unlucky.

Here is the myth that tells of this change in calendars, and of how the *'New Math'* came into being.

Nut is the **Night Sky**. She was one of a pair of ruling *Twins along* with her brother *Geb* (the Earth). Against their Father's wishes, *Nut* and *Geb* ran off and got married.

Fig. 56

The crescent Neomenia (New Moon) setting shortly after sunset

Sibling marriages have long been considered... well, different. When their father, Ra, learned of it he was outraged, and sent Shu (Air) to separate the newlyweds. Ra ordered that Nut not be allowed to bear a child in any month of the year.

The Egyptian timekeepers decided they needed to switch from a 360-day based *Lunar calendar*, to a 365-day *Solar* model, five days needed to be added. Fortunately, *Plutarch* tells us just how this extension was accomplished.

Thoth (the clever god) had pity on *Nut*. Playing draughts with the *Moon*, he won a seventy-second part of the *Moon's light* with which he composed five new days.

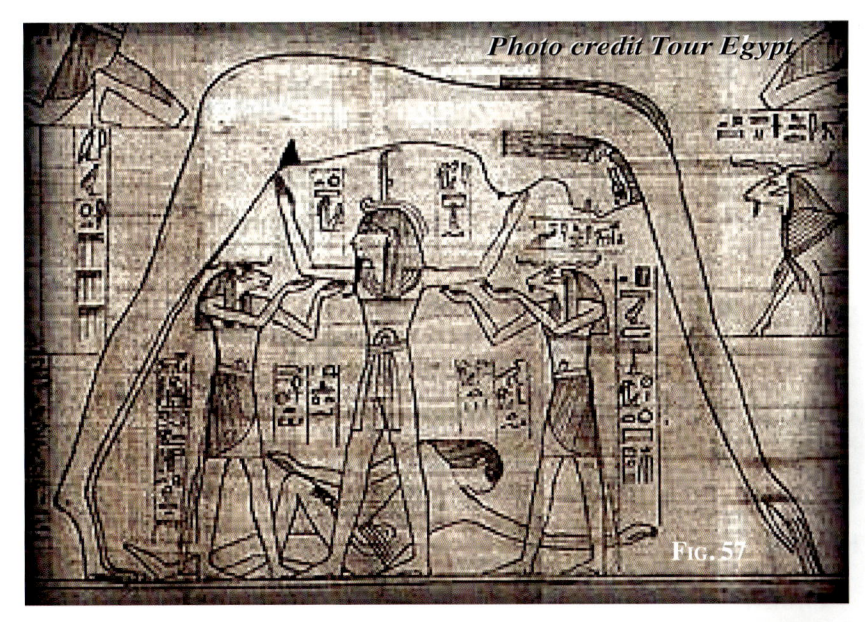

Photo credit Tour Egypt

Fig. 57

Nut (Sky) arching high overhead, supported by Shu (Air)

Athena's Web 25

Center and Circle

As these five intercaleated days had not belonged to the official (older) calendar, *Nut* was able to give birth to five children, one on each of the five 'left-over' days. These children were *Osiris*, *Horus*, *Set*, *Isis* and *Nephthys*.

Having adjusted to this daylight *Solar cycle*, the Egyptians continued to also use the *Lunar cycles* of the *Moon* in various religious ceremonies. The *Egyptians* used two calendars; one for the *Sun* and one for the *Moon*.

The *Spring* themes are being woven together in yet another dual-theme story format. *Twins* marrying each other, because they are the ones giving *Birth to Life*, to *All* born beneath the Skies as the *Vernal Equinox* passes through the constellation *Gemini*. We can hear the stories of Creation being told through a *Sibling* format.

It is what Heaven ordains.

Fig. 58

**Thoth, the Egyptian Moon
Photo credit Tour Egypt**

Ra is the original pharaoh. He gave birth to *Shu* and *Tefnut*.

They were *Siblings* who married each other and ruled together.

Then *Shu and Tefnut* gave birth to *Nut* and *Geb*.

They were *Siblings* who married each other and ruled together.

Geb and *Nut* give birth to *Osirus* and *Isis*.

They were *Siblings*, who married each other and ruled together.

While the *Vernal Equinox* moved through the constellation of the *Twins*, pagans felt all *Creation* was born of the *Twins*. It was a repeating theme, with an annual drumbeat marking cadence with its mythological baton every *New Year's Day* for centuries. Up and down, a-*One* and a-*Two*-ah.

Both of what we officially designate as *Sumerian* and *Egyptian cultures* date to the later passage of the *Vernal Equinox* through the constellation of the *Bull*; but the roots of each of these civilizations were already in place and at work, observing the rhythms of the sky and translating Earth's cycles from Heaven's point of view. The genesis of these stories, especially those involving the *Royal Number Two*, *Twins*, *Air*, *Birds*, and *Eggs* are being *mythologically animated* so as to give Birth to Life itself. These pre-Egyptian peoples living in the Nile Valley would tell variations on the *Tale of Two*, just as the pre-Sumerian peoples living in the Tigris-Euphrates River Valleys were looking at the same Skies, and telling slightly different versions. The myths reflect their worldview, as seen through the eyes of the local environment.

It was their *Sky*.

This world was spiritually split in *Two*. Today, we might think of this as being divided.

At that time they thought of it as what made you whole.

The Sun is used as an instrument to mark time; the Moon too.

Both *Circle the Serpent*.

Say hello to the image of the *Caduceus* (Fig. 43), the reading of Heaven's *Two* streams.

Two streams of life, each on their own track, pivoting about the same *Axis* Night and Day.

In an era when Sun, Moon, or stars could be used to mark time, harnessing Heaven to coordinate social or national activities was considered extremely important.

Photo credit Scott Hanko

Fig. 59

She is one and she is two...

These themes were observed, wondered about, and applied the full range of their collective imaginations for centuries. These indigenous peoples of the pagan world believed themselves to be looking up into the *Face of God*. They were attempting to determine '*His Will*'.

Center and Circle

Photo credit Tour Egypt

FIG. 60

Sirius or Isis

It gives a whole new meaning to being 'watched over'.

One reason the Egyptians first developed a *Solar Calendar* was because of *Sirius*, our brightest star. As we previously stated in Egyptian history, the **heliacal rising of Sirius** was used to date the *annual flooding of the Nile*. This flooding was the single most important event in the Egyptian year, and anything that could help determine its precise return was considered to be of huge importance. The agricultural future of the country was at stake. Close observation of these star positions helped them to better understand the motions of Heaven and the *Solar Year*, and in 4236 BC (according to Egyptologist J. H. Breasted) they were the first to officially institute this new 365-day year.

It was not until the 8th century BC that some other Mediterranean nations would follow suit and add five days to their 360-day year.

The question remains, if the Egyptians were prepared to make a change in their way of calibrating time in the 5th millennium, how long had they (or their predecessors) been using a calendar of 360 days before making the switch to 365 days?

We have no idea.

What the mythic record indicates is that by 4200 BC the skies were being closely monitored and had been for some time. *Nabta Playa* alone supplies archaeological evidence that serious celestial observation was going on back to at least 4800 BC, and possibly as early as 5500 BC.

In various cultures throughout Africa, Europe, the Indus River Valley, China and beyond, themes of *Center and Circle* can be found. Where this is true, we might suppose that we are dealing with cultures who are wrestling over the gnarly problems of the calendar.

These were the astronomical keys, hidden in mythological garb, needed to unlock the vault.

The *Sumerians* settled in the Tigris-Euphrates Valley and learned to work the rivers, as the Egyptians did on the Nile. The earliest Sumerian records stem from the mid-4th millennium down through the end of the 3rd millennium BC, but we know there were people living in these areas prior to the *Sumerians*. In the Tigris-Euphrates Valley, the Uruk period extends to 3900 BC, and the Ubaid period to 5300 BC. According to our astronomical reckoning, these are the peoples whose origins go back into the *'time of the Egg'*. This is when these tales would have made sense to the people on Earth looking up at the sky.

Enlil was one of the chief deities of the Sumerian religion. His name is thought to be translated as **'Lord Wind'** [18], consistent with a world view 'under the Egg'. He was Lord of *Wind* and *Sky*. He was also *Child of Heaven and Earth*.

Time always is.

Photo credit Nina Aldin Thune

FIG. 61

Pyramid of Khufu

27

Center and Circle

This notion subsequently evolves in *Greek mythology* with *Cronos* (Time), identified as the child of *Uranos* (the starry-Sky) and *Gaia* (Earth). Where the *Circle of the Sky* (Ecliptic) meets the *Belly of the Earth* (Equator), is where the '*cut*' of *Time* (sickle, spear, etc.) is made. Astronomy registers this point as the *Right Ascension Midheaven*, or RAMC.

The *Sumerian* culture (or their forbearers) remembers a time when AIR was the *spiritual medium of choice*. They tell of being born of the '*exhausted breath*' of *An* (God of heaven) and *Ki* (goddess of the Earth) after sexual union. [19]

Just as the ancient astronomical authors spoke of the *Twins* being associated with AIR, so they tied the stars of the *Bull* with the element EARTH. In moving from the *Twins* to the *Bull*, the metaphorical imagery is moving from AIR to EARTH. This 'shift' is reflected in the following **Egyptian myth**.

Shu, together with his sister *Tefnut* (our *Siblings*), comprised the *First Couple* of the Ennead. **Shu is the God of the AIR**. [20] He's the '*Atlas*' of *Egyptian mythology*. He holds up the *Sky*. When *Shu* was old and wearied of ruling, he abdicated in favor of his son, *Geb* (the Egyptian EARTH God; from AIR to EARTH) and took refuge in the *Skies* after a terrifying tempest which lasted nine days. [21]

In a number of myths, we see images of *Twins* giving way to *Bull*. *Shu* and *Enlil* represent Egyptian and Sumerian 'AIR' Lords who had gone before. *The Tale of Two Brothers* remembers the transition, as *Ba'al* opens a fissure above the face of the *Twins*.

The *Bull* rises before the *Twins*. There's a mythic story-line which generally runs as follows. In the beginning, there was *One*, unformed though he may have been. After growing lonely and/or bored, He sacrificed Himself to become *All of Creation*. *His Blood* became the *Oceans* and *Rivers*, *His Breath* the *Sky* and *Air*. We are *All One*, living within this *Giant Being*. Since the advent of Christianity, this '*One World Image*' has been translated as the *Body* and *Blood of Jesus*.

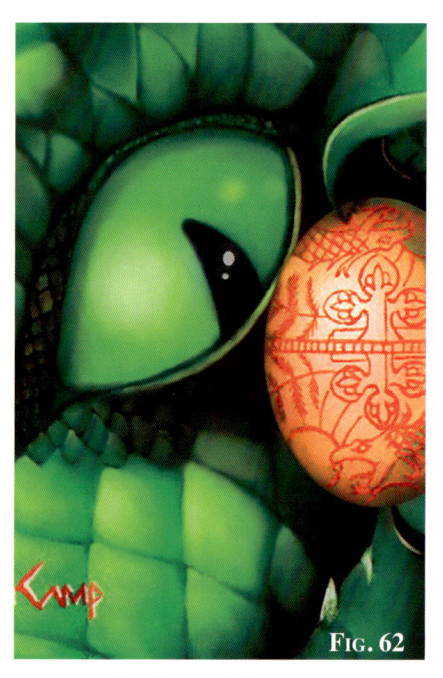

The Egg of P'an Ku
Photo credit Van Camp

An Asian example which bears the same mythic markings of our *Duality*, speaks of **P'an-Ku**, a *Dwarf,* **born from the Egg** (Fig. 62), which some say is represented by the *Yin-Yang* symbol he holds. This '*Dwarf*' grew at the rate of ten feet a day, filling the gap *between the Earth and Sky*. What we see is an eastern example of the same *Sky motif*. *One Being*, born of the *Egg*, holding as his *raison d'être* the essential symbol of *Duality*.

Are these cultural memories which pre-date known Chinese history? Or are they part of a *World-Wide awareness of precession*, the Skies, Time and the calendar? This Asian perspective only works if the Chinese thought that the stars of the *Twins* incorporated some form of *Duality*.

According to *Star Names*, Gemini was known to the Chinese "*as Yin Yang, the Two Principles,*"[22] which they show as a male/female pair of twins.

The mythic trail is leading us to the celestial signatures we are looking for, whether *East* or *West*. The question has always been, how far back do they go? The Skies may provide the answer.

This *Cosmic Egg* represented the *Great Bubble of Creation*, a self-contained biosphere, envelope and package which neatly contained the known and unknown universe. It's the inner side of the upper half of this *Great Cosmic 'Egg'* that we use as our model for planetariums today, the inner dome of the sky.

The changing face of the sky became the *Face Divinity assumed*, whether for each passing moment or millennium.

If *Time* is born of the union of *Heaven* and *Earth*, then days, weeks, months, years, decades, centuries and even millennia are the metaphorical children of this union. Observing *Spring* move across the face of the sky was simply one more '*hour hand*' on the face of their *Cosmic Clock*. We know the Egyptians worked with the month and year with consistent observation. Their temples were constructed for this reason.

Center and Circle

We know the *Sumerians* were developing *units of time*; the *seconds*, *minutes* and *hours* we use today. [23] These units endure; but they also used one which has fallen out of vogue. This '*hour hand*' was the *Vernal Equinox* marked by the paths of heaven, the birth of *Spring* each year.

During this epoch of *Divine Duality* associated with the element AIR- Creation, Time and Life were born again, an '*Egg laid daily by the Celestial Goose*'. But as the *Sky picture* changed, the image shifted, and the Sun, giver of life, love and consciousness gave birth to a new day now '*born every morning of the celestial cow like a sucking calf.*' [24] While *Spring*, marked by the Sun on this first day of the year, shifted from *Twins* to *Bull*, the image of the '*Face of Creation*' changed, this '*Body*' of the original Sky-Being who gave himself so that we might live.

These pagans who believed themselves to live '*inside*' the '*Body of God*' now saw themselves being transformed into a new mythic reality: living inside the horns of a *Great Celestial Bull* (*Apis*), or later, Cow (*Hathor*). For now the *Celestial Bovine* would rule supreme, but its roots ran deeper than dynastic Egypt. *Nabta Playa* provides us with that clue.

We look back to a time when people believed that we were all made in the *Divine Image*. Pagans of various persuasions took a giant step forward and saw the *Divine in their Sky*. This is who (they thought) the '*Divine*' wants us to be, as we are *His Reflection*. This is their *Celestial Choreography*, their *Divine Story-line*.

Please tell us, O Lord, what parts you would have us play today?

Each culture demonstrates their 'memory' of what times were like as they endeavored to *dance to a tune sung by the stars*.

As *Spring* passed through the stars of *Gemini*, the *Divine Will* was reflected by our being a perfect likeness of God, as expressed by the symbolism of *identical twins*. We must reflect God's will. The *Twins* were the archetype through which this vision was modeled.

It was their religion.

Fig. 63

The incredible edible egg

In our *Greek version*, one *Twin's* immortal, one mortal; but they cannot be separated long. For those watching the stars, first *One Twin* rises, then the other. As *One* sets, so the other soon joins Him.

Mythology was authored by those observing the Heavens. The astronomer/priests generated new stories as the Sky shifted, telling some of the same story lines for centuries, and then simply shifting gears. We will observe a number of these transitions in later myths.

One way to harness the myth, to crystallize its memory in the minds of people was by choreographing the ritual enacted on New Year's Day in theater.

Creation begun anew.

On this first day of the year the celestial reckoning was checked for accuracy. Exactly where did the *Vernal Equinox* now fall against the backdrop of the stars? This was the time of the year where they made sure they had it right.

For millennia our hunter and gatherer forbearers monitored the Skies. It has long been known that migrations move with the phases of the *Moon*. Stories of the Goddess Diana and the hunt are legendary. The *rhythms of Nature* obey the *Seasons*, the migrations follow their tides, the geece their flight, the bears their hibernation. How long did it take our ancestors to guage Lunar Mansions and their seasonal correspondences?

When did they first observe that through the course of a single evening, the Moon overtakes the next group of closest stars?

We don't know.

As we leave behind these tales of *Serpent and Egg*, the pattern begins it's Metamorphosis. Our theme is one of *Center and Circle*. Although the *Twins* will yield their place to the *Bull* on the *Circle*, the *Serpent* will not loosen its coils from the *Axis of Heaven*; at least not yet.

Through the 4th and 5th millennia BC, the *Celestial Snake's alpha star* wraps his coils in ever tightening circles around the *North Celestial Pole*, maintaining a firm grip on the flagpole of the *Highest Heaven*, come fair weather or fowl.

Megalithic Island

Photo credit BoyneValleyTours.com

FIG. 64

Now we journey north across cold seas to where the Atlantic wraps his chilly arms around the *Emerald Isle*. Northwest of Dublin the *Boyne River* meanders past a series of *Megalithic Monuments* Time forgot. Folklore offered clues, but exact knowledge of their whereabouts had been scattered with the quail among the thicket. Together with the nearby *Loughcrew Mountains*, these manmade temples may contain the greatest concentration of *Megalithic Art* anywhere.

The *Boyne River Valley* consists of three sites: **Newgrange**, **Knowth** and **Dowth**, each located on ridges just north of the *Boyne*. Radiocarbon dating puts their age to between 3700 and 3200 BC,[25] some of the oldest chronology for megalithic buildings in the world. *Newgrange* (Fig. 64) has been shown to align with the *rising Sun* on the *Winter Solstice*. Yet like so many earthen mounds in both Europe and the Americas, these sites were thought to be merely giant tombs for the dead. Increasing evidence suggests the *primary function* off these sites for *astronomical purposes*; to better understand our place in the world and the *Wheelings of Heaven* above.

Even after it was recognized that the long passage within the mound at *Newgrange* admitted a narrow beam of light at the Winter Solstice, some authorities have resisted both the *astronomical evidence* and the *folk tales*. They refuse to believe these *pre-Celtic peoples* had the sophistication to develop such highly engineered efforts.

One simple testimony to their abilities is many of these sites continue to accurately mark the *Seasonal Ingress* to this day.

FIG. 65

The passageway
Photo credit BoyneValleyTours.com

No one knows when the first *staff* was used to cast a shadow to observe the *Sun's* passage through the Sky, but these *Irish* sites use both *Sunlight and Shadow* to determine the *Solstices and Equinoxes*.

It's possible these chambers may even contain some of the *oldest known writing* in the world.[26]

Center and Circle

Fig. 66

The Ring of Power—The Ring towards the Loch of Harray

Photo credit David Barlow

Three Rings for the Elven-kings
under the sky,
Seven for the Dwarf-lords
in their halls of stone,
Nine for Mortal Men
doomed to die,
One for the Dark Lord
on his dark throne,
In the land of Mordor
where the shadows lie.
One Ring to rule them all,
One Ring to find them,
One Ring to bring them all,
and in the darkness bind them,
In the land of Mordor
where the shadows lie.
— J. R. R. Tolkien

Most folks today are familiar with Tolkien's *Lord of the Rings* trilogy. The author dipped his brush deep into the wellspring of both tradition and legend to write this series.

One of his central motifs was the **Ring of Power**. According to the story, it was from this ring that authority over all the others derived. The essential roots of this belief can be found in *Irish tradition*.

In the manuscript **Senchus Mor**,[27] the Irish divided the firmament above Earth into seven layers or divisions of Moon, Mercury, Venus, the Sun, Mars, Jupiter and Saturn. About the stars they believed, '*as a shell is about an egg, the firmament is around the earth*'.

Should we be surprised?

The emblem of the *Sun* has been called a *Circle* or *Ring*. The modern Gaelic term for dawn, *fainne an lae*, has been translated as '*the ring of light on the skyline at day break*'.

One power of the Sun-God was the ability to heal disease, and here we can understand some of the importance attached in ancient times to the wearing of *Rings* or other *Solar emblems*, which were primarily amulets and only secondarily ornaments. Sometimes the Ring became a miniature wheel, suggesting both the shape and path of the Sun (Fig. 67). Hence we find the word *roth*, "wheel", applied to a kind of circular brooch. The *Sun* became a metaphor for the *Great Celestial Wheel*; compare *roth greine*, the Sun, like Lucretius' *solis rota* (Sunwheel). We find God referred to not only as *ard-Ri grene*, 'supreme King of the Sun', but also as *ard-Ruiri ind roith*, 'supreme King of the wheel' which means the same thing.'[28]

One Ring... '*...and in the darkness, bind them.*'

And while the *Sun* rests, the Moon and stars continue to ride the celestial *Ring* at night.

Bind them, indeed.

Fig. 67

The Earth's Golden Ring
Photo credit NASA

Fig. 68

Counting Celtic cadence
Photo credit ColinsCamera

Athena's Web   31

Center and Circle

Sunlight and shadow

The image below (Fig. 71) is found on the top of a rock illuminated by a beam of light at *Knowth* on the *Equinox*.

On this day of the year, the shadow cast by a pole or stone at Sunrise falls due West.

As the *Sun* continues to rise it reaches high noon, when it marks due South. This is its greatest height and illumination. The shadow now falls to the North. The journey continues as the Sun glides to the West. Only on the *Vernal Equinox* does the Sun set due West, casting its shadow 'backwards' to the East.

On other days of the year, the shadow falls above or below this arc. The image of the cross within the circle is the basic concept behind both the *Celtic Cross* and the ground plan of these megalithic mounds. The quarters of time, Spring, Summer, Fall and Winter were sundered again in a division of eight, used by both *Celts* and *Etruscans*.

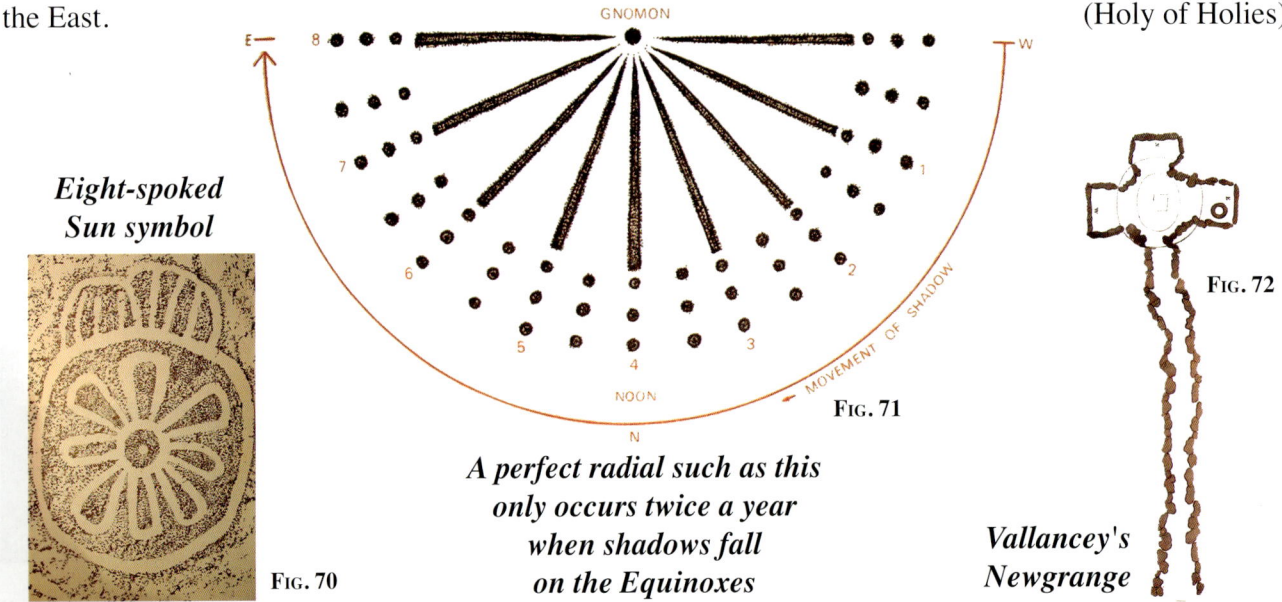

There is a picture (Fig. 70) of the carving illuminated on the Solstice. The image below (Fig. 72) is part of a drawing by Vallancey showing the tunnel leading into the Newgrange *sanctum sanctorum* (Holy of Holies).

FIG. 69

Photo credits this page
Inner Traditions

Eight-spoked Sun symbol

FIG. 70

FIG. 71

A perfect radial such as this only occurs twice a year when shadows fall on the Equinoxes

FIG. 72

Vallancey's Newgrange

32

Athena's Web

Center and Circle

It has been said that there is no human instrument more ancient or interesting than the *sundial*. It is literally a visible map of time. The original *sundials* were simply *standing sticks* or *stones*, set up to cast *a shadow in the Sunlight* and determine, by the angle and position of their shadow, the time of day or season of the year.

A **gnomon** is the upright on a *sundial* which projects a shadow onto its (usually) flat, circular face. When oriented to the Heavens correctly, it determines the time of day. If a *Maypole* is used as a *shadow stick*, then *the shaft of the Maypole* becomes the *gnomon*.[29]

The root of the term, '*gnomon*', is derived from the Greek, *gignskein*, "*to know*." A *gnomon* can then be translated as '*one who knows*.' It is from this word that the term '*gnome*' is derived; those fanciful creatures which can be found cast in clay and hiding out in English gardens everywhere. According to the dictionary, a *gnome* is, "*One of a fabled race of dwarflike creatures who live underground and guard treasure hoards*."[30]

It would seem they still do.

When *Jack Roberts* and *Martin Brennan* set out in 1980 to verify that most of the mounds in the *Boyne Valley* complex were aligned to the *rising or setting positions of the Sun* at critical times of the year, they were delighted to discover that beams of light crafted by forgotten builders cascaded into an inner chamber and illuminated specific images carved in stone, as if *"spelling out messages in an archaic code."*

Investigation and analysis of one of the greatest collections of megalithic art in the world revealed a dogged preoccupation with *Solar* and *Lunar symbolism*.

FIG. 73

Equinox light at Loughcrew
Photo credit Sean Rowe

As Mr. Brennan suggests, if one were to enter a church, one would expect to find religious themes and artwork depicted there. In similar fashion, if one enters a *site devoted to astronomical observation*, one might not be surprised to find artwork depicting what is in the Sky.

The simplicity of this analysis is further confirmed on the *Solstices* and *Equinoxes* when *carved Solar symbols*, the same used by astronomers today, are illuminated by carefully shaped beams of Sunlight, entering deep into artificially constructed tunnels, falling on stone slabs situatied at the innermost recesses.

Designed by megalithic man, the light of their wisdom is yearly *illuminated by the rays of the Sun* in a continuing annual ritual.

But these astronomical confirmations are not limited to inside these narrow chambers. Outside *Knowth* and *Newgrange* standing stones cast shadows on carved curbstones (called 'kerbstones' by the Irish), calibrated with lines to mark the shadow on corresponding dates, just as the light penetrates the inner recesses on *Solstice*, *Quarter-day*, or *Equinox*.

Time is quartered by *Equinox* and *Solstice*. This quartering of the year was split again to yield *an annual division of eight*. While the *Equinoxes* and *Solstices* mark the balance and extremes of the Sun's journey, the *Cross-quarter Days* provide a secondary cadence by announcing additional times for planting, harvesting and migrations. But *the cycles of the Moon* were also observed, and an annual weave of *Twelve* was interwoven with a theme of *Eight*. What goes around, comes around, in other words, the ultimate Celtic knot, TIME itself.

Photo credit Shira

FIG. 74

The reconstructed artificial mound (Nature didn't do it) at Newgrange

Athena's Web

Center and Circle

The combination of shadows on the outside, inner light penetrating darkened man-made tunnels, and *Solar* and *Lunar symbolism* illuminated by slender shafts of light make a convincing argument for primitive, and yet keenly sophisticated, astronomical observation.

Newgrange has only slowly yielded her secrets. It began in the summer of 1699 as farm laborers were pulling stones from a mound which were to be used for roads and fences. When they uncovered a large stone with engraved spirals, work stopped, word went out, and after some investigation, *Newgrange* had been discovered.

But it was not until *Gen. Charles Vallancey*, a British Army officer of the Royal Engineers came to Ireland in 1750 that some of today's modern theories concerning the astronomical possibilities of the site were given any kind of serious consideration.

Vallancey, a professional surveyor, was accomplished in cartography, engineering and physical geography. He called many of the megalithic sites, *Antra Mithrae* or '*Caves of the Sun*' and considered *Stonehenge* to be a *Solar Temple*.[31] He anticipated *Newgrange's* orientation with reference to the rays of the *Sun*. He was among the first to point out the importance of the *Cross-quarter Days*.

"The names of some of the ancient festivals are handed down to us by the mouths of the common people... Thus Cormac... says the four great fires of the Druids were in the beginning of February, May, August and November...

"The fires of the Druids lighted on the Neomenia (the New Moon) of the four quarter month's, and 'high mountains were assigned for their astronomical observations... Their festivals were generally governed by the motion of the heavenly bodies, was it not necessary that the people should be warned of their approach?'"[32]

Photo credit BoyneValleyTours.com   Fig. 75

Newgrange Gnome: "Forget somethin'..?"

It is under this same lunar phase (*New Moon*) that today's observational astronomers prefer to venture out with their telescopes, to best see Heaven's stars, unobstructed by Lunar glare. It is during the *New Moon* that the deepest skies can be seen and the greatest detail revealed.

A researcher working on these sites early in the 19th century wrote,

"Such artificial high places are generally situated in an eminence, frequently upon the tops of hills and mountains; and these stations were so chosen as to form a chain of connexion (sic) with each other in such a manner, that on the festival days, the 1st of May and the 1st of November most especially, the fires of Bel were seen from one to the other over the whole country.'"[33]

This was a series of holidays that embraced both the people and the land, keeping it in tune with the time. It's ferver, excitement, tradition, fellowship and competition all combined in a festive gathering eagerly anticipated.

Regarding stone circles she related, "it is thought that they expressed periods of time or astronomical epochs."[34]

In the last decades of the 20th century we have begun to realize that *many megalithic structures* built in the late stone age in *Ireland* and *northern Europe predate* other major sites found in the Mediterranean.

While some fault *Vallancey* for his belief the *Celts* and their priests, the *Druids* were authors of these sites, this was the belief of his day. We know these sites predated the Celts, but they probably inherited some of the older tradition, combined with some working knowledge of their capabilities.

* Birds, Eggs and Twins-----------------	* Nabta Playa ---	* Bovine -----	--- worship --* Knowth, Newgrange --
6000 BC	5000 BC	4000 BC	3000 BC

Photo credit Luxo

FIG. 76

A nautical world, centered on a highway running over the "Middle of the Earth" (Mediterranean)

In his work, *Druidism Revived Vallancey* claimed, "...the highest degree of Druidic order studied astronomy and divined by the aspect of the sun, moon and stars." [35]

To fault *Vallancey*, one would also have to fault classical writers who made similar claims.

FIG. 77

Eire off England- NASA

Was the complex outside Dublin built merely focused on a solitary annual ritual, marking but one day of the year? Or was it part of a system used to track the calendar, Sun, Moon and stars? Brennan's research makes sense of this ancient question.

Navigation's role in this growing stellar awareness should not be underestimated. Myth tells us the *Phoenicians of Tyre* settled *Crete* and *Thebes*, siring both *Minoan* and *Greek heroes*. And then, of course, there's *Carthage* and colonies in *Sicily*, *Corsica*, *Sardinia* together with *Spain*. The mechanism for diffusion is established.

Our earliest European tale speaks of the glories of a master mariner *Odysseus*, caught in the maelstrom of *Poseidon's wrath* for ten long years. The sea's briny waters are transformed into his prison's unbreakable bars. Ten long years is the key that will set him free.

He is a king of his kingdom as he captains his troops on the broad sea's back, with his men pulling together as one under sunny skies. But even as the tables turn, *Odysseus* remains the self-made man. He hews from the living tree a raft strong and true, *Master Mariner* still; just one not strong enough to overcome Poseidon's focused fury.

Homer sets the stage by painting the textures and hues of an island world together with the peoples who walked their shores. He sings the ageless song of hero and god, locked forever in their eternal watery embrace of *Man and Sea*.

FIG. 78

The Master Mariner Myth—overcoming the seduction of the Sea

| * Stonehenge, Phase One | * 1600 Stonehenge, final | Caesar visits Britain- 54 BC* |
| 3000 BC | 2000 BC | 1000 BC | 0 BC |

Center and Circle

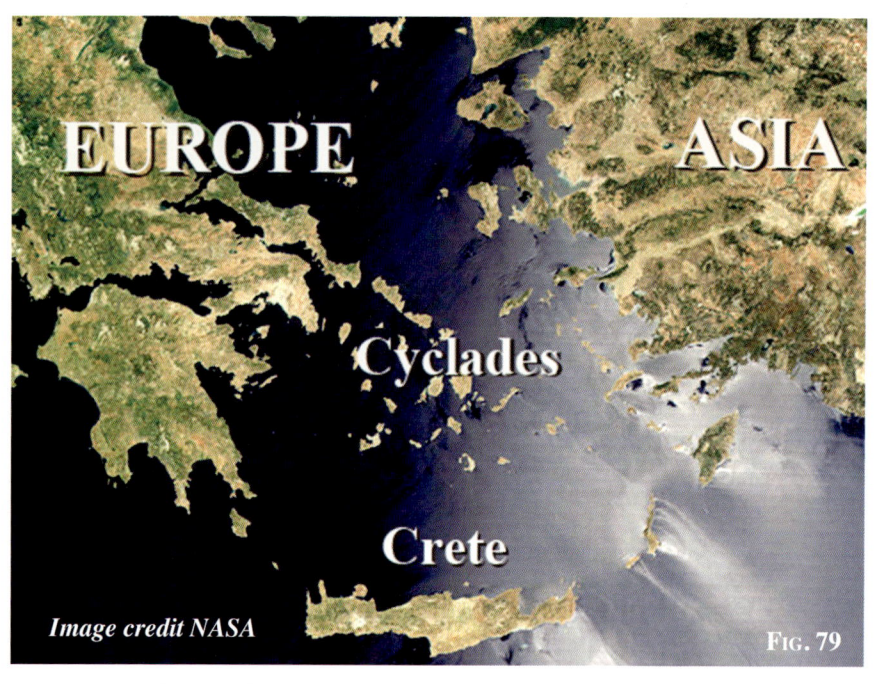

The Cyclades: Bridge between Asia and Europe

How long had this maritime folk tradition been evolving before it reached this pinnacle of prominence, penned by the first author of the western literary tradition?

Long before the era of Odysseus, as migrations spread to the Aegean around 7,000 BC, Europe's first farmers and herders appeared in Macedonia, Thessaly, Crete and the Peloponnese.

A thriving trade network was established around the central island group, the Cyclades, controlled by a rich and powerful elite. Delos, the sacred island of the old gods, stood at its spiritual and regional center. The oral mythic traditions remember that Apollo and Artemis (the Sun and Moon) were both born here. [36]

Time is calibrated in the ancient world by both Stonehenge and the Pyramids; but in the islands there are temples older than either of these. Radiocarbon dates for Newgrange (3700-3200 BC) have been mentioned. Malta's megalithic temples were used from 3600-2500 BC. Each site demonstrates considerable astronomical awareness, consistently tracking both solstice and equinox.

Located in the middle of the Mediterranean, just south of the Sicilian coast is Malta, a small island which serves as a stepping stone from Africa to Europe. The peoples who built these megalithic Maltese temples did so in two phases, during the Ggantija (3600-3200 BC) and Tarxien (3150-2500 BC). [37]

There is no memory of the people who built the temples found on Malta through these centuries. Although no written records indicate their purpose, archaeologists have inferred from ceremonial objects that they may have been used for sacrifices. What is evident from our perspective is that they were made for astronomical reasons, to help determine the nature of time by having a fixed framework from which to measure the motions of heaven. Islands provide perfect opportunities for visibility, with level horizons in the east, west, north and south.

At **Mnajdra** (Fig. 85), whose oldest temples date from the earliest construction on the island, one temple was aligned so that on the equinoxes, Sunlight passes through the main doorway and lights up the major axis. Meanwhile, on the solstices, Sunlight illuminates the edges of megaliths to the left and right of this doorway.

Heads tipped back, arms folded, and doing... what?

36

Athena's Web

Center and Circle

Image credit Heritage Malta

TEMPLE ROOFING
Note how each of the horizontal courses above the upright slabs protrudes over the one beneath. These courses are the remains of the base of a corbelled roof, the type of construction that may have originally been used to roof over parts of the temple.

SAQAF TAT-TEMPJU
Il-ġebel mimdud fuq il-ġebel wieqaf f'dan l-apsidi jisporġi 'l ġewwa. Dawn il-filati huma l-fdalijiet ta' bażi ta' saqaf bis-saljaturi, it-tip ta' kostruzzjoni li aktarx saqqaf partijiet tal-maqdes.

FIG. 81
Ħaġar Qim

At Ħagar Qim in Malta, an example of the roofing

Both Newgrange and the sites on Malta:
• *were begun within a century of each other.*
• *are both islands best reached by other islands.*
• *co-existed for four centuries (3600-3200 BC).*

As both land and sea routes developed, the stars became an increasingly importance means by which one could determine the points of the compass.

The stars could answer their navigational questions:

How can one see beyond this curvature of blue?

How do you determine north, navigating by the pole star?

How do you plot east and west by the risings and settings of the stars?

The advantages of mastering this knowledge was obvious- island hopping augmented by stellar navigation. Routes could be shortened, making longer journeys to out-of-sight locations. Trade could be increased, with greater profits realized. Human cargo, whether for adventurer, traveler or slave, was safer.

When we look to the art of the *Cyclades*, we are peering into the *heart of Greece's earliest culture*. One repeating theme is a strangely futuristic design of stylized figurines with outlines of the body and face, but with heads tipped slightly back, as if looking up (Fig. 80).

Now this is early Greece, contemporary with the *Old Kingdom of Egypt*, the *Sumerians* and the *Tarxien culture of Malta*. There is avid astronomical interest in locations surrounding them at this time. Might not these early maritime Greeks be also interested in the stars?

Their arms folded, they stand in a relaxed posture, facial features absent, looking up.

At what?

Are they looking at the *Children of the Sky*? Immortals to them. Are they honoring their wisdom, whether for the calendar, navigation, or guidance? These clay figurines suggest they are observing.

Is this their religion?

Are these their Gods?

But if these early worshippers are looking to the stars, wouldn't they build temples like those in Erie or on Malta? Wouldn't they want their own source of time-keeping closer at hand?

They did and they are called a **"Drakospito"**, or *"Dragon House"*. They're located throughout the islands, composed of the shale found in abundance in this rocky terrain.

FIG. 82

Image credit Heritage Malta

An ancient clay model of a temple superimposed on the Malta site

Athena's Web   37

Center and Circle

Evia, Greece's largest island after Crete, lies north of Athens and has three *Drakospitos* with which we are familiar, with one perched high atop *Evia's* southernmost summit. This spot was perfectly positioned to observe the skies, and was part of an ancient *Signal Fire System* running along the coast, just as we had previously observed at Newgrange.

How far back these *Drakospitos* extend, no one knows.

Dry stone can't be dated.

FIG. 83

The roof of one Drakospito

FIG. 84

The Drakospitos roof construction matches the architecture of the earliest Maltese temples, with solid stone corbeled ceilings working their way in towards the apex.

We see culture in the form of architecture, agriculture, and mythological information being passed along these navigational routes.

Another repeating artistic theme found throughout these islands is the *spiral*, perhaps derived in part from the *Spiraling Serpent* of European and African origins. Is this a symbol for Heaven's spiral, of time moving on its daily, monthly or yearly axis, and of its various layers of implications?

This *Spiral* winds its way in a watery weave through the islands of Greece, Malta, and Eire, all locations best reached by way of other islands.

Nabta Playa's construction and use dated from 5500 to 4800 BC, Newgrange from 3700 to 3200 BC, the temples at Malta date from 3600 to 2500 BC.

The *World Heritage Site* sign in Malta (Fig. 85) highlights the *Equinox Sunlight* traveling through the entryway into the central chamber located at the back of the site, fully filling and illuminating it. The thinner, crossing lines show the maxima and minima of the Sun in its annual pilgrimage. The smallest of these three 'temples', located on the lower right, dates back to 3600 BC.

Are these symbolic shorthand clues of an early maritime culture? The effort to construct temples here, as in Eire, would require considerable social intent and administrative coordination on the part of the builders. It would seem that projects requiring such planning, preparation and political motovation would require a high level of priority to warrant such efforts on the part of their leaders.

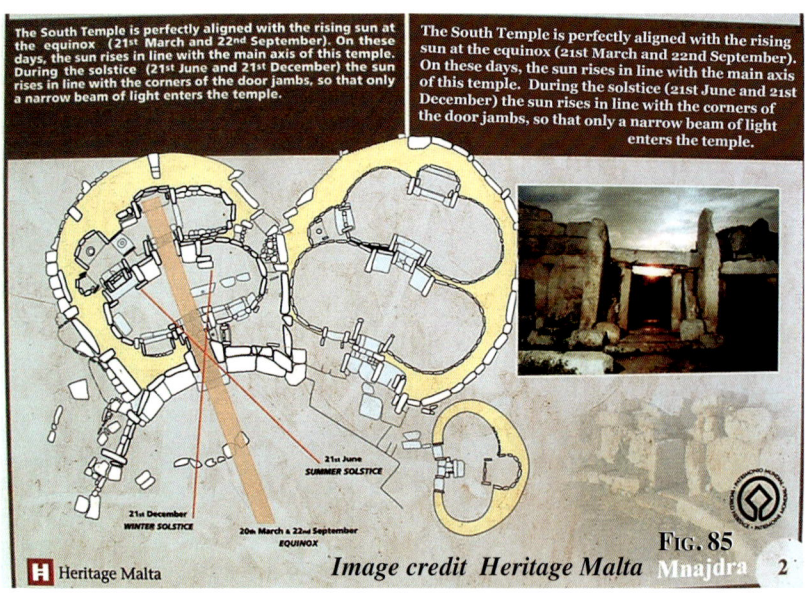

FIG. 85

Image credit Heritage Malta

Sunlight on the Equinox illuminates the back wall of the temple

Athena's Web

Center and Circle

In the valleys of the Nile, Tigris/Euphrates and Indus rivers, a keen interest was being developed in watching the heavens, of coming up with new ways to tell time, to mark the calendar for agrarian interests. As farming spread through Europe from the Middle East, so did the star lore tied to it. In *Homer* we find star lore crafted in maritime garb. In *Hesiod*, it plays out through farmer's fields in his **'Works and Days.'**

*"When the **Pleiades**, daughters of Atlas, are **rising**, begin your harvest, and your ploughing when they are going to set. Forty nights and days they are hidden and appear again as the year moves round, when first you sharpen your sickle. This is the law of the plains, and of those who live near the sea, and who inhabit the country..."* [38]

It is the law everywhere... The stars are being observed, and it is what we see in the *Cyclades*.

The *Art of the Cyclades* is contemporary with *Egypt's Old Kingdom* and the *Sumerians*. As noted, many of the figurines stand in symbolic silence, their heads tipped upward, as though watching or praying to, the stars.

But if the first settlers arrived in Greece in 7000 BC, how far back might one expect this maritime culture to go. What is the edge of the maritime world?

Where do you fall off?

Say hello to the *Maritime Archaic* and the *Red Paint People*.

The *Maritime Archaic* period lasts from approximately 7,000 BC to modern times, since it describes the lifestyles and communities of sea-mammal hunters in the subarctic.

FIG. 86

Another Drakospito on Evia

The **Red Paint People** of Maine were a subset of the **Maritime Archaic**, who thrived from about 5500 BC to 2000 BC. Their ability to fish is remarkably impressive. Remains of swordfish and other deep sea fish, as well as plummets, gouges, slate lance points and toggling harpoons confirm that these were seafarers of considerable skill. We know they lived in longhouses, and there's evidence of long-distance trade.

They left behind images of whales and stylized bird heads on combs and pendants. Their burials involved the use of *red ochre*, and their cemeteries were inevitably placed on high hills overlooking the sea.

But the *Red Paint People* may have been the last of an even older tradition.

From *Atlantis in America-Navigators of the Ancient World* by Ivar Zapp and George Ericson:

Photo credit MatthiasKabel

FIG. 87

Nautical legends remembered across the sands of Time

Athena's Web

Center and Circle

"James Tuck and Robert McGee of St. John's Memorial University uncovered a rectangular stone chamber of upright stones on the coast of Labrador that closely resembled similar stones found on the island of Teviec just off the coast of France. Both were burial sites where the dead were covered with red ochre, and dating of charcoal pieces from ceremonial burnings at these sites have been carbon dated as being 7,500 years old.

"The graves, like pyramids built in Mesoamerica, were oriented to reflect light at the time of the rising sun on one day only, at the time of the summer solstice. And instead of a red ochre burial we find at these sites an urn containing cremated ashes, obviously a special treatment for an unusual person, a shaman or tribal chief."

"The use of ground slate, a material inferior only to metal, in harpoons and bayonets in both northern Scandinavia and the northern shores of the Americas may not by itself reveal a shared maritime culture 7,500 years ago. But the use of red ochre, the similarity of designs and engravings, the use of bamboo in tools, and a similar use of oil lamps, all point to a shared culture across the North Atlantic." [39]

Photo credit BoyneValleyTours.com

Newgrange spirals above

Equinox Entrance: Waiting for the Sun

Our maritime tales, like the myths we chase, are elastic. The *Spiral* becomes a common thread, weaving together sites designed to help determine the mysteries of Heaven. But we also find potential for *trans-Atlantic communication*; a network bridging the cold North Atlantic, and for the sharing of information and trade. These are indigenous peoples who braved the *Atlantic Ocean*.

But this watery web is ethereal; difficult to prove and short on hard evidence. Even the environment they lived in, where sea meets shore, is one of the Earth's most abrasive. Not only do successive storms accelerate the deterioration of earlier habitations, but Man contributes to the erosion, as people move in on top of older sites, building upon the rubble of times gone by.

We are getting ready to resume our mythological trail with the approach of the *Heavenly Bull*, a time well honored in the rituals and traditions of the day.

What's not hard to grasp is the travel-aid the stars provided as man wrestled with *nautical navigation*. The Mediterranean was the autobahn of the ancient world. Coastal cities cultivated this lifeline for both support and security.

And the spirals of Malta below

Taurus

Are these stellar reflections? Do the dots represent Orion's Belt and the Pleiades?
—Lascaux Caves 17300 years ago

Center and Circle   *42*

Images this section courtesy of Starry Night Education

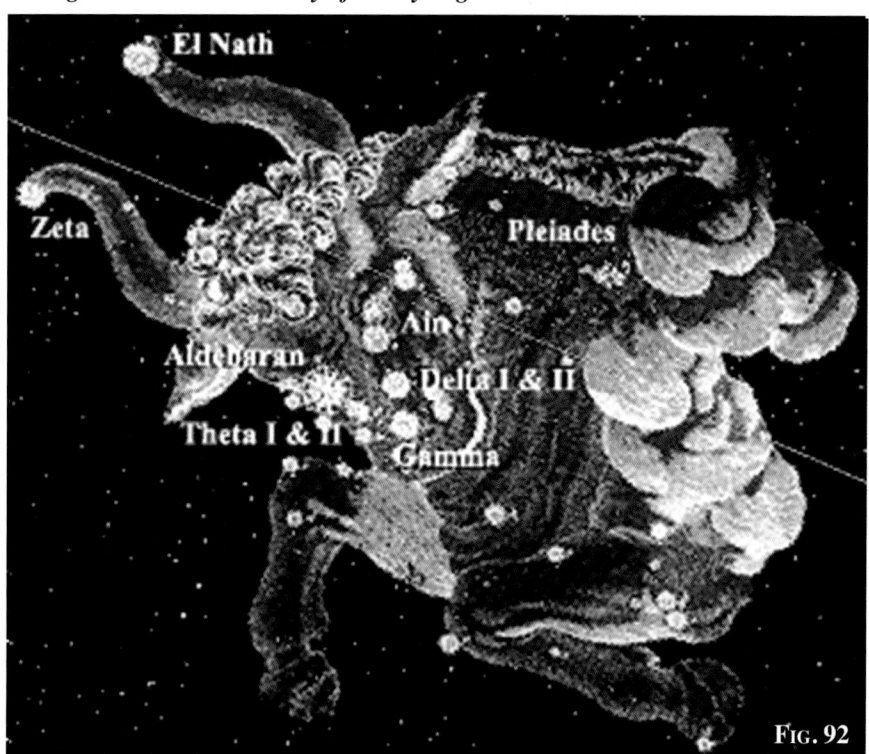

We pick up our mythological tale once again, having left behind the *Time of the Twins*.

In a quick recap, *Creation* was reflected in the start of the *New Year*, usually designated as *Spring*. *Spring* is when life is reborn, the world made new. In *Spring*, the flowers and trees bloom.

The pagan population believed that we live inside the body of *One*, great, *All-Encompassing Being*; but it's divine essence kept changing form with the changing face of the *Sky*. With this in mind, hear *Hesiod's* invocation to Zeus, the *Sky God*. If we translate the word '*Sky*' for '*Zeus*' each time it is invoked perhaps we might better grasp how it was perceived. [40]

"*Muses of Pieria who give glory through song, come hither, tell of Sky your Father and chant His Praise. Through Him mortal men are famed or unfamed, sung or unsung alike, as great Sky Wills. For easily He makes strong, and easily He brings the strong man low; easily He humbles the proud and raises the obscure, and easily He straightens the crooked and blasts the proud, Sky, who Thunders aloft and has His dwelling most high.*"

Makes a little more sense, don't you think? In our story line, the *Sky* is about to put on the mask of a *Bull*; having just discarded that of the *Twins*, an earlier time when *Two* had been *Numero Uno*.

Our oldest Greek myths speak of elemental powers ruling the Earth, when Titans married their *Siblings*. This tradition was passed to the Gods, as Zeus married *His Sister*. The pharaohs did it. For each it was *Heaven's Will*; but *Heaven's reflection* from an earlier time.

FIG. 92

Toro Time

Those who monitored the stars saw *Spring* moving from these dual themes towards the *horns of the Bull*. For almost 400 years, the *Vernal Equinox* advanced from the 'feet' of Gemini and approached the *horns of the Celestial Bull*. Anytime after 4300 BC the *Vernal Equinox* could arguably be said to be within the design of the *Celestial Bovine* (see Fig. 92).

Although the archaeological evidence leads us to the banks of the Tigris and Euphrates, classical literature takes us in another direction; to the banks of the Nile.

The story of Egypt stretches across several distinct periods. First, there is the prehistory, the time before Egypt's unification around 3200 BC, then there follow the *Old*, *Middle* and *New Kingdoms of Egypt* respectively.

Various transitions take place through the final millennium, with *the Ptolemys of Cleopatra's time* ruling in Egypt's twilight. Thereafter Egypt is absorbed by Rome.

Both the *Old Kingdom of Egypt* on the Nile and the *Sumerians* within the Tigris/Euphrates river valleys bore witness to and leave evidence of this *Bovine Spring*. The *Bull* or *Cow* (the important distinction here is not gender specific) had obvious *agricultural associations*, but this was bumped to a new level when, circa 4300 BC, the *yoke* and *plow* were devised and a new epoch in *farm machinery* literally transformed the landscape. This simple *cultural evolution* was to change *the nature of food production* forever.

Invention of the *yoke* meant now the *Ox* could plough the fields.

4800 * BC- Nabta Playa	Yoke/first Solar calendar * 4236 BC	...agricultural development...
5000 BC	4500 BC	4000 BC	3500 BC

Cultivation was no longer limited to what might be turned by hand with a hoe, as had been the case then. As a result the *grain supply sharply increased*. This *agrarian bureaucracy* applied their talents to other projects, such as work on the construction of the pyramids and temples.

The Apis Bull had to be black...

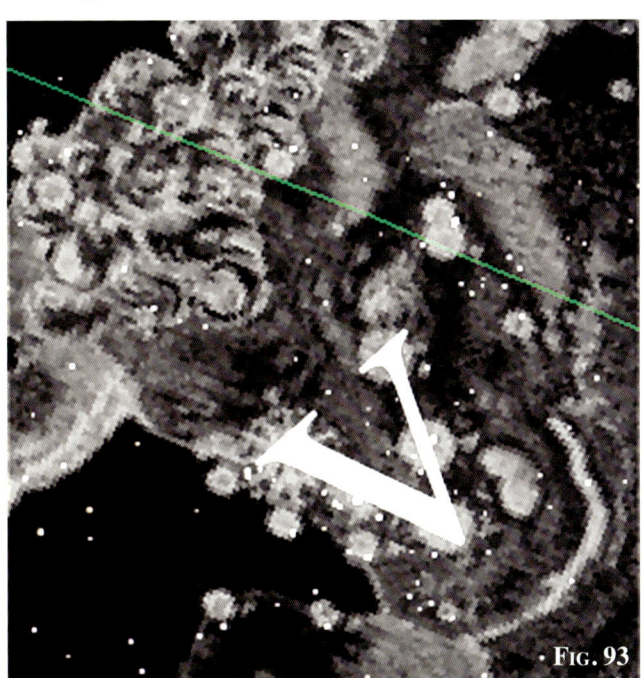

When he died, there was a huge search for a new *Apis Bull* [41] with special markings. These markings held great significance for the priests who searched to find this *new Bull*, and it is here that we find our first Heavenly hints of a *Bovine Birth*.

First of all, he had to be *black*. This may represent the night sky. In addition, he had to have a *white, inverted triangle* on his forehead. This could be a reflection of the *Hyades*, a distinctive stellar "*V*" in Heaven marking the *Head of the Celestial Bull* (Figs. 93 & 94).

On his flanks, there was to be found the mark of a *Crescent Moon*. The oldest astronomical texts say the *Moon has dignity in Taurus*, it is in its *Exaltation*, at that time marking the *New Moon* of *Spring*.

Both the *Bull* and *Moon* are symbols linked to agriculture. This is a *celebration of Life*, *Nature*, *agriculture* and the joy of a *New Year*, a *new Spring*; together with all its animated expectations.

The *Apis* celebration was searching for *a reflection of Heaven on Earth*. To think in terms of the earlier mythology, it was looking to find its '*Twin*' on Earth. It was important the *Apis* markings reflect what they now saw in the Sky. Once it had been *divine Twins*; now it was *Holy Cow*.

We live in a time when science is slowly emerging as the new 'religion'. It has not yet fully integrated its spiritual side into its discipline, but it will. Here we reflect on a time when art, and not calibrated accuracy, was what was striven for. We can convey *Truth* through either one, but each comes with a different set of rules for its application.

FIG. 93

...with an inverted triangle on his forehead...

Like North America sesquestered by oceans, Egypt is uniquely situated, protected on both eastern and western borders by natural impediments. This pocket of civilization, protected by arid desert, was allowed the time it needed to cultivate the crafts of a more advanced culture.

Around 3150 BC, Egypt unified. Memphis became its capitol. The **pre-dynastic Bull God** known as *Apis* emerged as an important deitiy during *Egypt's Old Kingdom*, just as *Spring* transitioned from '*between the horns*' to the '*head*' of the *Bull*.

There are many things we know about the *Apis*. There could only be one in the land, and this one was greatly honored.

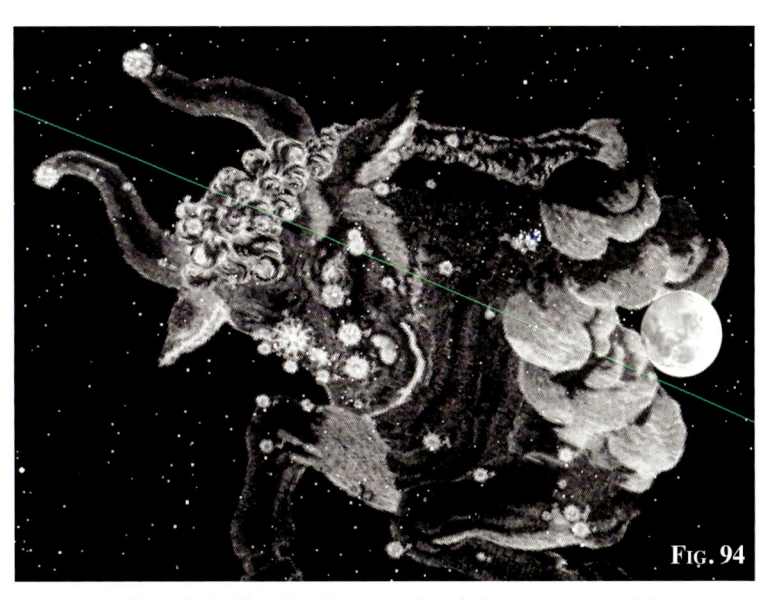

FIG. 94

...and on his flanks the mark of the crescent Moon

Unification * 3200 BC	Pyramid of Cheops 2560 BC *	...demise of the * Old Kingdom
3500 BC	3000 BC	2500 BC	2000 BC

Center and Circle

Fig. 95

Apis Bull
Photo credit Heidemarie Niemann

Any of these celestial figures may be rendered, not with astronomical precision, but artistic insight. This method of encryption does not deny its *'Truth'* because it is not scientific; it merely takes the image down a different path, illustrating alternate perspectives.

In myth, as in dreams, the important images stand out. The symbolism of the story-line may lend emphasis, or one might rely on size. For instance, Egyptians magnified their pharaohs in art, making them larger, stronger, and more powerful than all others. The gods in turn would be larger than the pharoph (Fig. 96, 101).

Similarly, on *Lakota leather star maps*, important groupings are drawn larger or more brightly than surrounding stars.

The symbolic clues found on the *Earthly Apis Bull* must reflect those of his *Celestial Bovine 'Twin'*. They sought Heaven's blessings. If Heaven approved their choice, prosperity and abundance filled the land. If the Nile floods proved either too great or too little, Heaven (the Gods) was (were) not happy.

Research at *Nabta Playa* suggests observation of the *Winter Solstice* back to at least 4800 BC, probably earlier. From 4300 to 2000 BC, the constellation of the *Bull* marked *the Vernal Point*, with its promise of growth and fertility.

Symbolically, celestially and practically, this essential image became the backbone and cornerstone of Egyptian civilization.

The management of the land and river, cultivating its fertility, spawned a mindful calibration of *Heaven's cycles* in order to better understand *Nature's rhythms*. This is what the Egyptians most desired, to be in tune with *Nature*.

Fig. 96

The power behind the throne
Photo credit Néfermaât

Photo credit Tour Egypt

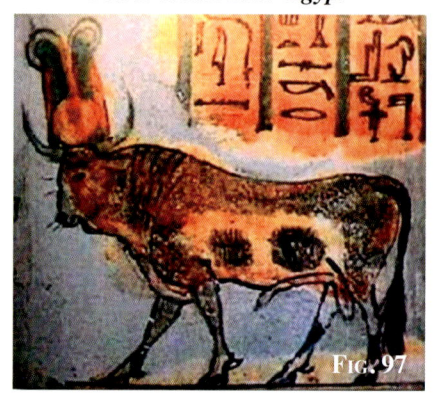

Fig. 97

The mark of fertility

Their temples were built to monitor what they earnestly believed to be the *Will of God*- the *Sky* itself. The purpose of all life was to reflect God's truth. They sought to understand the archetypal fingerprint and be in tune with *Destiny*.

The Sun's passage into *Spring* was synonymous with life, fertility, and the bounty of the Earth. As a culture, the Egyptians turned their *considerable administrative abilities* to the

Fig. 98

Pre-dynastic woman

preservation of their *agrarian cornucopia*, both for this life and the next. They say you can't take it with you, but the Egyptians sure tried, packing their funeral sites with all the *hedonistic luxuries* of life; food, beer, livestock, tools, material comforts and grooming items such as combs and mirrors.

When astrologers label *Taurus* an *'Earth Sign'*, they are talking about all the many, wonderful material creations that enhance life and make it enjoyable.

44

Athena's Web

Center and Circle

For a thousand years...

Images courtesy of Starry Night Education

They desired *Heaven's fertility*, so they touched Him. Rites were used to guarantee the ability to bear children, just as Egypt used her vitality to bear crops. This ritual was performed while the *Apis Bull* was on his way to Memphis to visit Hapy, the God associated with the annual flooding of the Nile and therefore Egypt's life-giving inundation. [43]

It was obvious to the Egyptians that *Apis* and *Hapy*, '*Heaven's Abundance*' and the '*Fertility of the River*', must be combined.

After the Apis visited with Hapy, He traveled by ferry to Memphis, where He took up residence in a temple with special quarters on the *south side of the building*. This is the direction one looks to see the stars at their culmination. It was here that the *Earthly Apis* would spend the rest of his life, watched over by the '*Bull of Heaven*'. [44]

All over the land there'd be rejoicing when a new *Apis Bull* was found. In one great national procession while the Apis Bull was led to His new home, people would line His path with questions. A positive response would be placed in a vessel on one side, with the negative response positioned on the other. [42]

The side of the path the *Apis* favored as he walked reflected *Heaven's Will*. This divine representative, this celestial mediator, could not only ascertain the *Will of Heaven*,

...He *was* the *Will of Heaven*.

Only women were allowed to approach *Apis* during this procession, which they did, naked (Fig. 98).

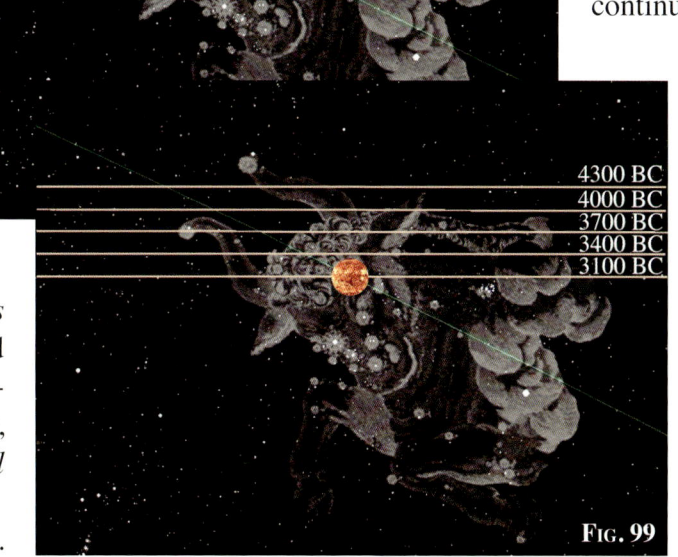

The *Apis Bull*'s most important role was fertility. On the pharaoh's accession, the two of them would circumnavigate the boundaries of the state, ritually and spiritually insuring the land's bounty. Cows were brought to the Apis to mate reflecting and insuring *Egypt's* continued *fertility*.

The *Apis Bull* was a stud, plain and simple. Be fruitful and multiply.

But if the *Bull* was seen as the *leader of Heaven*, then it would only be natural that the *Pharaoh*, as leader of his people be identified with him.

And he was.

Fig. 99

...Spring passed through the horns of the Bull.

Athena's Web

Center and Circle

The *power of Heaven* worked through *Heaven's appointed One*. The *Pharaoh*, like the *Apis Bull*, was *Heaven's chosen on Earth*.

This theme is repeated in various depictions with *Pharaoh as Bull*, lightly tossing his enemies aside with *His Horns*; *Pharaoh* standing before *Cow*, or embraced by the *Goddess* with the *Solar Disk* between *Her Horns* as *Hathor*.

For years Egyptian culture paid homage to the *Bovine* while using stylistic variations in an effort to distinguish it from other epochs. They had to stretch this *Bull* image over two thousand years as the precessed Vernal Point slowly made its way through the brightest stars of the *Celestial Farm*.

This image of *Solar Disk* between the *Horns* pervades Egyptian culture. Its astronomical origins lie in a mythic reality seen in the Sky for a thousand years and included the period of their unification (3200-3150 BC). Astronomically it was visible to the naked eye from about 4300 to 3100 BC.

If Egypt, the *Pharaoh*, and the future are all symbolically represented by the *Bull*, then you give this *Celestial Ambassador* your best, and they did; perfuming him and feeding him special foods daily.

All these *Bovine* images had a single, collective interpretation: *agriculture*, the ultimate source of power behind the throne (Fig. 96). *Egypt would* remain an *agricultural powerhouse* into Roman times.

*Egypt's fertili*ty was the end goal. Its cycle was measured by the stars, its volume by the river.

Photo credit Nathan Petersen

FIG**. 100**

'Father Nile' with his sixteen children

Egyptians were eminently practical. They knew the Nile was not without its fluctuations, so they set out to meter them each year.

"If the Nile rose no more than one meter above its normal level, the consequence was drought and famine. If it rose higher than usual, however, dams burst and fields were ruined. People therefore prayed for just the right amount of water, which meant sixteen cubits of water. This, incidentally, is the significance of the sixteen infants on the allegorical marble of the River Nile (Fig. 100) *now in the Vatican Museum."* [45]

The civilizations of both the Nile and Tigris/Euphrates river valleys were cultures precariously poised on deserts. The *angst* which this harsh environment evoked was driven by extremes between hot and cold, fire and water, feast and famine or life and death. A civilization that had to wrestle with drought, starvation, and plague on a regular basis attends to its needs, focusing all its skills on survival. *To unlock the secrets of Mother Nature was their quest*, by earnestly endeavoring to unravel the knots of celestial motion in seeking greater precision to the (they believed) answers in the Skies wheeling above them. Certainly, the obvious answer to famine was abundance and fertility. Once again, it was their equivalent of being fruitful and multiplying.

During Egypt's sojourn, both *Bull and Cow* would come to represent *Heaven's Cornucopia*.

In honoring the *Solar Disk* between the *Horns*, Egyptians were remembering when their civilization began, as Romans looked back to *Romulus and Remus*, or Christians recall the birth of *Christ*. Each evokes civilizations's birth, each is steeped in a state-authored endorsement of its Creation.

Give thanks to the life you live.

What was the eye of the needle through which these threads passed; of ritual, time, calendar, mythology, astronomy, astrology and religion? The influences of *Heaven on Earth* (or so they believed) all wrapped together as one in the *Spring*. Each holiday had its focus, but *Creation* was most shaped at the cardinal ingresses, *when Heaven and Earth aligned*.

Center and Circle

Creation begun anew.

Each year, we honor the *Child in the Manger*. So the Egyptians looked back to their *Genesis*.

For over a thousand years, Spring passed through the *Horns*. This image was to leave its imprint on Egyptian culture long after the astronomical event which authored it had come and gone.

The *Apis* celebration occurred at the end of this thousand-year period, during the early dynasties of the *pharaohs of the Old Kingdom*. The yoke had been invented a thousand years earlier and, according to J. H. Breasted, the *Solar calendar* had, too. *Agricultural abilities evolved for centuries* leading up to the *unification of Upper and Lower Egypt. Farming* did not commence with *Narmer's* unifying exploits. With the unification, an earlier way of life is passing, one of which we have only the barest of glimpses. From the unification on, there is additional archaeological evidence. Clues before unification are fewer.

The Palette of Narmer

FIG. 101

Photo credit Jeff Dahl

Photo courtesy Tour Egypt

FIG. 102

A Nubian procession, bearing African gifts

The Palette of Narmer (Fig. 101), once considered the oldest writing in the world (contemporary with late Newgrange) depicts *Bulls* overlooking a scene before them, *as if watching from Heaven above*. Note that it is *two Bulls*, reminiscent of the earlier dual epoch, Gemini.

The *palette* is dated *c*. 3200 BC, and is thought to represent *Menes*, (aka Narmer) the first pharoah of a unified Egypt.

What about the pre-dynastic phase, prior to 3200 BC?

Seventeen centuries after abandoning *Nabta Playa*, the *Upper and Lower Kingdoms*, independent until this time, fuse into a single nation state under Narmer. It's been assumed Egypt's *cultural impetus* arose by conquest from the Delta and North, rather than emanating from the sub-Saharan South.

Lower Egypt, the *Delta*, is a fertile region at the mouth of the Nile that fans out before reaching the Sea. It's called 'Lower' because it's closer to sea level. It was not until the Libyan Pharaohs (945-722 BC) overthrew their Egyptian overlords that the capital was moved from Thebes back to Memphis (Cairo). This was when Egypt began to open her shores to the eastern Mediterranean world.

Upper Egypt lies well to the south of Cairo along the Nile River. This is a hilly region- so named for the highlands, not its cardinal direction.

What transpired during the nearly two thousand years since the desert reclaimed *Nabta Playa*?

William Finders Petrie is considered the *Father of Egyptian Pre-history*. It was his conviction that there was a "peaceful", if not united, rule over all Egypt and Nubia during the entire pre-dynastic period.

If we journey south, to where the White and Blue Nile meet, we may uncover some of Egypt's earlier cultural roots.

Athena's Web   47

Center and Circle

Photo credit Mark Dingemanse

FIG. 103

Long before 'Kush' it was known to Egyptians as '*Ta Seti*', '*Land of the Bow*', from the fame of its archers. Research suggests that the line of kings living in northern Nubia were at least as early as the first pharaohs of Egypt. A cemetery of large tombs contains evidence of wealth and representations of rulers and their victories. They were found in northern *Nubia* in the *Kingdom of Qustul*, earlier than the dynastic period. It was in one of these thirty-three graves that a curious mystery was found.

It was *an incense burner*, dated no later than 3400 BC, showing incised figures with three ships in procession. They are sailing toward a palace with the first ship carrying a lion, and the second a king, sitting and equipped with a long robe, flail and White Crown, which are all later symbols of Pharaonic rule. At that time there weren't supposed to have been any pharaohs or palaces. Yet this censor was found nearly 200 miles into Nubian territory.

This discovery of the Nubian *Qustul Incense Burner* is considered one of the earliest certifiable uses of incense by a culture.

Taken as a whole, these clues hint at a *Nubian cultural impetus*, sowing the seeds of what later becomes *Egyptian civilization*. Originally linked through nomadic migrations to central Africa, this suggests a black African source for what blossoms as Egyptian civilization. While Spring passed through the length of the horns of the Bull, African culture gave birth to *Qustul* (3800-3100 BC).

In *Egypt* it was *Spring* for a young *Black Bull*.

Welcome to *Nubia* (Fig. 103). With climate changes and the drying up of the waters which had sustained *Nabta Playa*, the seasonal link between the Nile Valley and deeper Africa was severed. The local population may have trekked out of the Sahara towards the Nile below the second Cataract where they settled. The domestication of animals was begun and farming made its mark. The archaeological evidence leads south, not north.

Thus **Nubia**, Egypt's southern neighbor with its own civilization, preceded ancient Egyptian civilization. Known in the Bible as the *Kingdom of Kush*, the number of pyramids in *Nubia* are 223, double the number in *Egypt*.

48   Athena's Web

The Sumerians

FIG. 104

FIG. 105

FIG. 106

If we look at Egypt's early history, taking the pre-dynastic and Old Kingdom periods together, we are looking at an era contemporary with their more easterly neighbors in the Tigris and Euphrates river valley, the **Sumerians**. Each of these cultures awoke yearly to a Bullish Spring.

The *Sumerians* are often loosely grouped together under '*Babylonian history*'.

Babylonian history may be broadly divided into *three periods*. *Babylon per se* was one of several bustling cities on the *Tigris and Euphrates Rivers*. They were in what is today Iraq and Kuwait. This was the soul of the *Fertile Crescent*, an area which spawned agricultural development and, with it, the beginnings of town and village life. This is one of the original cradles of civilization. Much of the early history of this area is lost. It wasn't until the 19th century AD that clay tablets containing a nearly complete record of the *Epic of Gilgamesh* were unearthed.

After *Hammurabi* established *Babylon* and unified the lower valley of the Tigris and Euphrates around 1700 BC, the region begins to be thought of as *Babylonia*, home of the Semites. This is only after a continuous struggle for 600 years between Sumerians and Semites. As Rome absorbed Greek culture after conquering them, so the *Semites* adopted *Sumerian culture*, eventually surpassing their teachers in art and seal-cutting (Figs. 104 & 106). It was the fusion of these two cultures which produced what we think of as the first great period of *Babylon*. However, this poses a problem. Because the *Semites* adopted *Sumerian culture*, it's difficult to tell which influences were *Babylonian*, and which originated with the *Sumerians*.

Our understanding of the beginnings of *Sumerian culture* is sketchy. As early as 3500 BC these people came into the '*Plain of Shinar*' and began to drain the marshes. We know from excavations at *Ur* that by 3000 BC they had reached a fairly high level of civilization, developing a number of independent city states. Much of their early culture was built upon baked brick from the mud of the rivers.

It was this baked mud to which *Gilgamesh*, the *King of Uruk* referred when he said, "*I have not established my name stamped on bricks as my destiny decreed.*"[46] It is also from these baked clay tablets that we first see wedge-shaped markings. Just as the *Egyptians* were developing *hieroglyphics*, so the *Sumerians* were creating their own form of *cuneiform* notation.

The *Sumerians* also bequeathed to us our '60' time count, from which come the number of seconds in an minute and the number of minutes in an hour.

They helped frame *Time*.

Photo credit Stewart Cherlin

Head tipped back, hands folded, and doing... what?

---------- Sumerians --------- Akkad -- * ---* Babylonian culture ---------- Assyrian*Chaldean Empire

3000 BC   2000 BC   1000 BC   0 BC

Center and Circle

So while the *Sumerian civilization* was not what we traditionally call Babylonian, *Sumerian culture* helped shape *Babylonian beliefs*. We're not really sure where one leaves off and the other begins. There's evidence which suggests **Gilgamesh** is a **Sumerian myth**.

After the 600-year period of intermittent warfare between the *Sumerians* and *Semites* (2350-1700 BC), the city of *Babylon* arises together with the first 'proper' *Babylonian culture*.

Today, much of what is left of these brick cultures is rubble, bearing testimony to the words,

"From dust ye came…"

In the 8th century BC there emerges from the city of *Ashur*, a culture which was to gain supremacy over not only the Tigris and Euphrates, but the entire *Fertile Crescent* region, including a brief conquest of Egypt.

This is the *Assyrian Dynasty*.

Coming from the highlands of the Tigris-Euphrates, this area possessed quarries of limestone, alabaster and other hard stones. Together with the use of iron learned from the *Hittites*, these raw materials propelled this people to overpower those around them. Many records of earlier periods, including the *Epic of Gilgamesh*, come from this, the *third stage of Babylonian development*.

The destruction of *Nineveh* in 612 BC marked the end of Assyrian supremacy, but the *Chaldean Empire* followed, and again made its capital at *Babylon*. From this epoch comes *Nebuchadnezzar II*, the period of *Hebrew* captivity and the *Chaldeans*, a term synonymous with *astrologers*.

Like the Semities and Sumerians, *Nebuchadnezzar* copied much from the *Assyrians*.

It was during this period that **Babylon** reached its greatest glory as a city with their **hanging gardens** listed as one of the *Seven Wonders of the Ancient World*. In this epic the greatest God was *Marduk*, and his temple in Babylon was one of the main city-centers.

What is so important about this *final chapter* of *Assyrian* and *Chaldean* influence is that *"again and again the annals of the Assyrian monarchs confirm, elucidate, or supplement the Hebrew chronicles of Judah and Israel, while the creation and flood stories of the Babylonians as well as the Code of Hammurabi abound in striking parallels to the corresponding portions of the Old Testament"*.[47]

—*The Babylonian Genesis*, by Alexander Heidel.

When we study the myths of *Babylon*, there remains some question as to which period the original source material belongs.

This **'Fertile Crescent'** of the *Middle East* has given rise to many civilizations, from *Sumerian* to *Sadam*. They were all interested in the stars. With their low-moisture and high-visibility *Skies*, the *Heavens* would have appeared brighter and closer to those looking up at the night sky.

Abraham was born here, and later will fulfill his destiny by journeying with his family to a land of milk and honey.

Stories about the stars permeate the *Fertile Crescent*, and as we shall see, the children of Abraham were no strangers to their tales.

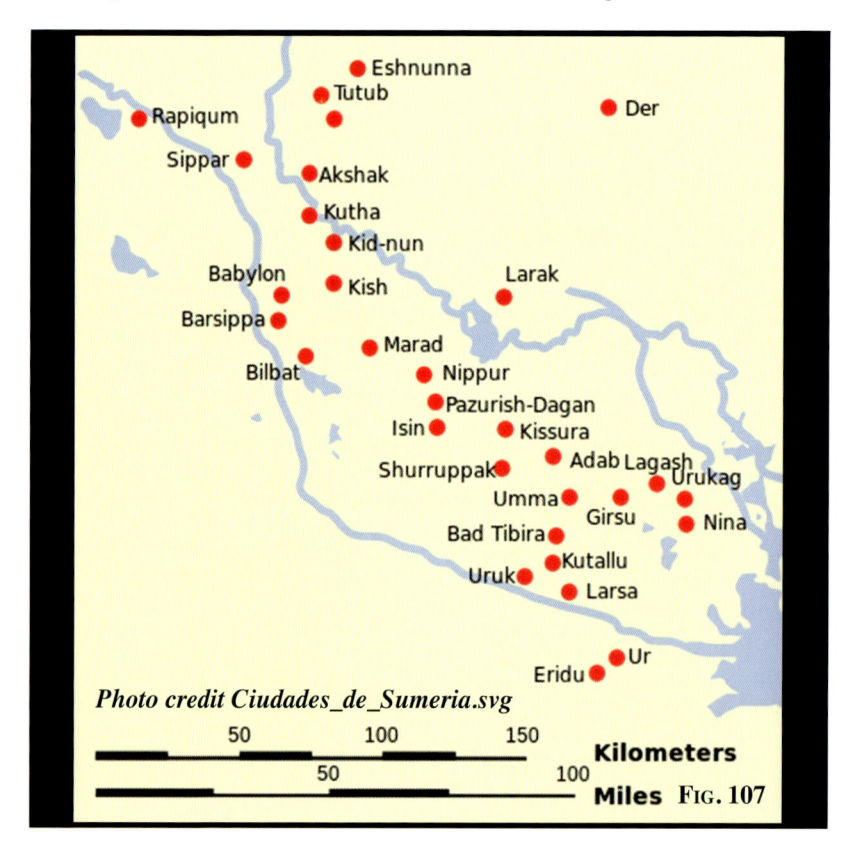

FIG. 107

River life was a slow evolution of learning how to drain and tame the marshes, maintaining enough water for the year round

Center and Circle

The trail of the Dragon has many tracks. We pick up one set beneath a night sky on a cool, desert evening about 5,000 years ago...

You are the teacher. Before you is history. What metaphor do you use to effectively convey the motions, depth, and vastness of space? As you ponder the infinite, endless possibilities ebb and flow through your mind. A *Void* might be an effective medium; or perhaps a *Sea*. But if you live inland, then ocean images would be out of the question. In that case your Creation might record a wilderness, forest, garden or even desert with distant fires burning. Inaccessible mountains, pools, caves, the body of a huge warrior, God or Goddess. Over the course of time, each of these and more have provided the backdrop for *Creation's Stage*.

The *Epic of Gilgamesh* paints the framework for one such *Sky Story*. In it, our hero embarks on a series of adventures illustrating the principle 'landmarks' of ancient heavenly observation.

Gilgamesh begins his quest by laying the 'groundwork' for the entire system by establishing its *Center*. He wrestles with an image whose source is the *Celestial Pole*, then grapples with another based on the *Vernal Equinox*. Having mastered these challenges, he has reached the culmination of his career. He has tasted victory and been showered in glory; he's established his name and claim to fame. *Destiny decreed* his deeds be recorded for posterity on the *bricks of Uruk*, where the names of the famous are written. But even as he basks in the glow of this new found attention, *disaster strikes*.

His best friend, Enkidu, falls sick and dies, shocking our hero into the realization of the frailty of the human condition. Urged on by anguish, he feels compelled to start another quest, but one with a far different purpose in mind. It is now not fame and glory that Gilgamesh seeks, but *eternal life*.

This is a man who has stood at the pinnacle of all *Creation* has had to offer. He is king and hero.

FIG. 108

Gilgamesh, King and Hero- whose name was destined to be written on the mud-baked bricks of the City-State Uruk in which he lived. The lion he holds is a royal symbol of his kingship.

What's the purpose of *Creation*; what point is there to success, victory, honor and glory if a transitory existence is but a brief brush stroke dabbed upon the canvas of life?

As his inner tumult rages over the loss of his best friend, *Gilgamesh* becomes *Hell-bent* to discover the answer and is willing to move *Heaven* and *Earth* to find it.

He sets out to discover the riddle of *Immortality*.

Gilgamesh's two celestial victories symbolize the glory of life. The *North Celestial Pole* is always visible to those living in the northern hemisphere. No matter what season of the year you chose to look, it's there. The Sun's passage over the *Vernal Equinox* is the traditional beginning of *Spring*. Seasonally, *Spring* represents the lengthening of the hours of the day over those of the night. It gives birth to Summer, and together these two points represent the visible manifestations of light and life throughout *Nature*, with all its warmth, vitality and produce. These two celestial markers symbolically reflect the full waking consciousness of life along with all its accomplishments. The struggle between *Gilgamesh* and these two personified points is mirrored in what he has achieved.

Such is life's bright side.

But now his path takes a wrenching turn. Grief stricken though he may be, he must now look to a *new horizon* to chart his course. He is devastated by the loss of his friend. If there is a way out, it is through *death's dark door* that *Immortality* must lie.

Gilgamesh is on his way to find out.

Athena's Web

Center and Circle

Gilgamesh, Enkidu and the 'Bull of Heaven' (Fig. 109)

If the *Vernal Equinox* symbolizes all that we have outlined, then its opposite point must do likewise, but in reverse.

The **Autumnal Equinox** marks the beginning of *Fall*, when Nature's essence retreats below ground into the roots. Like the *Vernal Equinox* (VE), the *Autumnal Equinox* is thought of as a doorway to a new vibration.

The *Sun* is going down.

It marks the entrance into a *Land of Shadows*.

It is here that *Gilgamesh* must now travel, *celestially scored* by the **constellation Scorpio**. He approaches the mountains in grief and despair, but with the same fierce determination that characterized his earlier conquests. The *Scorpions* standing guard over this western entrance to the Underworld are impressed by his courage, yet he asks only their advice.

On their counsel, he must follow the path of the *Sun* beneath the *Earth* through the "*12 leagues* (presumably hours) *of darkness*." There he will find the answers he seeks. There's to be no victory here; and no bright *Sun* of day. No honor, no glory, and no long-lost friend. Even the small amount of wisdom he is able to obtain on this quest winds up being stolen.

What's important is that he is leaving for us, in story form, a map illustrating the personal, terrestrial, and celestial reflections of his time, clothed in their contemporary garb.

It is in the first two conquests of this saga that our chief interest lies, for these illustrate an awareness of the correct positions of the *Celestial Pole* and *Vernal Equinox*. The mythological clues given as to the identity of *Humbaba* ('Hugeness') mark him as our Dragon; here depicted as a '*Watchman of the Forest*.' [48]

Why should we think of him as a Dragon? Listen to the warning *Gilgamesh* receives from his close companion, *Enkidu*,

"O my lord, you do not know this monster, and that is the reason you are not afraid.

"I who know him, I am terrified.

"His teeth are like dragon's fangs, his countenance is like a lion, his charge is the rushing of the flood.

"With his look he crushes alike the trees of the forest and reeds of the swamp." [49]

At the time of the telling of the *Epic of Gilgamesh* during the end of the 4th and 3rd millennia BC, the stars marking the position of the *Vernal Equinox* are *Taurus*. The *precession of the equinoxes* and *Celestial Pole* are mathematically linked like a *Circle* to its *Center*. The position of one cannot move without affecting the position of the other.

They are two and they are one.

In metaphorical fashion we are told that Humbaba,

"...guards the cedars so well that when the wild heifer stirs in the forest, though she is sixty leagues distant, he hears her." [50]

Having conquered the Dragon our two heroes now wrestle with the "**Bull of Heaven**."

In *mythological mathematics*, having determined the *Center*, we now use the *circumference* to find our *Circle*.

Center, then *Circle*.

Enkidu compels *Gilgamesh* to kill this *Bull* by thrusting his sword "*between the nape and the horns*." [51]

The two principle groupings of stars in Taurus are the **Hyades**, marking the head, and the **Pleiades**, marking the shoulder. Celestially speaking, *this* is the mythological mark where the correct astronomical cut must be made.

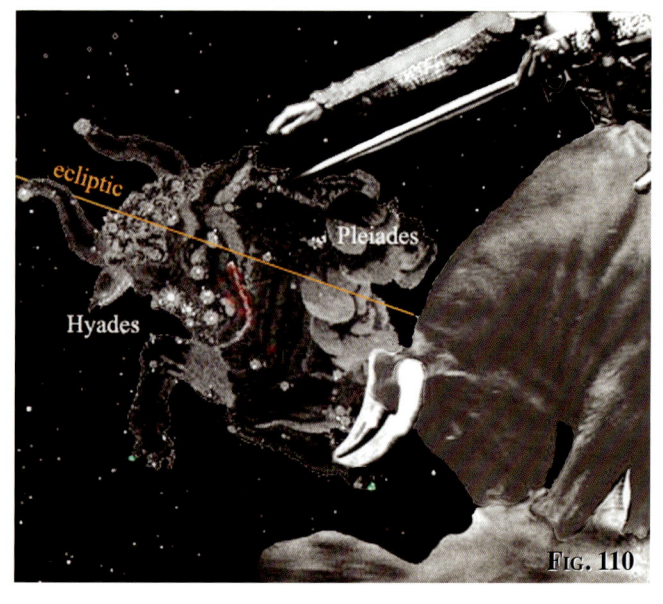

The 'Cut' between 'Head & Shoulder' (Fig. 110)

Center and Circle

Fig. 111

Gilgamesh myth flood tablets

From 2600 to 2400 BC the *Vernal Equinox* passed along the **Neck of the Bull**, marking the time we believe the oral tradition of *Gilgamesh* to have been chanted along the banks of the Euphrates.

Having overcome the "*Bull of Heaven*," the gods are angry. As a result *Enkidu* falls sick and dies.

Bewailing this unexpected turn of events, *Gilgamesh* sets out for the mountains,

"*...which guard the rising and the setting of the sun. Its twin peaks are as high as the wall of heaven and its paps reach down to the underworld. At its gate the Scorpions stand guard, half man and half dragon; their glory is terrifying, their stare strikes death into men...*" [52]

As *guardians* of the *Western Gate*, these creatures have watched the shadows come and go for years. It is the *Scorpions* who tell him that if he would seek *Immortality*, then the way ahead is through the 12 leagues of darkness.

"*Then he called to the man Gilgamesh, he called to the child of the gods: "Why have you come so great a journey; for what have you travelled so far, crossing the dangerous waters; tell me the reason for your coming?"*

"*Gilgamesh answered,*

"*For Enkidu, I loved him dearly, together we endured all kinds of hardships; on his account I have come, for the common lot of man has taken him.*

"*I have wept for him day and night, I would not give up his body for burial, I thought my friend would come back because of my weeping. Since he went, my life is nothing; that is why I have travelled here in search of Utnapishtim my father; for men say he has entered the assembly of the gods, and has found everlasting life. I have a desire to question him concerning the living and the dead.*

"*The Man-Scorpion opened his mouth and said, speaking to Gilgamesh,*

"*No man born of woman has done what you have asked, no mortal man has gone into the mountain; the length of it is twelve leagues of darkness; in it there is no light, but the heart is oppressed with darkness. From the rising of the Sun to the setting of the Sun there is no light.*

"*Gilgamesh said,*

"*Although I should go in sorrow and in pain, with sighing and with weeping, still I must go. Open the gate of the mountain.*

"*And the Man-Scorpion said, "Go, Gilgamesh, I permit you to pass through the mountain of Mashu and through the high ranges; may your feet carry you safely home. The gate of the mountain is open.*

"*When Gilgamesh heard this he did as the Man-Scorpion had said, he followed the Sun's road* (from his setting) *to his rising, through the mountain.*" [53]

The *constellation Scorpio* is opposite *Taurus*, located 180 degrees from each other on the *ecliptical path*. It marked the *Autumnal Equinox* current in the earliest oral traditions of the telling of this myth.

Thus, we have *Gilgamesh's* conquest and establishment of the *North Celestial Pole* (his 'victory' over Humbaba), his 'cut' for the *Vernal Equinox* (his victory over the Bull of Heaven), and his journey to the *Autumnal Equinox* (Scorpio). This is the entrance into the *Underworld* as he follows the Sun as it passes beneath the *Earth*.

The most complete version of this story comes from the library of the *Assyrian Empire in the 7th century BC*, but there's evidence it existed as early as the beginning of the 2nd millennium BC. The stars tell us this story is *a thousand years* older still.

The path of our hero has taken us on a journey around *Creation*, illustrating a *Celestial Framework* in terrestrial terms through one *Sumerian's* eyes.

Imagine that.

Athena's Web

Center and Circle

> At the approach of Marduk Ti'amat is the only one to stand her ground in the midst of her army. Marduk then spreads his net, encompasses her, and defeats her by shooting an arrow through mouth, belly and into her heart. Following the victory, and after a great feast, Marduk organizes the calendar, fixes the polestar and permanently appoints the seven gods of 'destinies' or decrees.

FIG. 112

Ti'amat is the Serpent, Marduk the God riding Her Back

Returning to our theme of *Center and Circle* in *Dragon and Bull*, another contemporary myth which parallels many of the essential features of the *Epic of Gilgamesh* is known as the **Babylonian Creation Myth** and describes the epic battle between **Marduk** and **Ti'amat**.

At the approach of **Marduk**, **Ti'amat** is the only one to stand her ground in the midst of her army.[54]

This was an apt astronomical observation for Draco, rooted as she was in the *Center of the Celestial Field* (Fig. 114).

During this period of mythological development, *Thuban* was the pole star. *Thuban* is considered to be the 'Heart' of the Dragon because it represents the 'Center' of the constellation. The *Center of the Dragon* is reflected in the *Center of Creation*, the *Center of Heaven*. It's like staking a dog on a 6 foot rope in the back yard. The dog is rooted to the stake.

The Dragon is rooted to the pole.

Neither can wander any further than the 'rope' permits.

Marduk then spreads his net...
Marduk is the *Babylonian Sky God*. The Tigris-Euphrates River Valley lays archaeological claim to development of the 'Web' of heaven, a cross-mesh of lines, linked to both Earth and Sky, that allowed examination and calibration of the paths of the stars and planets.

Photo credit © 2005 Joshua McFall FIG. 113

The Babylonian Pyramid- The Ziggurat

Voyager v. 1

Center and Circle

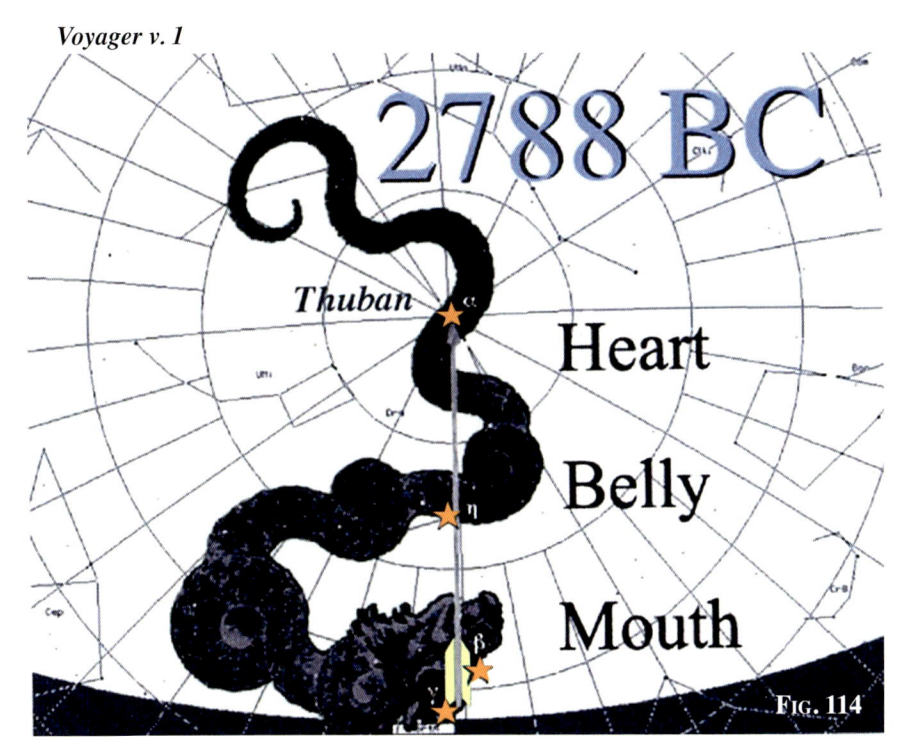

Marduk's Arrow

Fig. 114

'*Marduk's Net*' is what we think of today as lines of longitude and latitude, or possibly their forbearers, lines of azimuth and altitude. One system uses the horizon, the other the Equator. Look at the diagram above (Fig. 114). We've caught a Dragon in our net.

When *Marduk*[55] spreads his net, he encompasses her, and defeats her by shooting an arrow through mouth, belly and into her heart.

They're describing astronomical motion in mythological terms.

The *Pole Star* is at its closest approach in 2788 BC, and for centuries stargazers would have watched as this distant fire of *Heaven* slowly tightened its noose, coming closer and closer to the exact rotational pole of *Heaven* and *Earth*. Four stars in this myth point the way to both *Thuban* and the *North Celestial Pole* (NCP).

Just as today *Ursa Major* (Big Bear) is sometimes used to 'clock' times of night in different seasons, how much more easily ancient skywatchers might have used these four stars as a huge hour hand, with *alpha Draconis* (*Thuban*) at the very *Center* of the stellar field?

What a wonderful time to be studying the stars and calibrating the astronomy of the skies!

Our Dragon is a child of *Time*, and each culture that invokes the Dragon is wrestling with a deeper understanding of '*Time*' and how it worked.

The myth is recording for us some of their accomplishments. After years of effort, having finally determined Heavenly motion, there is a joyous celebration.

Following the victory, and after a great feast, Marduk organizes the calendar...[55]

Exactly. This is the chief focus of the fruit of their labors.

...and permanently appoints the seven gods of 'destinies' or decrees.[56]

The seven visible planets.

It's a *University of Learning*. They're figuring out *Time*, the calendar and astronomy, with new philosophies, mythology and an expanded department of mathematics.

Once a grid system is established, both stars and planets can be equally observed, calibrated, and probed in an effort to reveal the secrets of mathematical motion. The *Greeks* and *Romans* will come to lean heavily on both the myths and mathematics learned from these Eastern wise men.

Having 'encompassed her', *Marduk's Net* holds the Dragon still for closer examination. By taking careful aim, we shoot and penetrate Heaven's essence. We use the stars *beta* and *gamma* (the Jaws), and *eta* (the Belly). Together they point to *alpha* (the Center or Heart).

Ultimately, desire to penetrate Draco's secrets led to the development of *spherical trigonometry*. The myth is describing their relationship to and understanding of their *Sky World*. It's what they saw when they looked up at night.

The star map shows the positions of the stars circa 2788 BC when **Thuban** was at its closest approach to the *North Celestial Pole*. The stars suggest this myth is slightly older than the *Gilgamesh Epic* (2600-2400 BC).

While we're on precession's path, we should mention that once again the astronomical time line indicates this would be a Sumerian myth. *Hammurabi* and the *Babylonians* won't show up for another thousand years or so. The stars whisper it's proper title: it should be the **Sumerian Creation Myth**.

Athena's Web

Center and Circle

'Center and Circle' keep bobbing to the surface in variations on a theme. These 'power points', the North Celestial Pole and Vernal Equinox, were considered to be the *Lathe of Creation*; 'Life' contoured by the *Will of Heaven*.

This symbolic shorthand of '*Center and Circle*' keeps reappearing throughout our myths. Now we will hear a tale of *North* and *East* woven in an Egyptian weave.

Because the *Vernal Equinox* was so important, it's position and perceived manifestations were carefully monitored. Whenever the North or East points aligned with a star in heaven, it 'triggered' the influence of that star, releasing its 'vibration' on Earth.

Or so they believed.

In the following myth, the curiosity of the Pharaoh is piqued when he finds out that just such an alignment will occur during his reign, and he wants a front row seat. The essential elements of '*Center and Circle*' are very evident, but it turns out to be a little more than Pharaoh bargained for.

The *Uraeus* is our Dragon.

In working with the Dragon's evolution, the wings are a later adaptation. The earliest representations of the Dragon in Egypt are as the *cobra*, a wingless '*flame-spitting serpent*'. This *cobra*, known as the '**Uraeus**', and was often shown in an aggressive posture, with his cowl spread and tongue darting as if ready to strike. He was the tireless protector of the Pharaoh.

The *Uraeus* looks like the constellation Draco with his head reared back. *Serpents*, even mythological ones, share common biological characteristics.

This myth is a story about a *Serpent* and an **Eastern fort** on the edge of the frontier. The *Uraeus* marks *North*, the fort *East*.

North and East. Center and Circle.

Here's the tale:

Geb caused the golden box in which Ra's Uraeus was kept to be opened in his presence. Ra had disposed of the box, together with his cane and a lock of his hair, in a fortress on the eastern frontier of his empire as a potent and dangerous talisman.

FIG. 115

Photos this page courtesy Tour Egypt

Pharaoh with Uraeus

When opened, the breath of the divine serpent within killed all of Geb's companions then and there, and gravely burned Geb himself...

When he was restored to health Geb administered his kingdom wisely and drew up a careful report on the condition of every province and town in Egypt. [57]

Ra is the Sun. According to astrology, the *Sun* rules *gold*. Ra's *Uraeus* is kept in a *golden box*.

The 'talisman' inside the box is a star the *Vernal Equinox* is getting ready to trigger by precessional motion. Since the *Vernal Equinox* moves at the pace of about a degree every lifetime, it isn't often one of these alignments occurs and Geb has the time to ponder it and wants to see what will ensue.

When the Sun returns on its annual pilgrimage and 'triggers' (conjuncts, aligns with, parallels) the *Vernal Equinox* each *Spring*, the power of the star is discharged on the Earth. Unfortunately, the star's power was more potent than the Imperial Court realized, and when Geb gets too close to observe, this 'dark star' explodes

The Uraeus. Notice the lack of wings on the Cobra.

Athena's Web

Photo credit Tour Egypt

Fig. 117

Contemplating the mysteries of the past

(discharges), and blows up in their faces.

Black-faced Geb.

Buddies go bye-bye.

This interpretation may seem tenuous, but we will see some of these same patterns repeating again in other myths, (specifically with *Cadmus*). These are some very old mythological reflections indeed.

When the *Sun* returns to the *East* in the *Spring*, it aligns with the *Vernal Equinox* in an annual event.

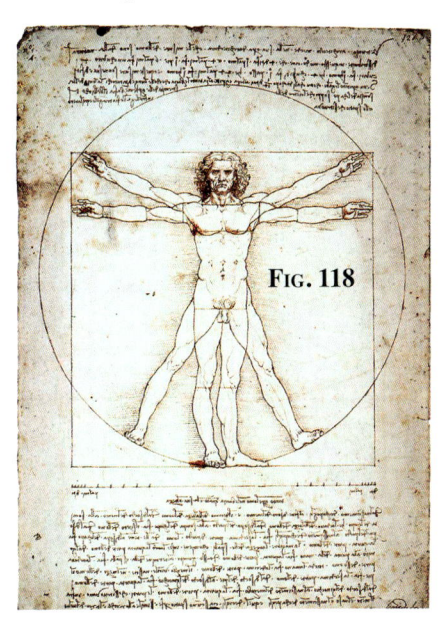

Fig. 118

Man.
Mirroring Heaven on Earth?

Athena's Web

If, however, there is also a 'star' aligning with the *Vernal Equinox* at that time, then the power of the star is triggered, either as the Sun aligns with the *Vernal Equinox*, or on the *Full Moon* following the *Vernal Equinox*.

This is perhaps best illustrated through *Passover*, when the power of a star aligning with the *Full Moon* of *Spring* propelled the *children of Abraham* out of Egypt and onto their *Exodus*. *Judaism* remembers this ritual passage on the *Full Moon* to this day.

Apparently, Geb learns a great deal from the experience because after he recovers he's a good boy and does all his homework, keeping a '*careful report on the condition of every province*'. That this experience is a sobering but edifying one is also a theme that we will see being told many centuries and several cultures later.

But as God made man in his *Own Image*, does man then reflect God within *his* soul?

Pagan culture believed the *Face of God* looked down on them from above.

Center and Circle

Like the *Seasons*, this face changed with the cycles of Heaven. This face could assume the form of a *great Goose*, a *beautiful Bull* or a *bare-breasted Goddess*.

And just as God made us in *His Own Image*, so we can see *His Image* reflected in us. This is why Pharaoh wore the *Uraeus* in the crown. It was a reflection of Draco at the highest point in the sky.

As above, so below.

As time went by Draco slipped from the pole, so the *Uraeus* moved down and out, from crown (Fig. 195) to forehead (Fig. 116).

Pharaoh was reflecting Heaven's stars in his crown and person. He was *Heaven's Will on Earth*.

For Egypt.

Fig. 119

Earth.
A dim reflection of Heaven?
Photo credit Dean Martin

Center and Circle

Photo credit Harold Moses

FIG. 120

Norse Dragons found in both Sea and Sky

Let's keep our eyes on the *celestial highway* with *Center and Circle* as we change lanes. We've been examing the signs of traditional astronomical interest in the desert river valleys of the Tigris-Euphrates and Nile, but our celestial symbolism is not limited to these state-sponsored 'universities.'

Like a good detective novel, we hunt for clues. During the time the *Vernal Equinox* is passing through the stars of the Bull, we are looking for new themes of *Center and Circle*, North and East, or *Serpent and Bull*. Various means of mythological mortar were used to bind the two.

In an effort to throw us off the track, our Serpentine figure has donned a disguise. He's wearing a horned helmet and has a handle-bar mustache.

Be on the lookout.

According to **Norse tradition**, *Jormungand*, the *World Serpent*, was one of the three children of *Loki*, *the trickster God* and the giantess *Angrboda*. When *Jormungand* was born, he grew at such an alarming rate the gods kidnapped and threw him into the sea around Midgard. Given this room the Serpent continued to grow until he encircled the Earth and lay with his own tail in his mouth.

The first time the red-headed **Thor** and *Jormungand* met was in the *Land of the Giants*. While there, *Thor* was challenged *to pick up a cat*. After a great struggle in which he was only able to lift the paw of the beast, the God admits defeat. It was all a magic trick. The 'cat' turned out to be **Jormungand** the **Midgard Serpent**.

How do you 'lift' the starry skies?

The next time *Serpent* and *God* meet was when *Thor* travelled to get a cauldron to brew ale for the gods. While visiting the giant *Hymir*, the two decide to go off fishing together. *Thor pulls the head off Hymir's largest Ox* (how the constellation appears in heaven- no hindquarters) and takes it as bait.

They set out in Hymir's fishing boat, but *Thor* rows out too far and Hymir becomes fearful and begins to panic, concerned that the *Midgard Serpent* will attack. Unconcerned, *Thor* takes the *Ox's Head*, fastens it to a hook and hurls the bait over the side.

Immediately, the *Serpent* strikes.

As soon as *Jormungand* bites the *Ox's Head*, the hook cuts into his mouth. The great *Serpent* gives the fishing line a tremendous yank, but this time *Thor* is ready. He exerts all his strength, braces his feet on the bottom of the sea and slowly draws the furiously writhing *Serpent* into the boat. Jumping up from where he had been hiding in the back, Hymir panics and cuts the line, allowing *Jormungand* to escape just in time. *Thor* throws his hammer, but misses. Furious, he turns and lashes out at Hymir, knocking him into the *Sea*.

World Serpent and *Ox's Head* bound by fishing-line.

Center and Circle.

The myth suggests that it is contemporary with, and possibly a little older than it's Sumerian counterpart (*head* vs. *neck* of the *Bull*). But whether mists of sea or sands of desert, they all shared the same sky.

FIG. 121

Thor and Jormungand

Photo credit Vishnu Avatara

Center and Circle

FIG. 122

Vasuki, King of the Snakes, wrapped around Mount Mandara, churning the Sea of Milk.

The prophecies say that there will be a third meeting of *Thor* and *Jormungand* in the final battle of Creation, *Ragnarok*. The prophecies also say that *Thor's* inability to defeat *Jormungand* is not a good omen for the future.

We moved northwest from Mid-East to sit with the World Serpent in chilly Baltic Seas. Now we travel east, to India, to discover what we can. Our focus remains *Center and Circle*, as we again work with themes of *Serpent* and *Bull*. Here's the tail.

Long ago, *Indra*, *King of the Gods*, began to lose his power.

Vishnu appeared to him, smiling, and advised him as follows. He was to take *Mount Mandara* and use it as a stick. Then the Gods and Demons were to take the snake **Vasuki** as a rope and churn the *Sea of Milk*. This action would produce the liquid of immortality along with other wonderful presents.

The coils of *Vasuki* wrapped this mountainous stick, and Creation was 'agitated' back and forth. To complete the vision, *Vishnu* had to assume the form of a tortoise so that *Mandara*, as the stick, had something solid on which to turn. *Vishnu's* energy also sustained *Vasuki*, *King of the Snakes*, so that everyone, Gods and Demons alike, saw him seated in glory on the peak of *Mandara*.

Here we find a wonderful allusion to the *Serpent* high atop the canopy of *Creation* in every cloudless night sky.

However, the *Snake* suffered from these painful labors. While the Gods pulled him by the tail and the Demons by his head, torrents of venom escaped from his jaws and poured down on the Earth in a vast river which threatened to destroy everything, divine and demonic alike. *Siva (the Destroyer)* was asked to drink the venom, which He was able to do.

Finally, a state of balance was established, and the marvelous fruits of their labors could be had.

The Sea of Milk produced **Surabhi**, *the Marvellous Cow*, Mother and Nurse of all living things.

All Living Things.

The VE was seen to move into the stars of the *Bovine* in India.

Athena's Web

Center and Circle

Mount Mandara reflects the path the Sun, Moon, planets and stars take as they rise in the East at the horizon, culminate in the South at their peak and set in the West level with the horizon once again. Their path traces out a rough 'bell curve,' like a mountain in the sky. Indeed, mountains are wonderful terrestrial markers against which the progress of the stars and planets might be mapped night after night. The Greeks will use this same image in the paths the Gods walk in their hallowed halls on *Mt. Olympus*. The summit of this mountain is where the coils of *Vasuki* wrap themselves (Fig. 122).

FIG. 123

Sunrise from Mt Olympus

Once *Surabhi*, *the Marvellous Cow* came into being, the *Tree of Paradise*, the delight of the nymphs of Heaven, scented the Earth with the perfume of its flowers. Then followed the *Moon* and *Lakshimi*, the *Goddess of Fortune*.

Indra was able to drink the *ambrosia* and became rejuvenated and regained his lost vigor.

The essential components of our *Bovine World View* are coming into focus. Creation rocks back and forth on Heaven's rhythms, whether night and day or summer and winter, and 'churns' Creation into Life. At that time it produced a *Sea of Milk*, which in turn yielded *Surabhi*, *the marvellous Cow* and the *Tree of Paradise*.

Indra, the King of the Gods becomes fatigued. Prior to drinking the ambrosia, he seemed to be wanning in power. It's what astrologers would call the end of an 'Age.'

An *Age* is the time it takes for the *Vernal Equinox* to pass through a single constellation of the *Zodiac*. Judging by the tone of the myth, the '*Age*' *of Taurus* is being born. There's no longer any *Wind* in the celestial sails of Gemini's imagery. *Vishnu*, who sees beyond the present, knows that a new epoch is getting ready to dawn.

The '*Age*' *of Gemini* is giving way to the '*Age*' *of Taurus*.

The astronomy does not give a sense of the energy winding down as it comes to completion.

The astrology does.

The mythology does.

Were the people who looked to the skies and crafted these images priests, mythographers, astrologers or astronomers?

Yes.

It was a collective effort, shared by those on the inner *Circle*. Many people put their heads together and discussed events before putting their mythological stamp of approval on it, sharing their myths with the populace at New Year's.

While we're here in India, let's look at another Indian myth of *Center and Circle*.

Serpent and *stick* were our images with *Vasuki* and *Mount Mandara*, but now our pair take on a different form. This time the *King of the Snakes* and a magical column make their mark.

Now his name is **Takshaka**.

FIG. 124

Takshaka, aka Taxaka-
Photo credit LR Burdak

Here's the story.

King Parikchit was out hunting one day, *exhausted* after chasing a wounded gazelle. During this pursuit, he unintentionally offended a *hermit* of the *highest virtue* who was observing a *vow of silence* in the *heart of the forest*. For this offense, the hermit's son placed a *curse* on the king, saying that before the week was up the *snake Takshaka* would *burn* the king with his *poison*, and *he would die*.

When the king heard the news, he built *a palace atop a column* which stood *in the middle* of a *lake*, secluding himself there as a precautionary measure. But *Takshaka* succeeded in *outwitting the guards* by changing some *snakes* into *wandering monks* who

60   Athena's Web

offered the king *water*, a *sacred plant*, and *fruits*. The king received the *monks* and their gifts, and dismissed them.

Since this was the last day of *the accursed week* and the *Sun* was getting ready to set, the king summoned his ministers so that they might celebrate his successful navigation of the *curse* by sharing the food brought by the ascetics.

Among the fruits there appeared a *strange insect* shining like *red copper*, with *glittering eyes*. Emboldened by the approaching sunset, the king picked up the insect and placed it on his neck.

"*The sun is about to set, and I have now no fear of death. Let the hermit's speech be accomplished, let this insect bite me.*"

Then *Takshaka*, for it was he, wrapped the unsuspecting king in his *coils, uttering a great roar*.

The ministers *burst into tears* and *suffered the keenest grief*; but *even as they fled* the monster's roar, *they saw the marvellous reptile rise into the air*.

The King of the Snakes, red as a lotus, had fulfilled the *curse*. The king *fell dead*, as if *struck by lightning*, and *the palace was wrapped in fire*.[58]

We begin to see some of the mythological patterns unfold as we move from myth to myth. There are common elements at work here, not all of which are astronomically derived; the poisonous

Photo credit James Thornhill

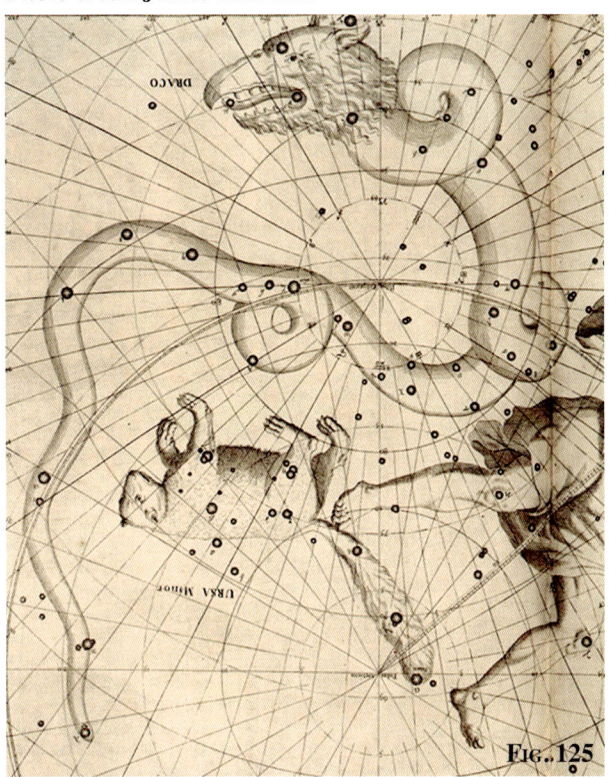

The two stars in Draco's head are the brightest

fumes released by the *Uraeus* when Geb opened the box, the poison that dripped onto the Earth as the Gods and Demons rock the *Serpent* back and forth, the poison of *Jormungand* and the poison King Parikchit dies from.

We have *Center and Circle* woven into the myth, but it's not strictly the combination we have been witnessing so far. In this myth there's an emphasis on the *North Celestial Pole* and the *Serpent*.

The *hermit* is in the *heart of the forest*. The mythological personality of the Dragon is being conveyed through the hermit. He, like the Dragon, is *a solitary creature*. He has used his time to study the *deeper mysteries* of life. He is both *wise* and *powerful*. This is a *monk* of the *highest virtue*. At this time, Draco is the *highest constellation* in the sky.

Center and Circle

Within a week, the King has a palace constructed *on top of a column*, right in the *middle* of a *lake*.

The column is the *Earth's Celestial Axis*, which turns at the *Center* (in the middle) of a body of water. This is a *microcosm of Creation*: Death and destruction- common mythological themes. The Dragon reaches out and embraces them all.

Draco is a vast constellation. Many people are familiar with the *Big Dipper*. It's fairly large for a stellar grouping and boasts particularly bright stars. The seven brightest stars of the *Big Dipper* comprise a portion of the larger constellation *Ursa Major*, and some think the Dragon neatly wraps both this and the smaller Bear in his coils as he winds his way through the nightly sky (Fig. 126).

Let me repeat.

Draco is a vast constellation.

While *alpha* of a series usually designates the brightest star in a constellation, in this case it doesn't. It's been said that *Thuban*, the 'Heart' of the Dragon must have been more brilliant at one time in the past, but there may be another, simpler answer.

It's more likely *Thuban* was given the designation *alpha Draconis* not for its brilliance as a star, but rather for its importance.

It marked the Center of Heaven.

Center and Circle

The two brightest stars in Draco are both located in, what most people agree to be the *Head of the Dragon*. They are currently designated as *beta* and *gamma Draconis,* the second and third stars of the constellation.

They are in fact the first and second brightest. As you look at the constellation, they stand out. But whether these two stars are the *eyes* or *jaws* of the Dragon is left to the imagination of the reader or tradition of the culture.

There is yet another myth in which *Serpent* and *Shaft* is 'highly' visible. Take note of the themes that begin to repeat themselves whenever the Dragon is around.

This Dragon myth from India involves a young Brahman named **Utanka**, who was asked to deliver a *pair of earrings* to his tutor's wife. The Brahman was warned to take care, as **Takshaka, the King of the Serpents** [59] had long coveted these jewels.

While on his way, the Brahman noticed a *naked beggar* who kept appearing and disappearing from sight. When *Utanka* stopped to *perform his ablutions*, he laid the earrings down. Quickly the *beggar sneaked up* and *stole* them.

After *Utanka* finished, he caught the *old man*, who immediately changed back into a *Snake* (it was *Takshaka*), and glided into a *cleft* in the Earth.

The Brahman *probed the hole* with his *staff*, but without luck.

Indra saw that he was *overwhelmed with grief* and sent his *thunderbolt* to help, which followed the *staff* and entered the *cleft, bursting open the hole*.

Having entered the *World of the Snakes*, *Utanka* found it full of *admirable establishments*, crowded with *porticoes, turrets, palaces* and *temples* of *different architectural type*. He chanted a hymn in praise of the **Nagas** (*the Devic Snakes*), but still they would not return the jewels.

Then *Utanka* entered into a deep meditation. A marvellous symbolic vision of nights and days, and of the years and seasons, unrolled before his inner eye.

(See *Preface*, page *xv*.)

Finally, with *Indra's* help, *Utanka* was able to get *Takshaka* to return the *earrings*.

The myth illustrates the power of the *Serpent* manifesting through the principles in the story. Draco represents (in part), those that live on the fringes of society, whether as monks, hermits or naked beggars. The Dragon can be and often is associated with hard work, old age, death and destruction. Remember that Yu worked so "*his hands grew worn and his feet thickly calloused, but in the end his labors were rewarded.*"

Photo credit Arthur Millner

FIG. 127

Vishnu: The god standing on seven intertwined snakes: Ananta, Sesha, Takshaka, Karkota, Padma, Mahapadma and Sankhapala

Some of the high side of Draco is also seen in the *Realm of the Serpents*. It is filled with...

"*admirable establishments, crowded with porticoes, turrets, palaces and temples of different architectural type.*"

All of these are buildings that can be easily used as a framework with which to study the stars.

And as we now know, in order to study Time, the calendar and religious order, you had to pay homage to the Dragon.

May I suggest that in this myth, it is the earrings that *Takshaka* seeks, because they are Draco's two brightest stars?

Takshaka is obsessed with trying to obtain them.

He chases them night after night.

FIG. 126

Draco and the Bears

62   Athena's Web

Center and Circle

The '*cleft*' into which the *Serpent* slips, and into which *Utanka* probes with his *staff*, is the correct spot to align *Heaven and Earth*.

We've got the *shaft*, and the *Serpent* is showing us exactly where to put it.

This notion of *Serpent and Shaft* is a culturally pervasive one, its imagery strong. If we return to the *Ancient Near East* in the 19th century BC, we find an Iranian bronze in the shape of a Dragon (Fig. 128). The 'socket' on the belly is the perfect spot for a pole, wand, baton or *Celestial Axis* to sit.

Perfect is good.

In the same manner a baton leads the parade, so the stars (Gods to them) of Heaven were seen to influence the destinies of kings and nations. To 'honor' them by following their motions dutifully was of the greatest concern.

While the *Vernal Equinox* is most often associated with *Spring*, the *North Celestial Pole* is visible throughout the year, and is not visually limited to a single season.

As *Commander of the Center-of-the-Circle*, the Dragon's inclusion was always welcome as celestially and astronomically correct.

Tradition shows it most frequently invoked at *New Year's*.

Because the term '*Draco*' is Greek, there is no collective term for 'Dragon' prior to their time. We recognize this creature by His (or Her) physical and mythological attributes.

Our *Iranian Dragon* is composed of a *Lion's Body*, a *Serpent's Head* and the *neck* and *claws* of a *Bird of Prey*. This is of the same breed of Dragon as those found on the *Gate of Ishtar* (Fig. 130).

While the *Gate of Ishtar* was originally built by **Hammurabi** in the 18th century BC, it was rededicated by *Nebuchadnezzar II* in the 6th century BC. What we see is one section from the *Gate*.

Center and Circle.

Serpent and Bull. Over and over again all over the *Gate*.

Think of it as *mythological Morse code*, dash and dot.

There are many myths that weave together *Center and Circle* in a dance with the stars, but each might also stand as a story on its own. Stories about *Serpent* and *Shaft* are examples. They both deal with the *North Celestial Pole*.

We will see many examples of a myth about one or the other, including one about *Moses*, but that's not for another six centuries.

Whatever the constellation, the *Vernal Equinox* is the sign of *Spring*. Spring *is* Creation. Creation is the starting point of Time. With twelve constellations (thirteen if you count Ophiuchus) falling along the path of the ecliptic, by definition there can be only One who is first.

First in Season.
First in God's heart.
First-born.
First time.
First-in-line.
First.

When *Taurus* formed '*the Body of God*', myth personified the *Sky Story* of the times in various *Bovine* combinations, whether as a *Bull*, *Cow* or *Calf*, *Apis*, *Ba'al*, *Bel*, *Ba'alberith*, *Nandi*, *Surabhi*, *Hathor* or any of a host of other incarnations.

FIG. 128

A socketed bronze Iranian Dragon c. 1900 BC

Athena's Web

Center and Circle

Here's **Inanna** (Fig. 129).

*My Father gave me the Heavens,
gave me the Earth.
I am Inanna!
Kingship He gave me,
Queenship He gave me...
The Heavens he set as a Crown
upon My Head,
The Earth he set as sandals
upon My Feet...
The Anunnake (the Gods in their
entirety) trundle along.
I am a splendid wild Cow;
I am Father Enlil's
splendid wild Cow,
his splendid wild Cow
leading the way!* [60]

Photo credit AmberinSea

FIG. 129

**Inanna or Ishtar
Goddess of Love and War**

This question of 'being first' was incredibly important judging by the amount of attention and devotion paid to it by these archetypes. The earliest astronomical writers come to associate *Venus* with the *stars of the Bull*, and here we are seeing some early indications of that relationship. It didn't matter whether she was *Venus* (Roman), *Aphrodite* (Greek) or *Inanna* (Sumerian); the loving, sensuous, fertile and alluring goddess translates through the cultures nicely, thank you very much.

The agricultural concerns of the Bull combined easily with the gifts of *Venus*. *Taurus* is a sign of nature. It has a close relationship with the *Moon*, as we have seen. With both *Venus* and the *Moon* finding favor here, *Mother Nature* nurtures us all from *Her breasts*.

This is a time steeped in the abundant produce of the Earth, whether domestic or wild. People born of the Earth are practical, conservative and thrifty. This is what it means to live under the influence of an Earth sign.

We now turn to one of the most prominent and powerful symbols from the celestial pantheon. Dipping into *Mother Nature's* rich bounty, we come up with an image that was functional and nearly omnipresent. Like many symbols in the mythological pantheon, it was to take various forms.

The image of which we speak is the Tree. It predates the temples and was set apart in sacred groves, on hilltops and high places.

It would come to embrace *All of Creation* beneath it's vibrant branches. It was known as the *Tree of Life*.

Center and Circle

Dragon and Bull

Gate of Ishtar

FIG. 130

Center and Circle

The World Tree

©1990/2000

Fig. 131

Yggdrasil—
World Tree by Jen Delyth

Most of what we know about *Norse Mythology* comes from *Iceland*, where early educational interests recorded and preserved the old myths. The cosmology left behind depicts the Earth as a *Circle* of land surrounded by *Ocean*. In this *Ocean* lies the *World Serpent*, while in the *Center* of the land is a mighty *Tree*: **the World Ash** that would come to be known as **Yggdrasil.** This *World Tree* is the 'foundation stone' upon which *Norse mythology* is built. It is the *Guardian Tree* of the Gods, for beneath its canopy they hold council. It's a symbol of universality, linking the various races of beings, and forming the *Center of the Nine Worlds*.

Fate, destiny and wisdom were all linked to this *Tree*. *Odin, King of the Norse Gods* voluntarily hung himself in agony from it for nine nights, to gain power over the *runes*, one of several forms of Norse divination. He died during this trial but *resurrected himself* with the knowledge gleaned. Before he died, *Odin* was pierced by the spear and cried out in pain.

After triumphing over death, *Odin* cut his great spear *Gujngnir* from *Yggdrasil* and carefully carved *magical runes* on its shaft. It never missed its mark in battle.

Yggdrasil is the most stately tree in the world. It was looked upon as the *backbone of the universe*.

Center and Circle

Its *branches spread out above the heavens* and overhang the nine worlds. It was supported by three great roots. Beneath these roots lay various treasures. At the end of one of these roots, *Mimir* guarded a magic well. *Odin* desired to drink from this well in order to gain its secrets. From this well could be learned "*many truths unknown to any other person.*"[61] To drink at this well *Odin* had to sacrifice one eye, which he agreed to do. His good eye was the *Sun*. His sacrificed eye, thrown into the deep waters of the well, was the *Full Moon*.

Under the branches of the *World Ash* lived the nine worlds of *Norse Mythology*.

Asgard was one of these worlds, the home of the *Aesir*, the gods that included *Odin, Thor, Loki,* and the more familiar Nordic deities.

Vanaheim was another, the home of an older generation of gods thought to be fertility spirits.

There were seven other worlds.

Jotunheim was the land given to the *Frost Giants* by *Odin* at the Creation.

Alfheim was the land of the light elves.

Muspelsheim was the land of the fire giants.

Midgard was humankind's home.

Svartalfheim was the land of the dark elves.

Hel was the realm of the unworthy dead, and retains an obvious association with today's *underworld myths*.

Finally, cold *Nifheim*, the abode of **Nidhogg** our Dragon, which lay beneath *Yggdrasil's* roots. These great tentacles lead to the cold, frost-bitten north.

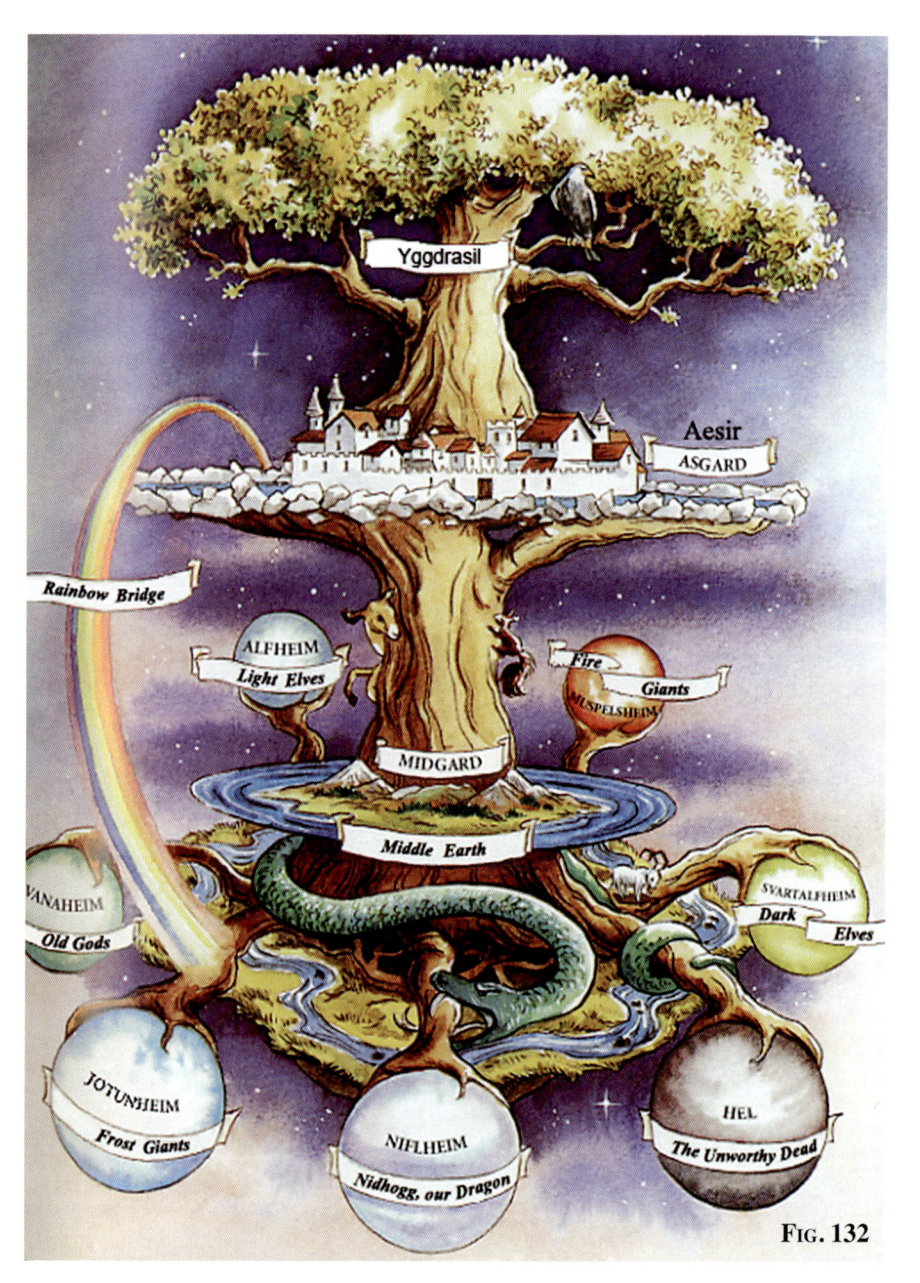

Fig. 132

Yggdrasil, the World Ash

Nidhogg lies beneath the tree, in the freezing mist and darkness of *Nifheim*, the lowest of the nine worlds. Here, he ripped apart corpses and hungrily devoured them. Neither fire nor flood could deter him from his ceaseless feasting on the vast and inexhaustible supply of the dead. When he tires of this morbid menu, he turns and gnaws on the roots of the *Tree of Life* itself.

Nidhogg is placed at the bottom rather than the top (where we see it astronomically) of the *Tree* to highlight the Dragon's degenerative nature. *Nidhogg's* gnawing on the roots earmarks his self-destructive *Draconic characteristics*.

What do we know about these traditions and where do they come from? Once again, there is an intermingling of cultures, this time of *Germanic* and *Norse* traditions.

Center and Circle

In the 1st century BC, Germanic-speaking peoples occupied an area between the Rhine, Danube and Vistula Rivers, south of the Baltic Sea. Between the 3rd and 6th centuries AD [62] they expanded into the Scandinavian countries of modern Denmark, southern Sweden and Norway, as well as southern and eastern England. Other tribes moved further afield, into southern Russia, Italy, Spain and even Africa, but these left less permanent impressions.

Between the 8th and 11th centuries, these same *Scandinavians* went through another period of expansion as the *Vikings*.

Germanic tribes on the continent closest to the Roman Empire were absorbed by and converted to Christianity at earlier dates, but tribes north of Denmark held onto the older pagan roots until the start of the 11th century AD.

As a result, much of our information on the myths of the '*Northmen*' (Norse Men- from the Dutch) comes from this period.

According to the *Prose Edda*, Norse myths foretold a *time of the end* wherein the *Tree of Life* would play a *Central role*.

It would be known as the great battle of **Ragnarok**, fought between the *Gods of Asgard* and the *Frost Giants*. During the battle the **Rainbow Bridge** would be finally destroyed. This bridge currently ends at the residence of *Heimdallr*, who eternally waits for that fateful day while guarding it against the *Frost Giants*.

Ragnarok is the *Twilight of the Gods*. In this myth *Loki*, *Fenrir the Wolf* and *Hel*, goddess of the Underworld and her *Army of the Dead*, unite to battle the Gods. During the battle, *Thor* and *Jormungand* meet for the third and final time. In this epic *tour de force*, *Jormungand* comes out of the Ocean and poisons the Sky itself. *Thor* confronts him, throws *Mjollnir*, his magic hammer, and smashes the *head of the Serpent*. *Jormungand* drips some of his poison onto *Thor*, who is only able to walk nine steps before he too, falls and perishes.

So what role does the *World Tree* play in this mythic battle?

It allows humanity to survive the ordeal. Using the Great Tree's branches as protection, two people survived to repopulate the world. *Lif* (*Life*) and her mate *Lifthrasim* (*Eager for Life*) shelter in the Sunlit branches of the *Cosmic Ash* at the end of the world while *Ragnarok* is fought.

After the Earth is purged by fire and water in this battle, the young couple climb down to renew the human race.

"*The bellowing fire will not scorch them; it will not even touch them, and their food will be the morning dew. Through the branches they will see a new sun burn as the world ends and starts again.*" [63]

After the Earth is purged by the fire and flood of this battle, a *New Age* dawns, emerging from the sea like a great volcanic island.

There are elements of this myth that bear interesting similarities to the *Book of Revelation* in the New Testament and of a great battle coming at the *End of Time*, marking a new epoch, a New Age, and a new spirit of cooperation.

Fig. 133

Heimdallr stands before the rainbow bridge blowing his horn

Athena's Web

Center and Circle

As the *World Tree*, we have encountered *Yggdrasil*. Her *Branches* support the stars and planets, held aloft by the strength of *Her Arms*. The *Trunk* of this *Great Tree* is the *Axis of the Earth*, forming the imaginary, but ever-so-powerful shaft upon which She turns. Strong arms indeed must carry this heavy load. Her *Roots* are the soil of the Earth and all that lies beneath.

This marvellous *Tree*, unlike any other, unites Heaven, Earth and Underworld. Pagans felt the *Tree of Life* to be the *Central Axis*, the *Celestial Spinal Column* translating *Heaven's Will* into Earth's reality. *Siberian myth* perceives the *World Tree* in this way. Like the *Serpent* in the crown and diadem of the Pharaoh, the *World Tree* is visible in the *Crown of Silla*,[64] Silla being one of the three *Kingdoms of Korea*.

Hungarian folklore[65] speaks of the *Sky-reaching Tree*. As *Cosmic Tree* it was important in *Latvian mythology*. It's also a common element in *Lithuanian* folk painting, and frequently carved as a design into household furniture.

Isis and *Osiris* were said to have first emerged from under the branches of an Acacia Tree. The *Egyptians* referred to it as the "*Tree in which life and death are enclosed*."[66] Translation?

Tree of Life.

Assyria[67] also named a '*Tree of Life*'. It was not of a woody, leaf-bearing fiber, but rather a curious series of nodes and criss-crossing lines. It is a tree in name only.

In *Greek mythology*, *Heracles* fought with *Ladon*,[68] a serpent-like Dragon who guarded the *Golden Apples* in the *Tree*.

Photo credit Jim Naureckas FIG. 134

A myth with muscle, Heracles and his Lion-skin

Our *Circular Coil* theme returns, this time wrapped around the trunk of a *Tree*. This last *motif* passed to the *Greeks* from *Near Eastern* and *Minoan* sources. In the *2nd century AD*, *Pausanias* saw an archaic cult image carved out of ceder wood in a temple treasury at *Olympia*—a depiction of *Heracles* and the *Apple Tree* of the *Hesperides* with the Dragon wraped around it.

The Tree is the *column*, the *North Celestial Pole of Heaven*.

During recent mythological periods, Draco has guarded the knowledge of where the point pierced his stellar body, in many cases having to take it right down the throat.

The *Bahai's* correctly interpret the *Tree of Life* as referring to the "*reality of the Manifestation of God in whatever Age he appears.*"

Exactly.

It represents the here and now, which is always changing.

It's measuring the current epoch, the contemporary choreography, the *Living God*.

It's just that *Time* marches on and their 'here and now' is our 'there and then'. Plugging the present time into the present reality is the only way to go, but if you leave the Image on the conveyor belt of Time too long it gets away from you. Old models were correct for their particular epoch, but not for ours.

It's the anguish of the Ages.

Yesterday's gone.

You can't put new wine in old bottles.

In *Chinese mythology* one carving of the *Tree of Life* depicts a *Phoenix and a Dragon*.[69] Not surprisingly, the Dragon represents immortality.

Dating from around 1200 BC, a recent discovery uncovered three bronze trees in China, one of them 4 meters high. At the base (like the Norse version) was a Dragon. There was fruit hanging from the lower branches.

Trees also play a prominent role in Germanic paganism. *Thor's* Oak, *Hercules's* Apple, *Osiris's* Acacia, *Buddha* did the Bodhi—each has represented the *Axis Mundi* for the Earth at different times in different geographical locations.

In embracing the *World Tree*, we have stepped outside the boundaries of our Bovine box. The *World Tree* was a representative axis for the Earth long before the East Point entered Gemini, long even before the Serpent seized the *Frozen Throne of the North*. It continues to play a role on the world's mythological stage.

Did you have a good Christmas?

Fig. 135

Michelangelo's God

Anything under *the Tree*?

The myth lives to this day.

Which leads us to the **Yule Log**.

These are all *New Year's festivities*. They throw out the old calendar and bring in the new. The *Winter Solstice* marks the *Sun's* lowest ebb in the sky, but it turns and begins to ascend once again after three days.

For three days following the Solstice, the Sun is in the grave, but is then reborn. The length of Sunlight during the day begins to increase, the Sun begins to climb higher in the Sky, and *the shadows of night* begin to recede.

Any of the holidays around the *New Year* are associated with the renewing of the calendar, whether on the 1st of January or the first day of Spring. The calendar is the *Child of Time*, and Time is born of the relationship between *Heaven and Earth*.

Cronos was the child of *Uranos* and *Gaia*.

Just like the myth says.

Those who opt for a calendar based on the *Winter Solstice* look to the birth of the Sun for the start of the year.

Those who opt for a calendar based upon the agricultural cycles start their year in Spring.

The tradition of the *Yule Log* radiates from the *Tree of Life*. The *Winter Solstice*, *Christmas* and *New Year's* are all holidays that deal with the beginning and end of the year. At *Christmas,* we reflect on all that has been, whereas at *New Year's* we look ahead to the promise of life made new.

The *Yule Log* ritual can be part of the *Winter Solstice* or the *Twelve Days of Christmas*, *Christmas Eve*, *Christmas Day* or even *Twelfth Night*.

Originally Germanic customs selected an entire *Tree*, inserting the root end into the hearth first, allowing the rest of the *Tree* to project into the living space.

Center and Circle

The first mention of the *Yule Log* in England dates to no earlier than the 17th century. Clergyman Robert Herrick called the tradition a "*Christmas Log*" [70] and said that it was brought into the farmhouse by a group of males, who were then rewarded with free beer from the farmer's wife. The clergyman claimed, in keeping with tradition, that the fire used to burn the new *Yule Log* be kindled with a remnant from the log that had been burned the previous year. One year emerges from the next.

In *Catalonia*,[71] a log is wrapped with a blanket and given special care several days ahead of *Christmas*. On *Christmas Eve* the log is hit to go '*cagar*'. The blanket is then removed to reveal the gifts that have been 'expelled' by the log.

Since the *Yule Log* represents the circular shaft around which the world turns, it is a symbol of the coming year ahead, and how it will fare. This is the birth of the year. The gifts 'expelled' by the log are the gifts the coming year will bless you with, starting right now.

If we understand this as the *New Year*, and if the *Tree* contains the 'essence' of what will happen to us over the course of the year ahead, then the following makes sense. The names have been changed, but the rituals remain.

Among the *Serbs*,[72] the *Yule Log* is known as a *Badnjak*, and it is a central feature of the *Serbian Christmas* celebration. Felled early on *Christmas Eve* morning in a solemn procession, the selected log, usually a young straight oak, is brought into the house and placed on the fire.

As the fire is lit, prayers to God are sent up so that the coming year may bring much happiness, love, luck, riches and...

You guessed it. Good food.

In *Bulgaria*,[73] a young man of the family was sent out in his best clothes to cut down an oak, elm or pear tree.

Before it was cut, a *prayer of forgiveness* was necessary or it couldn't be chopped down. The doorway to the beneficial realms had to be first opened. The *Tree* was to be carried on the right shoulder and forbidden to touch the ground. After arriving at home, the end of the *Tree* is bored out and filled with Chrism (a mixture made of wine, cooking oil and incense). The hole is then plugged, with its end wrapped in a white linen cloth, before being burnt on the hearth.

The *Yule Log* is said to have special healing powers. Leftover pieces are supposed to be lucky, while items carved out of the wood were believed to be blessed.

The branches of this *Tree* embrace the stars of Heaven while its roots nourish themselves on the fertile rivers of the Underworld. This central myth is found *around the world*. A *Winged Serpent* hides amongst branches of this Tree, gazing down at us from above and watching as day turns into night, as one Season rolls into the next.

Although we have journeyed across many miles in this outline, there's yet one more *Tree* that we should visit before we cease our tale. For now our quest takes us on another journey across a great sea.

Fig. 136

Image credit Marinell Turnage

The burning of the Yule Log was thought to be magical

Center and Circle

The Mayan World Tree

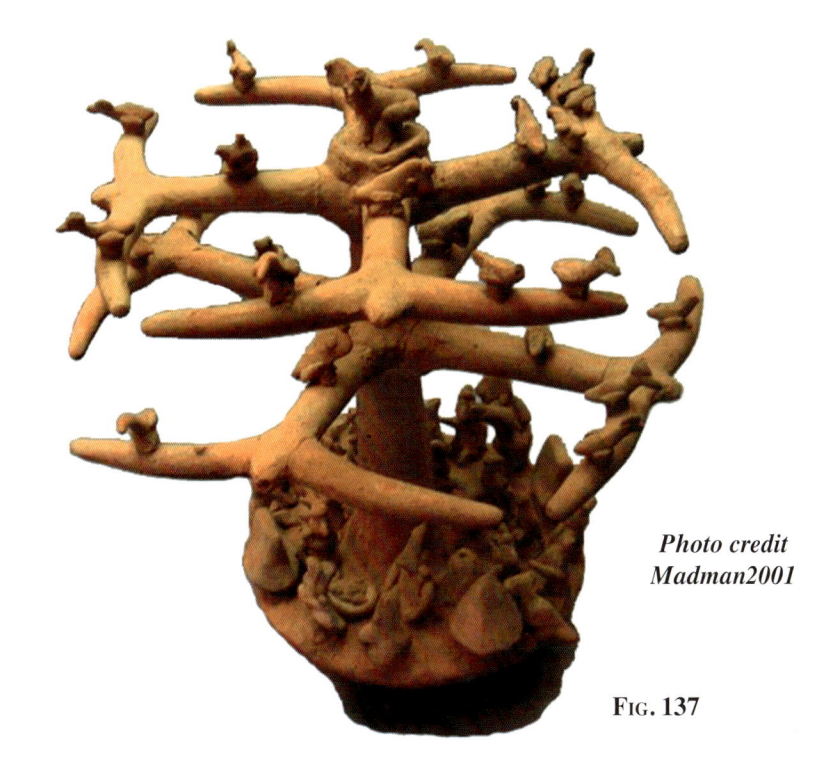

Photo credit Madman2001

Fig. 137

The Mayan World Tree is oriented to the four points of the compass

"This is the account of how all was in suspense, all calm, in silence; all motionless, still, and the expanse of the Sky was empty."

"This is the first account, the first narrative. There was neither man, nor animal, birds, fishes, crabs, trees, stones, caves, ravines, grasses, nor forests; there was only the Sky."

"The surface of the Earth had not appeared. There was only the calm Sea and the great expanse of the Sky."

"There was nothing brought together, nothing which could make a noise, nor anything which might move, or tremble, or could make noise in the Sky."

"There was nothing standing; only the calm Water, the placid Sea, alone and tranquil."

"Nothing existed."

"There was only immobility and silence in the darkness, in the night. Only the Creator, the Maker, Tepeu, Gucumatz, the Forefathers, were in the Water surrounded with Light. They were hidden under green and blue feathers, and were therefore called Gucumatz. By nature they were great sages and great thinkers. In this manner the Sky existed and also the Heart of Heaven, which is the name of God and thus He is called." [74]

The opening lines of the **Popol Vuh**, the "**Book of Community**," the *Bible* of the *Maya*.

When placed in a cosmic context, the words of the *Popol Vuh* spring to life. If we place ourselves out in the middle of a large open field, alone at night, and attempt to observe the material world about us, we might better understand this picture.

"In the beginning..." becomes the darkest night, without Moon or star to guide a weary soul. It is *Chaos*, *Void* and *Dark*, stripped of *Light*. Then we are given a string of names that existed while yet living within this *Void*.

They are all One.

Athena's Web 71

Center and Circle

Photo credit Matt Logan

FIG. 138

As Kukulkan, bringing the serpent energy to Earth...

But they appear as more than One, and we are given several of their different guises, the different masks, that this *One Essential Entity* may wear. Each is a different facet and manifestion of the same *Central Crystal*.

The Creator *is* the Maker, *is* Tepeu, *is* Gucumatz and the Forefathers.

'*The Forefathers*', '*E Alom*',[75] is translated as '*those who conceive and give birth*', while '*e Qaholom*' is '*those who beget children*'. These two have been translated collectively as '*the Forefathers*', but literally they are the *One who conceives and begets children*. Taken collectively, these themes can be further distinguished as the unmanifested *Heaven and Earth*. *One* begets *Two*, who together beget *Three*, etc.

Life's duality was the original birth of God. The Maya leaned heavily on this theme of duality. We will see it reappear in the successive layers of their mythological evolution in the *Popol Vuh*. From *One* (the collective combination of all that *is* and *isn't*, *was* and *wasn't*, *formed* and *unformed*) comes *Two*. The image the myth engenders is *Creation* pregnant with *Her children*, but they have not yet been born and are still 'within' Her being. In the Mayan model this 'pregnancy' lies hidden beneath *Green* and *Blue Feathers*, reflecting the essential hues of the Earth and Sky.

The ol' *One-Two*.

But at this point we stumble upon a common mythological misconception. This notion of '*the Sea*' pops up over and over again in various mythologies, often in Creation Myths, and it represents the homogenous uniformity of 'background noise' that existed before anything came into Being.

This is '*the Sea*' being spoken of at this point in the Mayan Creation Myth.

"...*and the expanse of the sky was empty.*"[76]

This is also the '*calm water, the placid sea, alone and tranquil*' defined by the Greeks as *Oceanus*, the *World Ocean*. It is not what we think of when we picture the Atlantic, Pacific or Mediterranean. Neither is it what comes out of the tap in your kitchen. It is both more and less than these. These geophysical bodies are the children of this essence, but they are only offspring, still tied to their Mother's apron strings, the Earth. We are speaking here of an even vaster *Sea*, one that is free and limitless. Homer calls *Oceanus* '*the progenitor of all the Gods*'.[77]

It is a *Sea* whose waves lap against the thinnest shores of consciousness. The Greeks believed that their version of this *Great Cosmic Sea* issued from the Underworld and flowed in a circular stream around the Earth. *Helios* and *Eos* lived on it's banks in the East (The Sun and the Dawn), and as Helios was carried in its current, He disappeared into its waves in the West at Sunset.

The Sun is carried on its currents through the day, and it sets with the waves in the West.

This Ocean encircles the Earth, land and sky, night and day. These *Waves* carry the tides of both *Heaven and Earth*. And it goes even deeper. It emanates from the Underworld. "*Nothing existed.*"

That Ocean. The waves of which continue to create the 'raw material' for, substantiate, nourish and 'fill' *our* Creation.

The seen and the unseen, the thought and the unthought, the dreams and the nightmares.

Center and Circle

Photo credit Hanneorla *Photo credit Travis Shinabarger*

Compare with Spiraling Serpent Fig. 35

...*or as the Quetzacoatl Pyramid (superimposed by a Spiraling Serpent)* FIG. 139

In contemporary language this could be thought of as '*stardust*.' As Aquarius dawns near, a new mantra will be heard across the land,

'*From Stardust you came, and to Stardust you shall return*.'

Now close your eyes and see it. These are the *Inner Waters*. It's a more accurate vibrational rendering for our moment in Time.

The *Union* of *Heaven and Earth* is a fundamental theme of pagan mythologies, especially in their Creation Myths. From this Union derives all of Creation, everything born under the Sun.

Including the Sun.

As we explored the *Gilgamesh epic*, we pondered how our imaginations might harness this concept, the 'story' of *Heaven and Earth*. Here in Central America, we are seeing a fresh mythological perspective, but also find our old friends the *World Ocean* and *World Tree* gainfully employed.

As we have witnessed in other mythologies, the two most important *celestial cornerstones* are North and East. '*They are one, and they are two*' to quote Dr. Gimbutas.

Here we are again witnessing that fusion, a *celestial mythological* and *astronomical shorthand* combining our now familiar *Center and Circle*. *Center and Circle*.

We're working from the inside out.

Even though different cultures use different divisions for their management of Time, we must all still deal with the same cycles of Heaven. *All* the astronomies of antiquity had to begin with the *Center* if they were ultimately going to understand and predict any of the long-term rotations of Heaven.

Athena's Web

Center and Circle

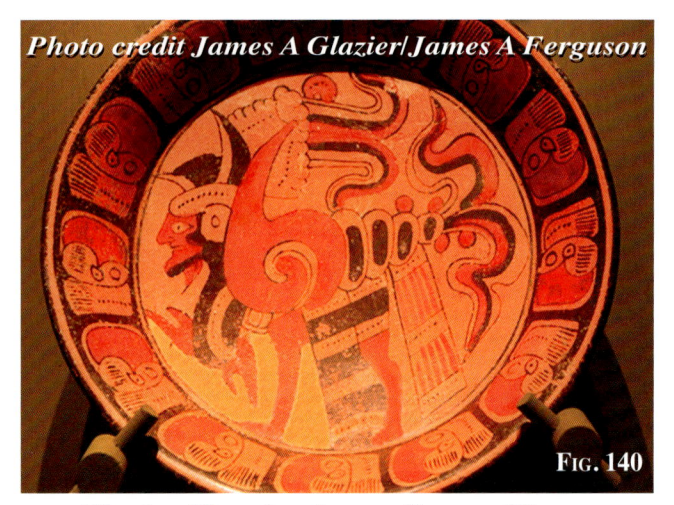

Fig. 140

The familiar circular attributes of Draco
(Photo credit James A Glazier/James A Ferguson)

It's evident Heaven runs on circles (if ellipses can be loosely included as 'out-of-round' circles). Circles are controlled by their center (or epicenters).

We are searching for the *Mayan Center*.

Once again, the *Tree* is the Center. In the pre-Creation overture act of our drama, the *Sea* is what the *Tree* is in the Center of. The *Sea* will eventually become the 'modeling clay' of all Creation, to 'circle' with all the other celestial children of the Skies around the pole after it has been sufficiently molded, worked and shaped.

We have seen it manifest thus far as *Twins* and *Bull*, but that's western imagery.

The *Popol Vuh* continues, and from the Oneness of Nothingness, there came Two, and from these Two came Union and Creation.

"*Then came the word. Tepeu and Gucumatz came together in the darkness, in the night, and Tepeu and Gucumatz talked together. They talked then, discussing and deliberating; they agreed, they united their words and their thoughts.*"

You can reduce that last paragraph to a single sentence:

"*In the beginning was the word.*" —John 1:1
Same thing.

Back to the *Popol Vuh's* Creation,

"*Then while they meditated, it became clear to them that when dawn would break, man must appear. Then they planned the Creation, and the growth of the trees and the thickets and the birth of life and the creation of man.*"

"*Thus it was arranged in the darkness and in the night by the Heart of Heaven who is called Huracán.*"

We have encountered '*the Heart of Heaven*' before, in the guise of the Dragon's star, *alpha Draconis*. At one time, it was His 'Heart' (*Thuban*) that marked the Center.

Throughout legends around the world, power over the elemental forces of nature have been associated with the Dragon, especially in Chinese, Japanese and Malaysian cultures, but they are found in the West as well.

Hercules must save a kingdom beset by a Dragon vomiting floods in one variation of a familiar story.

The Serpent's most famous character role in Greece was as *Typhon*, who chased the entire pantheon of Gods across the Mediterranean to the Nile, Tigris and Euphrates Rivers. While there, each deity disguised themselves in various forms in order to escape his fury.

In Mayan mythology, we are introduced to him as **Huracán**, [78] a term now associated with hurricanes and the tropical storms of the Caribbean.

So, say hello to our **Mayan Dragon**. The Maya knew him as *Kukulkan*, the Quiche as *Gukumatz*, the Aztecs thought of him as *Quetzalcoatl* and *Tlaloc* but it doesn't stop there. In Zapotec mythology, he is *Cocijo*. We are talking about a fairly small geographic area here, yet one filled with linguistic diversity. Other labels are *Tohil*, *Bolon* and *Tzacab* which brings us to an important point.

With every people, with every culture, you have different languages. With different languages, come different names and labels. With different names we might imagine different identities, unless, of course, it was the same image that kept coming up using different names. Even the myth gives us a whole heavenly host of names for the same spiritual entity. Most of them translate to '*Feathered Serpent*'. We have been told that this design was 'arranged' by '*the Heart of Heaven*' who is *Huracán*. They have determined the Center of the night sky.

"*The first is called Caculha Huracán.*
"*The second is called Chipi-Caculha.*
"*The third is called Raxa-Caculha.*
"*And these three are the Heart of Heaven.*" [79]

They are Three and They are One.

Center and Circle

If you consider the image and not the name, they're the same; *the flying Snake, the feathered Serpent, the Dragon.*

That's why He is known in other mythological traditions as the *World-Wide Serpent.*

One image,
One World,
One Snake.
Different names.
Many, many different names.

*Photo credit James A Glazier
James A Ferguson*
Fig. 141

The Feathered Serpent

In the late 19th century, Richard Hinckley Allen wrote this simple but telling conclusion at the end of his exploration of the constellation of the Dragon in his work, "*Star Names, Their Lore and Meaning,*" p. 212:

"*There seems to be confusion, and some duplication, in the nomenclature of Draco's stars, but their many titles show the great attention paid to the constellation in early days.*" [80]

Now we are looking at how the Mayan interpreted these '*stars of the Feathered Serpent.*'

In addition to the multiple handles being given to our Dragon, we once again find our familiar coupling of energies. While the archaeological evidence does not support the theory that the cultures living in the Americas extend back to Gemini (6300–4800 BC), the mythological record suggests otherwise. Once again, we have our celestial shorthand beginning to weave the Web of Creation.

"*Then Tepeu and Gucumatz came together; then they conferred about life and light, what they would do so that there would be light and dawn, who it would be who would provide food and sustenance.*" [81]

We are seeing the beginning of a pattern often repeated in Mayan myths, of taking counsel before dawn, while it was still dark.

They are seeking council by conferring with the Gods.

They are looking at the stars.

Of course, we are talking about the future, because at this stage of development in the myth the sky hasn't been created yet.

"*Thus let it be done! Let the emptiness be filled! Let the water recede and make a Void, let the Earth appear and become solid; let it be done.*

Fig. 142

**Hun-Hunahpú shooting
Vucub-Caquix in the Ceiba Tree**

"*Thus they spoke.*

"*Let there be light, let there be dawn in the Sky and on the Earth! There shall be neither glory nor grandeur in our creation and formation until the human being is made, man is formed.*

"*So they spoke.*" [82]

Without an audience, theater is not much fun. The Gods feed off the applause of the crowd and the 'juice' keeps them going.

"*Then the Earth was created by them. So it was, in truth, that they created the Earth.*

"*Earth! They said, and instantly it was made.*" [83]

This is the causal plane. The Mayans are demonstrating 'Mind over Matter'. You bring things into manifestation by thinking them into being.

Think it.
Speak it.
See it.

You bring it into being and 'see' the manifestation of your efforts.

"*Like the mist, like a cloud, and like a Cloud of Dust was the Creation, when the Mountains appeared from the Water; and instantly the Mountains grew.*

*Photo credit
James A Glazier
James A Ferguson*

"*Only by a miracle, only by magic art were the Mountains and Valleys formed; and instantly the Groves of Cypresses and Pines put forth shoots together on the surface of the Earth.*

Athena's Web

Center and Circle

"And thus Gucumatz was filled with joy, and exclaimed: "Your coming has been fruitful, Heart of Heaven; and you, Huracán, and you, Chipi-Caculhá, Raxa-Caculhá!"

"Our Work, Our Creation shall be finished," they answered.

"First the Earth was formed, the Mountains and the Valleys; the currents of Water were divided, the Rivulets were running freely between the Hills, and the Water was separated when the high Mountains appeared.

"Thus was the Earth created, when it was formed by the Heart of Heaven, the Heart of Earth, as they are called who first made it fruitful, when the Sky was in suspense, and the Earth was submerged in the Water.

"So it was that they made perfect the Work, when they did it after thinking and meditating upon it." [84]

The Mountains appearing from the Water is evidence of *Oceanus* as the background noise.

Magic brings the world into being. It's a mathematically elegant solution, both efficient and effective in its simplicity.

Works for me.

Do notice that there seems to be a distinction being made here between the *Heart of Heaven* and *Huracán*.

The *Heart of Heaven* is the *Axis Mundi*, the mathematical shaft that becomes the trunk of the *World Tree*. In its relationship to the Earth, it sits there, forever, or at least, as long as there's an Earth. It is rooted to the planet, and therefore to our vision of Heaven. It is what helps us form our frame of reference.

However, the *Heart of Heaven's* relationship to the sky and the stars *does* change. It is now (at the time of the writing of this myth) a Mayan Dragon that hugs the axis.

Huracán, on the other hand, is the mythological personality who guards this celestial shaft of Heaven and Earth and for thousands of years (Fig. 19) monitored this secret of Heaven.

Hence, these two terms describe the same thing; the *Center of the Circle*. Over time these serpentine stars have drifted from their endless vigil and finally laid down their burden. After literally thousands of years of flying overhead, the Dragon has set, and slept. As the Egyptians saw it, he must now contend with his own mortality.

We hear similar themes derived from different times and locations, in hidden pockets of populations who remember walking these paths in their stories. Their whispers are still on the wind.

Huracán was a *God of Wind*, *Storm*, *Fire* and *Lightning*. He lived in the windy mists above the floodwaters (where the rains emanate from in the Skies). In iconography He is depicted as a deity of lightning with only one human leg.[85] The other leg is shaped like a serpent.

Are you surprised?

The *Heart of Heaven* is the *Tree*.

Huracán is our *Serpent*.

There's a *Serpent* in our *Tree* of Creation.

Say hello to Draco, Eve.

FIG. 143

The Mayan World Tree
Photo credit Madman2001

This is **the story of a maiden** named *Xquic*. When *Xquic* heard the story of the **Great Tree** and the ripe fruit, she was amazed and asked her father, a great lord named *Cuchumaquic* about it.

"Why can I not go to see this Tree which they tell about?" the girl exclaimed. "Surely the fruit of which I hear must be very good." Finally she went alone and arrived at the foot of the Tree planted in Pucbal-Chah.

"Ah!" she exclaimed. "What fruit is this which this Tree bears? Is it not wonderful to see how it is covered with fruit? Must I die, shall I be lost, if I pick one of this fruit?" asked the maiden.

Then the Skull which was among the branches of the Tree spoke up and said:

"What is it you wish? Those round objects which cover the branches of the trees are nothing but skulls."

So spoke the Head of Hun-Hunahpú (the spirit of the Tree) turning to the maiden.

"Do you, perchance, want them?" it added.

"Yes, I want them," the maiden answered.

"Very well," said the Skull. "Stretch your right hand up.

"Very well," said the maiden, and with her right hand reached toward the Skull.

In that instant the Skull let a few drops of spittle fall directly into the maiden's palm. She looked quickly and intently at the palm of her hand, but the spittle of the Skull was not there.

"In my saliva and spittle I have given you my descendants," said the Voice in the Tree.

"Now my head has nothing on it any more, it is nothing but a skull without flesh. So are the heads of the great princes, the flesh is all which gives them a handsome appearance. When they die, men are frightened by their bones.

Gourds of the Calabash Tree
Photo credit Polyparadigm.

"So, too, is the nature of the sons, which are like saliva and spittle, they may be sons of a lord, of a wise man, or of an orator. They do not lose their substance when they go, but they bequeath it; the image of the lord, of the wise man, or of the orator does not disappear, nor is lost, but he leaves it to his daughters and sons. I have done the same with you.

"Go up, then, to the surface of the Earth, that you may not die. Believe in my words that it will be so," said the Head of Hun-Hunahpú and of Vucub-Hunahpú.

And all that they did together was by order of Huracán, Chipi-Caculhá, and Raxa-Caculhá."

After all of the above talking, the maiden returned directly to her home, having immediately conceived the sons in her belly by virtue of the spittle only.

And thus Hunahpú and Xbalanqué were begotten.

And so the girl returned home, and after six months had passed, her father, Cuchumaquic, noticed her condition. At once the maiden's secret was discovered by her father when he observed that she was pregnant.

Then the lords, Hun-Camé and Vucub-Camé, held council with Cuchumaquic.

"My daughter is pregnant, Sirs; she has been disgraced," exclaimed Cuchumaquic when he appeared before the lords.

"Very well," they said. "Command her to tell the truth, and if she refuses to speak, punish her; let her be taken far from here and sacrifice her."

"Very well, Honorable Lords," he answered. Then he questioned his daughter:

"Whose are the children that you carry, my daughter," And she answered,

"I have no child, my father, for I have not yet known a youth."

"Very well," he replied.

"You are really a whore. Take her and sacrifice her, Ahpop Achih; bring me Her Heart in a Gourd and return this very day before the lords," he said to the two owls.

The four messengers took the gourd and set out carrying the young girl in their arms and also taking the knife of flint with which to sacrifice her.

And she said to them:

"It cannot be that you will kill me, oh, messengers, because what I bear in my belly is no disgrace, but was begotten when I went to marvel at the Head of Hun-Hunahpú which was in Pucbal-Chah.

Center and Circle

Skulls in Tree photo credit Benjamin Jakabek

Fig. 145

The approach to the Calabash Tree

"So, then, you must not sacrifice me, oh, owl messengers!" said the young girl, turning to them.

"And what shall we put in place of your Heart? Your father told us: 'Bring the Heart, return before the lords, do your duty, all working together, bring it in the Gourd quickly and put the Heart in the bottom of the Gourd.' Perchance, did he not speak to us so? What shall we put in the Gourd? We wish too, that you should not die," said the messengers.

"Very well, but my Heart does not belong to them. Neither is your home here, nor must you let them force you to kill men. Later, in truth, the real criminals will be at your mercy and I will overcome Hun-Camé and Vucub-Camé. So, then, the blood and only the blood shall be theirs and shall be given to them. Neither shall my Heart be burned before them. Gather the product of this Tree," said the maiden.

The red sap gushing forth from the Tree fell in the Gourd and with it they made a ball that glistened and took the shape of a Heart. The Tree gave forth sap similar to blood, with the appearance of real blood. Then the blood, or that is to say the sap of the Red Tree, clotted, and formed a very bright coating inside the Gourd, like clotted blood; meanwhile the Tree glowed at the work of the maiden. It was called the "Red Tree of Cochineal," but [since then] it has taken the name of Blood Tree because its sap is called Blood.

"There on Earth you shall be beloved and you shall have all that belongs to you," said the maiden to the owls.

"Very well, girl. We shall go there, we go up to serve you; you, continue on your way, while we go to present the sap, instead of your Heart, to the lords," said the messengers.

When they arrived in the presence of the lords, all were waiting.

"You have finished?" asked Hun-Camé.

"All is finished, my lords. Here in the bottom of the Gourd is the Heart."

"Very well. Let us see," exclaimed Hun-Camé. And grasping it with his fingers he raised it, the shell broke and the blood flowed bright red in color.

"Stir up the fire and put it on the coals," said Hun-Camé.

As soon as they threw it on the fire, the men of Xibalba began to sniff and drawing near to it, they found the fragrance of the Heart very sweet.

And as they sat deep in thought, the owls, the maiden's servants, left, and flew like a flock of birds from the abyss toward Earth and the four became her servants.

In this manner the Lords of Xibalba were defeated. All were tricked by the maiden.[86]

Let's do a little comparison shopping, shall we?

"*In the beginning God created heaven and earth. Now the earth was a formless void, there was darkness over the deep, with a divine wind sweeping over the waters.*" Genesis 1:1-2

Remember our *Waters*? These are the mystical waters of *Oceanus*. Compare the above opening lines with the opening lines of the *Popol Vuh* (see p. 71).

"*God said, 'Let there be light,' and there was light. God saw that light was good, and God divided light from darkness. God called light 'day', and the darkness he called 'night'. Evening came and morning came: the first day.*"

"*God said, 'Let there be a vault through the middle of the waters to divide the waters in two.' And so it was. God made the vault, and it divided the waters under the vault from the waters above the vault. God called the vault 'heaven'.*"

Odin just split his Egg.

"*Evening and morning came, the second day.*"

And so it goes through the course of seven days, with the Earth, vegetation, etc. coming into being. For our purposes the fourth day is interesting, so let's look at that:

"*God said, 'Let there be lights in the vault of heaven to divide day from night, and let them indicate festivals, days and years. Let them be lights in the vault of heaven to shine on the earth. And so it was. God made the two great lights: the greater light to govern the day, the smaller light to govern the night, and the stars. God set them in the vault of heaven to shine on the earth, to govern the day and the night and to divide light from the darkness. God saw that it was good. Evening came and morning came: the fourth day.*" [87]

The Mayan Gods took their council in the evening, and then took action in the morning, creating man with the dawn.

In *Genesis* chapter two, God rests after his labors, and then makes Eden, Adam and Eve.

In Genesis 3:1 we meet the **Serpent in the Tree**.

"*Now the snake was the most subtle of all the wild animals that Yahweh God had made. It asked the woman,*

"*Did God really say you were not to eat from any of the trees in the garden?*" *The woman answered the serpent,*

"*We may eat the fruit of the trees in the garden. But of the fruit of the tree in the middle of the garden God said,*

'*You must not eat it, nor touch it, under pain of death.*'"

Then the serpent said to the woman, "No! You will not die! God knows in fact that on the day you eat it your eyes will be opened and you will be like gods, knowing good and evil."

'Then Yahweh God asked the woman,'

"What is this you have done?"

"The woman replied,

"The serpent tempted me and I ate."

FIG. 146

Adam-Eve, Serpent-Tree, Creation Myth

'Then Yahweh God said to the serpent,'

"Because you have done this,
Be accursed beyond all cattle
and all wild beasts.
You shall crawl on your belly
and eat dust
every day of your life.
I will make you enemies
of each other:
you and the woman,
your offspring and her offspring.
He will crush your head
and you will strike its heel."
'To the woman he said:'
"I will multiply your pains
in childbearing,

Center and Circle

*you shall give birth
to your children in pain.
Your yearning shall be
for your husband,
yet he will lord it over you."*

'To the man he said,'
"Because you listened to the voice of your wife and ate from the tree of which I had forbidden you to eat,
*Accursed be the soil
because of you.
With suffering shall you get
your food from it
every day of your life.
It shall yield you
brambles and thistles,
and you shall eat wild plants.
With sweat on your brow
shall you eat your bread,
until you return to the soil,
as you were taken from it.
For dust you are and
to dust you shall return."*

The man named his wife "Eve" because she was the mother of all those who live. Yahweh God made clothes out of skins for the man and his wife, and they put them on.

Then Yahweh God said,
"See, the man has become like one of us, with his knowledge of good and evil. He must not be allowed to stretch his hand out next and pick from the Tree of Life also, and eat some and live for ever."
"So Yahweh God expelled him from the Garden of Eden, to till the soil from which he had been taken. He banished the man, and in front of the garden of Eden he posted the cherubs, and the flame of a flashing sword, to guard the way to the tree of life."

—Genesis 3:1–24

We have two Creation Myths remarkably similar in many of their details. In one tradition it is *Huracan*, the one-legged, snake-legged God, in the other, it's the *Serpent* that wraps itself around the branches of the *Tree*. Each in turn speaks for the *Tree*, with Woman as the principle mediator.

It's a simple formula, but we tend to forget from time to time.

If you are born, you must die.

The *Tree* is the measure by which the Sky can be read. In times past the World once used both *Tree* and *Mountain* as their calibration for the Earth's Skies. As people began to pursue this knowledge more intently they built *Newgrange*, *Stonehenge* and the *Pyramid of the Sun* as their observatories. The *World Tree* represents part of the 'hour hand' of history, or at least mythology, that Time forgot.

God warns *Adam* and *Eve* not to touch the fruit under the pain of death, while *Xquic* curiously asks,

"*Must I die, shall I be lost, if I pick one of this fruit?*"

"*Pain of death...*" says God.

The correspondences are too striking to be mere coincidence. The *World Tree*, like the Dragon, is part of a once worldwide system of communication based upon symbolism rather than phonetics. If your doorway of understanding lies through *phonetics*, the names are different; they *sound* different. If your doorway of understanding is *symbolic*, then the images *look* the same. It's that easy. Are you *looking* or *listening*?

The Dragon and *Tree* are linked together, behind various masks, around the planet. We find *His tail* visible for all who looked up in the night Sky and were told that a *Great Serpent* flew around the pole, shaft, blow-gun (*Head of Hun-Hunahpu*), column or *Tree*.

Now. You've been very patient. Would you like to learn how to lance a Dragon?

Expelled from the Garden

Fig. 147

A Moment in Time

Center and Circle has been our mantra, and we return to the banks of the Nile to see what they saw as the heartbeat of life quickened as *Spring* with its *Crimson Ball*, blossomed over the desert.

The following four pairings of images are snapshots in time, taken at the same instant, one thousand years apart. By observing the *'hour hand of Heaven'* we can tell what 'Time' it is.

We are looking *West* and *North*, waiting for the Sun to set so the skies can get dark enough and we can 'see' the stars being 'born' to their daily motion.

The *Equinoxes* have long been used to tune the *Celestial System*. *Spring* is associated with the *East* because it births the year. *Dawn* is the start of the day.

The two are linked.

We're using *Sunset* as our marker because that's when the skies darken. Like the owl, the stars awaken and begin their sojourn through the quiet night. It's after sundown when the stars of the Chinese Dragon first appear in the fading light. This is important for our storyline. In it, we are given seasonal markers for three out of our four seasons (Figs. 25-29). We are monitoring the precise motion of the stars, carefully making our 'mark' from Earth.

The stars we are watching are seasonal stars: the fruit of knowing the *Celestial Center*. The motion we are using to best observe these 'eastern' stars is the daily rotation of the Earth—even though they are momentarily setting in the west.

Ever since the *Vernal Equinox* started moving through the stars of the *Twins*, themes of *Duality* have been woven together on a legendary loom.

Those who work with the *Sun* as a marker work with the day. Those that use the *Moon* as a marker work with the night. This is an important distinction between *Solar* and *Lunar calendars*. One of the many 'messages' the *Twin Serpents* are telling us is that we need to learn to use both.

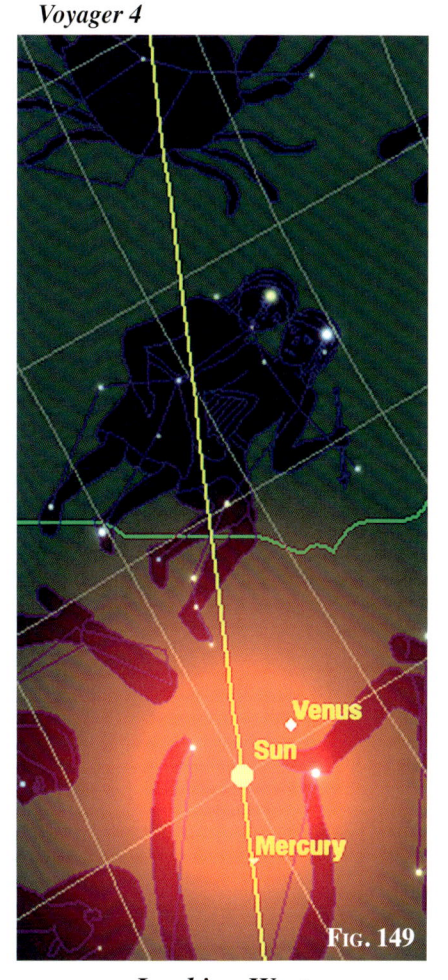

Center and Circle
Voyager 4

Fig. 149

Looking West after the Sun has set and the stars come out at night in 4000 BC. The Vernal Sun is just entering the horns of the Bull.

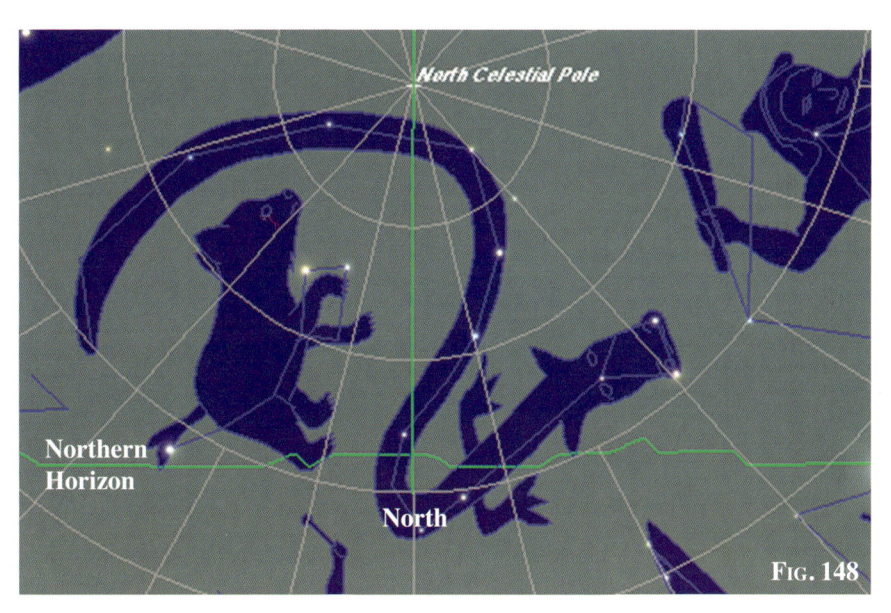

Fig. 148

Looking North at the same moment, on the 4000 BC Equinox

Athena's Web

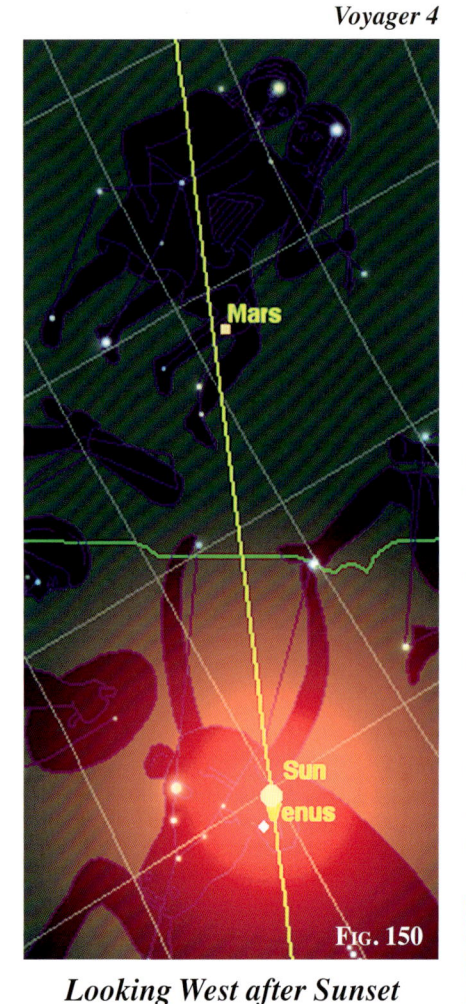

Center and Circle

One of the important distinctions between the *Center and Circle* in these illustrations is the speed with which the stars move along the *Circle of the Ecliptic*. By 1000 BC, the *Vernal Equinox*, marked by the first day of *Spring* falls in the back of the *Ram* (Fig. 155).

When working with precession, there's more motion close to the middle as opposed to near the center. Fortunately, the first 'bend' in the neck of the Dragon is abundant with stars (the cowl of our cobra?), providing a host of visual markers against which polar precession might be calibrated.

Where the astronomical 'cut' is made serves to help mark the *Center of Heaven*, symbolized by *Apophis* (Fig. 152) bending his neck to the 'sacrifice'.

FIG. 150

Looking West after Sunset one thousand years later in 3000 BC.
The Vernal Sun is cresting the Hyades. The yellow horizontal line is the horizon. Here, the tips of the horns are just setting.

Apophis

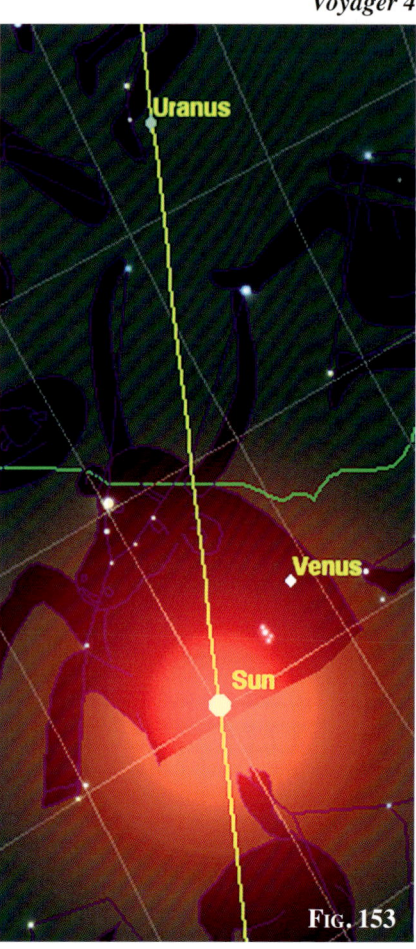

FIG. 153

Looking West after Sunset in 2000 BC.
The Vernal Sun has just left the shoulder of the Bull. Notice how little Draco moves, being closer to the pole, compared to the Equinoxes.

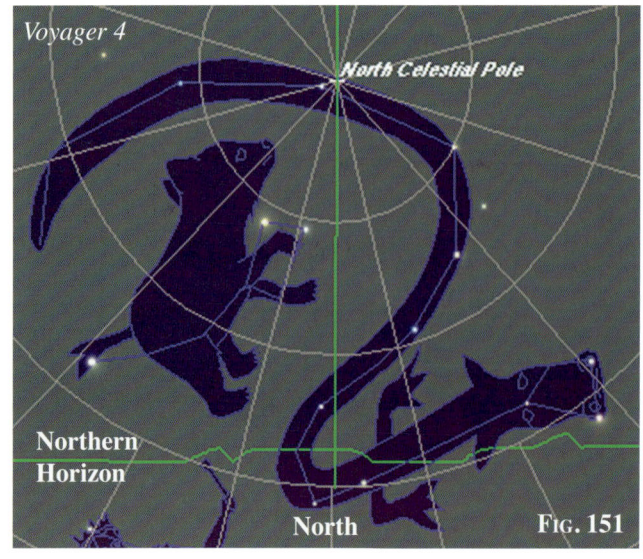

FIG. 151

Looking North on the 3000 BC Equinox

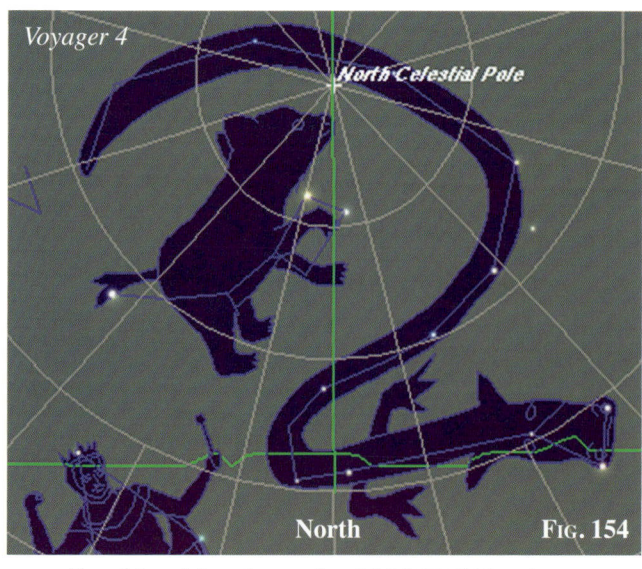

FIG. 154

Looking North on the 2000 BC Equinox

82

Athena's Web

Center and Circle

Voyager 4

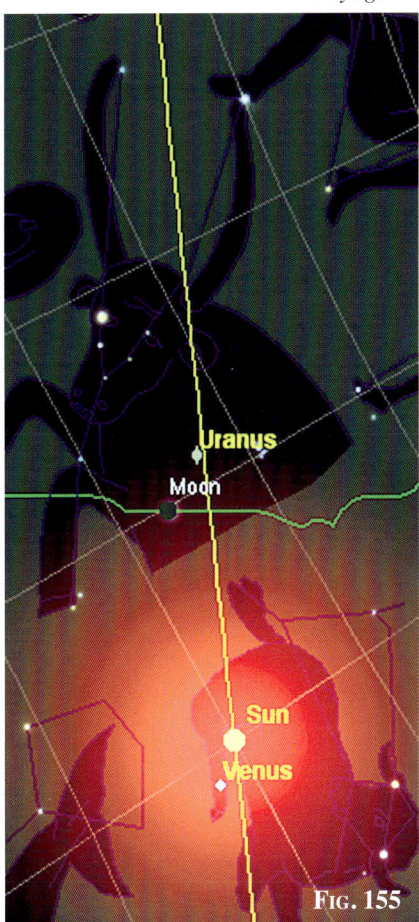

Photo credit Brother Childs

Let's take a closer look at this Egyptian serpent, *Apophis*.

We are reminded (p. 8) that *Apophis* is the sworn enemy of the *Sun God Ra*. Further, the king's enemies,

'*shall be like the snake Apophis on New Year morning.*' [88]

The *Serpent* has been run through because another year has gone by and it's time to see how far the Heavens have moved '*Out-of-Center.*' If, as we suspect, the line running *due North* is our *Tree of Life*, then it is what astronomers call the **Celestial Meridian**. This line that runs through your zenith (right overhead) and then '*threads*' the *North* and *South Poles*.

As we look South, we witness the majority of Heaven's celestial activity. Stars and planets that rise in the east culminate along this line. Once they pass this 'high-point', they begin to descend as they head toward the West.

The circumpolar constellations never rise and set because they never cut the plane of the horizon. But they cross the *Celestial Meridian* and here lies the essence of our myth.

Fig. 155

Looking West after Sunset in 1000 BC.
The Vernal Sun has precessed along the vertical yellow line (Ecliptic) to the rear of the Ram.

Fig. 157

If you know where to stand, any tree can be used to help find Heaven's Center.

Trees are *Nature's Meridian*. They can be used to determine the point around which the Heavens rotate, and then, by simply turning around and maintaining the same line of sight, one can observe the *Meridian* as it extends from the *Northern Pole* over your head and arcs around the Earth through the *Southern Pole* in a huge circle.

When the Sun aligns with the *Meridian*, it's *high noon*.

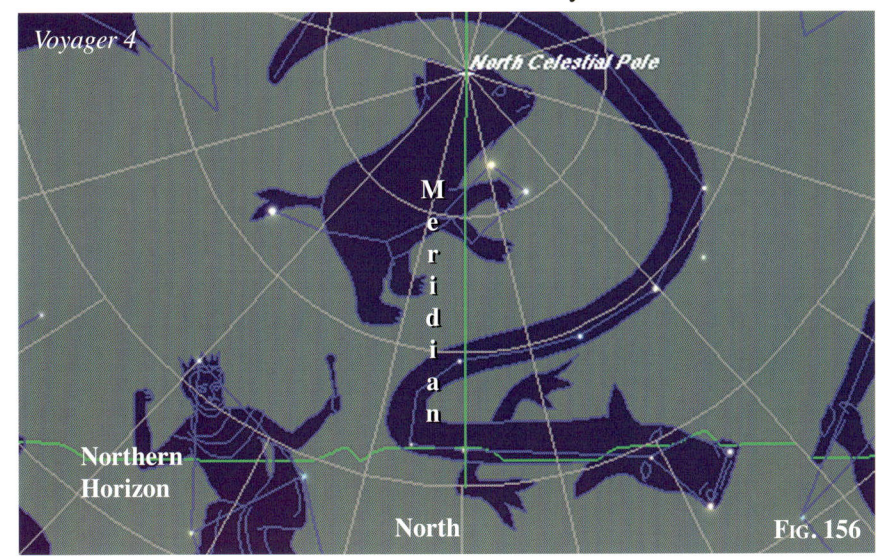

Fig. 156

Looking North at the same moment, on the 1000 BC Equinox

Athena's Web

Center and Circle

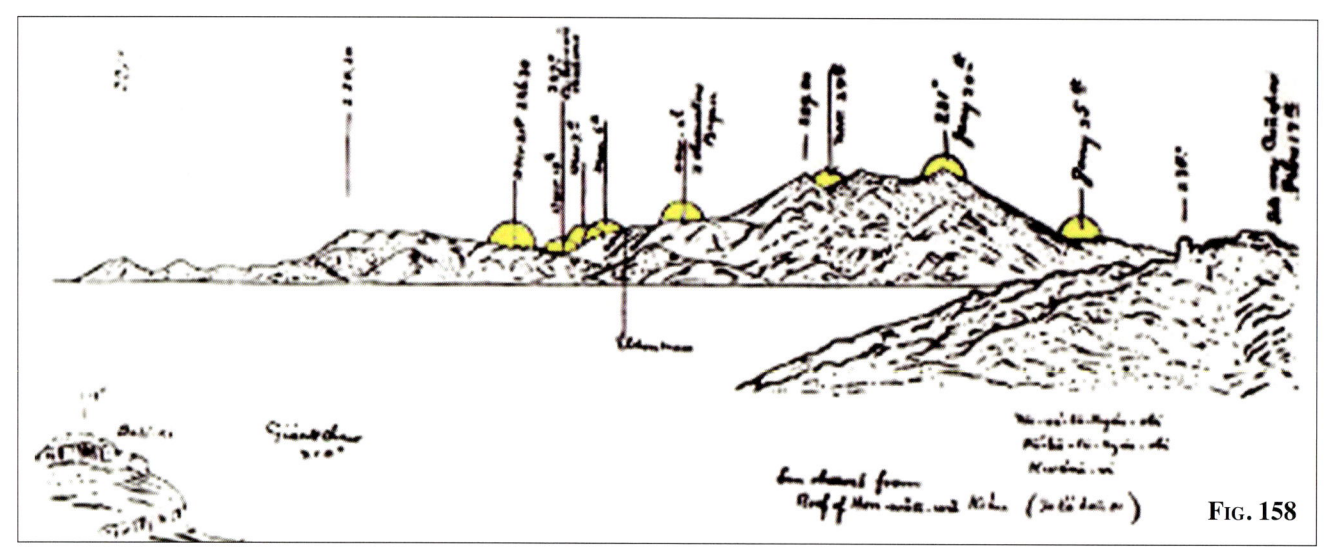

FIG. 158

Hopi Horizon—noting the position of the Sun along the skyline through the year as explained by native informants to 19th century anthropologist Alexander Stephen

The *Meridian* is our *Tree of Life*. It is rooted to the Earth and is less impacted by the 'drift' of precession. It anchors our gaze.

The Mayans took this concept one step further. Rather than simply employ one *Tree*, they used *five*. Their *World Tree*, in this case a *Ceiba* (*Ceiba pentandra*) was placed in the *Center*, with four additional *Trees* being aligned with each of the *four cardinal directions* of the compass. Once you have two visual points to work with, the process becomes similar to that of utilizing the foresight and backsight of a rifle. You use the two points together to align your focus, your attention, and in this case to align the *crosshairs* with your target.

And if you really want to do it right, you introduce a third point at some distance, like the *Dogon* (Fig. 42) and *Hopi* did.

You include a third point on the Earth's *horizon* (Fig. 158).

FIG. 159
Uhu

Of course, these early *Mother Nature* friendly sighting sticks evolved over time. Cut and dressed timber would eventually take the place of living trees, while mountainous terrain, islands, and natural stone would also be pressed into service. Finally, dressed marble and polished sandstone would be employed as more finely calibrated markers of Heavenly motion.

Solar and Lunar sites still work. If you know how to read them (knowing where to stand helps) they continue to 'tell time'.

However, stellar sites begin to lose their focus the day after they're built since precession slowly drags the stars along the horizon.

You need the priests (astronomers) to help re-calibrate the temples (Fig. 22).

FIG. 160

Notice how the Serpent lays his head along the ground just as Draco does when the stars emerge in Spring (Fig. 29)

Photo courtesy Tour Egypt

Fig. 161

Egyptian Serpent cut by the Tree

As we just mentioned, the *Ceiba* was placed in the *Center* of four other *Trees*, aligned on the four cardinal directions.

Genesis tells us the *Tree* stands "*in the middle of the garden...*" [89] Wherever you go, there you are.

The *Tree* works anywhere, on any continent, in any land. From a single *Center*, the order of *Creation* can be maintained. From *Chaos* to *Creation* and back again.

In the same manner that *Marduk* took careful aim in order to shoot his arrow through the jaws, stomach, and into *Ti'amat*'s heart, so the Egyptians are 'cutting' the neck of the *Serpent*, just as *Hun-Hunahpu* shoots *Vucub-Caquix* from the *Ceiba Tree* (Fig. 142).

Photo credit G. Licence

Fig. 162

The Serpent opened their eyes

Center and Circle

Photo credit Rodrigo Alvarez-Icaza

Fig. 163

Sunlight and Truth—Fruit from the Tree of Knowledge

To be born on this planet is to eat fruit from the *Tree*. For those who knew how to discern *the Will of God* from the *face of the Sky*, this was your first test. If you can't get the *Center* right, you don't get the *Circle* right; and if you don't get the *Circle* right, you're out of touch with *God*.

Any of them.

As we climb back down out of the *Tree* and plant our feet on the ground, it's time for us to leave this garden behind, for during the time that the Vernal Equinox was passing the *stars of the Bull*, the Earth's fertility was emphasized and Nature, whether wildcrafted or domesticated, was what was seen as preeminent. This lush fertility would give way to another epoch, introducing a new series of celestial frames, but the patient Bull had one last act on center stage yet to play.

So let's put the pieces together. We have the *Tree*, rooted in the Earth. Against the shaft of the *Tree* we can observe how the stars move at night, the rotation of the *Heavens* and the geometry of space.

The *Tree* and the *Dragon* are teaching us about our World. Not only are we learning about science, mathematics, astronomy, and history, we're also seeing art and mythology. Why did cultures focus so much attention on the *Bird*, *Eggs* or the *Bull*? They were tracking the stars. In addition to the disciplines mentioned above, use of the calendar instilled a better understanding of agriculture, not only by timing such events as when the Nile would flood, but also by providing agricultural star tips that were old when the ink on Hesiod's parchment was still fresh. When we hear of agriculture spreading through into Europe, know that the star lore went right along with it.

Athena's Web 85

Center and Circle

More Mayan World Tree

Fig. 164

The Ceiba, World Tree of the Maya

The **World Tree of the Maya**[90] is a *Tree* to be proud of. One of the largest *species* in the *Central American Rainforest*, it's light gray bark is capped by a unique flat top crown. It's huge buttressed roots provide shelter for bats, who also help fertilize the tree and keep the insect population down. The *bat* is one Mayan symbol for the Underworld. The powerful subsoil root system supports a trunk that represents the "*middle ground*."

The Norse call it *Middle Earth*. Anteaters populate this zone, feeding on the large protein-rich termite nests which lodge among its lower branches.

Center and Circle

FIG. 165

Eye on the Sky

The *Tree's* upper branches radiate out horizontally, providing an excellent haven for the dome of Heaven to rest upon. One earthly star that shines forth from here is the Harpie, the largest of all eagles, who finds this tree's canopy a perfect roost. If we take another look at the *World Tree* and how it was depicted by the Maya, a few additional concepts begin to come into focus.

This *'Tree'* is a living observatory to monitor, calibrate and chart the Heavens. Here we can see a contemporary *'horoscope wheel'* superimposed on top of the *Mayan World Tree* image. A *'horoscope'* is nothing more than a tool, a lens with which to observe the Heavens and witness a moment in time. The name means *'hora'* the time or the hour, and *'skopus'* the observer; *"to observe the hour."* The *horizontal line dividing the chart in half* represents *the horizon* (Fig. 165). Anything above that line is what you can see in the Sky. Anything below that can't be seen because the Earth gets in the way. The blue *'slices of pie'* are the Sky above, while the green *'slices of pie'* are the Earth below. East, where the Sun and planets rise is on the left side of the wheel, while West is on the right. For those in the northern hemisphere, South is at the top of the wheel because we have to look South to see the planets and most of the stars.

Now, we notice some interesting things if we superimpose the *World Tree* back onto this image (Fig. 166).

First of all, the 'step' configuration marks 'units' along the *Tree* that can be used for *longitudinal observational calibration*. Secondly, if you look along the central beam that falls along the horizon, you will notice some cut-out notches. These were used to *calibrate celestial latitude*- how close, or far, stars rise or set from the *Celestial Meridian*. The vertical line climbing to the top of the wheel represents due South, our 'high point' of Heaven. Astrologers remember this traditional seat in their nomenclature, the *Medium Coeli* or MC, Latin for *'middle of the Heavens'*. We've been focusing on the Dragon and its guardianship over the North Celestial Pole, and for that one must look North.

How do you find due North (the Pole star) if you're looking at the planets? Turn around.

In order to accurately portray the *Skywheel*, you'll notice that the Earth is pictured 'upside down' (Fig. 165) from what we are used to (North on the top), but again, this simply represents a matter of focus. Were they paying attention to the activity in the Southern Skies or the Northern? The answer, of course, is that they were interested in both.

FIG. 166

Americas meet the Old World

Athena's Web

Center and Circle

Naturally, as the *Sun*, *Moon* and stars pass below the western horizon, they pass out of sight, into the various 'layers' of the 'Underworld'. The Egyptians also believed that where the planets ventured below the horizon, they were in the Underworld.

The Mayans are expressing a similar concern, and once again *Death and the Underworld* celestially derive from being buried '*beneath the Earth*'.

The ability to 'see' (mathematically deduce, observationally understand, accurately predict) beneath the Earth is remembered by the Maya (or their predecessors) in mythology. Once again, we return to the *Popol Vuh*.

Vucub-Caquix [91] was very proud of himself, and as *Creation* thinks he's pretty hot stuff.

"I shall now be great above all the beings created and formed. I am the sun, the light, the moon," he exclaimed. *"Great is my splendor. Because of me men shall walk and conquer. For my eyes are of silver, bright, resplendent as precious stones, as emeralds; my teeth shine like perfect stones, like the face of the sky. My nose shines afar like the moon, my throne is of silver, and the face of the earth is lighted when I pass before my throne.*

Vucub-Caquix represents one of the previous Mayan four 'Ages' and is reflecting the state of being of the World. Obviously, *Vucub-Caquix* feels as though Life is good. He is the *Sun* and the *Moon* and all the other celestial bodies, because he is *God of the Sky* for their time. Like most indigenous peoples everywhere, they felt as though God "*made man in his own image*" to quote one source. That is *exactly* what is happening here. This God has vision.

"So, then, I am the sun, I am the moon, for all mankind. So shall it be, because I can see very far."

But it is a time before they have figured out planetary motion below the horizon; out of sight, out of mind. Therefore the 'power' to understand fully half of the information is not yet readily available, and the myth 'remembers' this limitation.

"*So Vucub-Caquix spoke. But he was not really the sun; he was only vainglorious of his feathers and his riches. And he could see only as far as the horizon, and he could not see over all the world.*"

Their astronomical abilities had not yet come fully up to speed.

The face of the sun had not yet appeared, nor that of the moon, nor the stars, and it had not dawned. Therefore, Vucub-Caquix became as vain as though he were the sun and the moon, because the light of the sun and the moon had not yet shown itself. His only ambition was to exalt himself and to dominate. And all this happened when the flood came because of the wooden people. [92]

What these, and all the other *Stargazers of Time*, were after is a greater understanding of the cycles of Heaven, for herein lies their future. *Vucub-Caquix* has learned (or more appropriately, the people of his time learned) how to use oppositions to read the face of the night Sky. He can see far, but as noted, only to the horizon.

To people of the past, the whole point of this operation was to understand *God's Will* and to be able to look ahead and warn the people about the future. This is why *Vucub-Caquix* feels his splendor is great, because he does have some power. He is the Sky, with his eyes represented as the two luminaries, the *Sun* and the *Moon*.

FIG. 167

Vucub-Caquix
Photo credit J. Elizabeth Clark

Center and Circle

His teeth (the stars) shine like perfect stones. Here they mean precious jewels, *'like the face of the sky'*.

Vucub-Caquix is vain because his power is incomplete. He does not know as much as succeeding generations will know once they mathematically determine the astronomy of below the Earth, or as they see it, of *Death* and the *Underworld*. Learning how to use *these* to chart the future could be helpful, however, and it could teach you how to pick the right 'time' to *'walk and conquer'*. We know that later the Aztecs will make extensive use of celestial correlation to wage battle. What we are witnessing here is the time when these abilities are still in their infancy.

"...the face of the earth is lighted when I pass before my throne."

Visible and available.

The *Popol Vuh* even gives us an indication as to when *Vucub-Caquix* yielded his reign, during the last Mayan 'Age'.

The current Mayan Age is *Tonatiuh's Face* (Figs. 168 & 173). The Maya tell us that it began on *August 13, 3114 BC*, therefore the previous Age must have ended at the same time. This was when the flood came *'because of the wooden people'*.

The previous Mayan Age was *Atonatium*, the *Sun of Water*. We are told that at the end of the *Fourth Epoch* (Fig. 174) everything perished because terrific storms and torrential rains covered the Earth. Reaching the peaks of the highest mountains the Gods changed men (the *wooden people*) into fish to save them from the *Great Flood*.

One tale, all speaking with one voice. The names have been changed to protect the innocent. We hear of the *World Tree* wherever we go, about stories of a *'Great Flood'* which destroyed civilization on the planet. That's Old Testament, Babylonian, Greek and now even Mayan myths each describing a huge *World Flood*.

Are they speaking of *different* floods (that covered the entire World)? Are they wrong, or do they reflect a reality no longer remembered by history, only legend?

And this brings us to one of the major distinctions between the *Skywatchers of the Americas*, and those from the *Old World*. They both involve what has come to be called *'the Great Year'*.

But first a little background.

In the *Old World*, those learning to discern the motions of the Sky looked to the two brightest lights of Heaven to base their wisdom, the *Sun* and the *Moon*. The *Sun* was made to rule over the day, and the *Moon* to rule over the night.

Genesis told us so (p. 79).

Photo credit Rafael Saldana

FIG. 168

*The Aztec Calendar Stone's 'inner Circle' depicts Five Mayan Ages, with our current 'Age' (**Tonatiuh's**) in the Center*

Looking back, we know that the *Moon* was an early 'calibrator' of Time. So far as we can tell, the Egyptians were the first to switch to a *Solar calendar* in 4236 BC (Time Line, pp. 34-35). The point is, many civilizations used both as depicted by the *Caduceus*, with a *Serpent* for each *Luminary* in the calender; one ruling the day, while the other rules the night.

Center and Circle

The *Dynamic Duo*.

Naturally, there are *Twelve Lunar* cycles to each *Solar* cycle.

Sort of.

The problem is, *Twelve Lunar* cycles don't actually equate to *One Solar* cycle; they '*almost*' equate. And here's the rub. As a result, we have to watch this calendar and effect periodic repairs in the form of *Leap Years*. The mathematics of *12 to 1* comes close to working, but the fit is not exact. Every four years we must 'insert' another day into the calendar in February, giving it a 29th day instead of its 'normal' 28. This is because our *Solar Year* is 365 *and a quarter days* in length. Every four years this extra '*quarter day*' adds up to another full day, so we insert February 29th, accommodating the leftovers. On *Leap Years* therefore we have 366 days in the year. A year that is not a leap year is called a common year.

But even these computations aren't exactly right, as the year is actually 11 minutes short of 365 and a quarter days, and therefore, still more adjustments must be made. Every 400 years we add yet *another* day in an effort to keep pace.

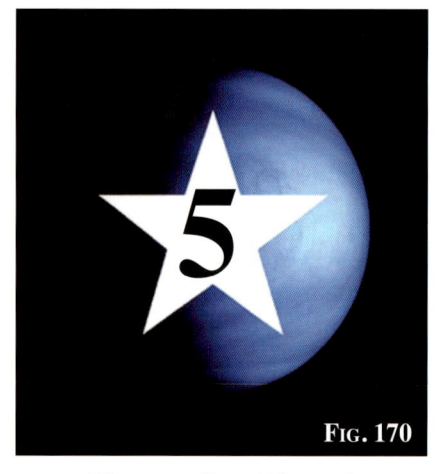

FIG. 170

Venus — Star Planet?

Even *this* correction is not entirely enough. We've determined that somewhere in the future we'll have to add still another day not currently figured into the system in order to keep seasonal pace between the *Sun* and *Moon*.

It's as though we were rolling through Time on this rubber tire and every fourth year there's a distinctive 'thump' as the patch hits the road.

Thumpity, thump.

Naturally, this *Moon-Sun* relationship is reflected in our units of time, which is why we have twelve months (from the Old English, *monath*, related to the Dutch *maand* and the German *Monat*, from the '*Moon*') to the year.

This issue is also a common one with *Lunar calendars*. We know the *Greeks*, *Romans* and *Hebrew* cultures used *Lunar calendars* as their mainstay. Many *Arabic* cultures still do. The *Moon* has a cycle that's difficult to miss (and being the brightest object in our night skies doesn't hurt either). But since the *Lunar month* is only 29 and a half days, twelve times that is 354, which falls some eleven days short of the *Solar year* (365.25 days).

Because of this shortfall, after two 'years' you're 22 days behind, after three years, 33. This is where the priests come in. It was their job to monitor the *Sun-Moon* relationship and make certain the pair didn't get very far out of seasonal alignment with each other. As a result, every two or three years anyone who uses a *Lunar calendar* must make corrections, inserting an *intercalary month* (instead of an *intercalary day*) to keep pace with the correct seasonal cycles. Caesar threw out the *Moon* (beginning each month with the *New Moon* as an official calibrator of Time) when Rome's priests were found to be selling these extra months to the highest political bidders. *Caesar* knew of the corruption because, as a former *Pontifex Maximus* (chief priest), he had been guilty of this infraction himself. For a price, you could hold power longer while still in office.

The Mayan calendar never ran into this problem. They didn't try to 'make' the *Moon* fit the cycles of the *Sun*, nor the *Sun* to the *Moon*. They didn't try to make *Twelve* equal *One*, when in reality, it's *Twelve and a fraction equal to One*.

FIG. 169

Harvest Moon

FIG. 171

Sun

Athena's Web

The Mayans used **Venus** as their tabulator. Since they were working with only a single planet, and not the *Luminaries*, they had to choose another numerical ratio to work with to become the Mayan model of Time. In attempting to 'honor' the essence of *Venus*, the Mayans meditated on her motion thoughtfully, as they did with each of the planets. While they contemplated *Venus*, they realized that in its passage around the Earth over the course of eight years, *Venus* makes five *inferior conjunctions* with the Sun, and these five conjunctions form a nearly perfect *Five-pointed Star* when measured against the *Circle of the Ecliptic*. The superior conjunctions do the same. This pattern is unique to the transits of *Venus*.

The *Old World* used the brightest luminaries in the Sky to establish their relationship of *Twelve to One*, choosing this as their model.

The Mayans used *Venus*, the *third brightest object* in the Sky after the *Sun* and *Moon*, and selected '*Five*' as the basis for their model of Time.

The **Great Year** is the amount of time it takes for the '**Precession of the Equinoxes**' to transit just once around the *Circle of the Ecliptic*. At present, it takes 25,765 years to complete a *single cycle*.[93] Of course, such a time scale presumes that the 'wobble' currently being experienced by the Earth remains a consistent figure over the course of 25,765 years, which I would personally tend to doubt.

The *Old World* took this 'unit' (*The Great Year*) and divided it by *Twelve*, producing '*Ages*' of over two thousand years (2147) apiece.

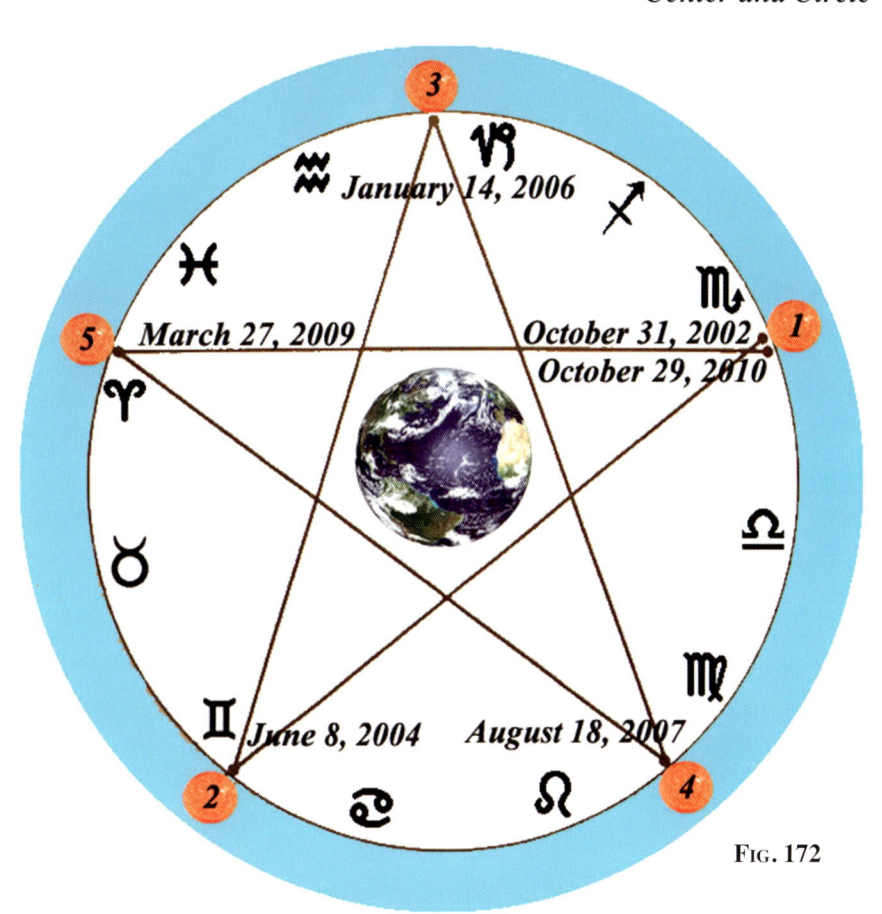

FIG. 172

Venus in eight years with five inferior conjunctions

The Mayans took this 'unit' and divided it into *Five,* generating Ages of a little over five thousand (5135) years.

The Mayans (or their predecessors) honored '*Five*' in a tip of the hat to *Venus*. The *Old World* honored '*Twelve*' in deference to the *Sun* and *Moon*. The Mayans divided the *Great Year* by *Five*, while the *Old World* divided the *Great Year* by *Twelve*.

One is not right, nor the other wrong. They are simply using two different '*foundation stones*' upon which to build their framework.

They're two different systems. Because the Mayan did not use *Twelve* as their base, their units of time seem strange to us, with 20 days to their 'week', and 18 months to their 'year', with the five unlucky days at the end of the year (just as the Egyptians thought their five end days were unlucky) representing 18 'and a fraction' months.

The Mayan myth tells us the last epoch was a period of great storms and rains, apparently leading to extensive flooding.

FIG. 173

The Current Age: Tonatiuh
Photo credit Joseph A Ferris III

Center and Circle

This is when *Tonatiuh*, the Lord of the current Mayan Age, began on *August 13th, 3114 BC*. We first see him being honored by the *Toltecs*, then the *Maya* and finally the *Aztecs*. His hair is *blond*, representing the light the *Sun* gives off. The *wrinkles* on His Face were thought to show great maturity and age. The *obsidian knife* piercing the tongue helped to open the doorway, to communicate (speak) with the Gods. Next to the *Maya*, the *Aztecs* developed a most sophisticated *Solar calendar*. It was felt that *Tonatiuh* was responsible for the movement of the *Sun*, and that *Life* demands *Sacrifice*. The Maya thought that the current Epoch would be destroyed by *Earthquake*. This is the Epoch that ended on *December 21, 2012*.

Julian Days [94] are a method of dating time that counts, one by one, the days that have elapsed since **January 1st, 4713 BC**, just as our *Vernal Equinox* was entering the tips of our *Bovine's horns*. If we subtract the number of *Julian Days* between *December 21st, 2012 AD* and *August 13th, 3114 BC* we come up with *1,872,024 days*. If this is a consistent interval with the other *Mayan Epoch*s, then all we should have to do is go back in time this number of days and we will discover the starting dates in terms of the *Gregorian calendar* (our calendar).

If this simple assessment is correct, then the date of the start of the *Fourth Epoch* would have been *April 17th, 8239 BC*.

Give or take a day.

FIG. 176

Ehecatonatium
Sun of Wind

This is the glyph for the *Lord of the Second Epoch* (Fig. 176). In this era, humanity (or the Gods attempt at humanity) was destroyed by strong *Winds*. The Gods transformed human beings into *Apes* or *Monkeys*, so they might cling to the trees and not get carried away by powerful *hurricanes*.

The *Wind Epoch* began *August 24th, 18,490 BC*.

Photo credits this page Joseph A Ferris III

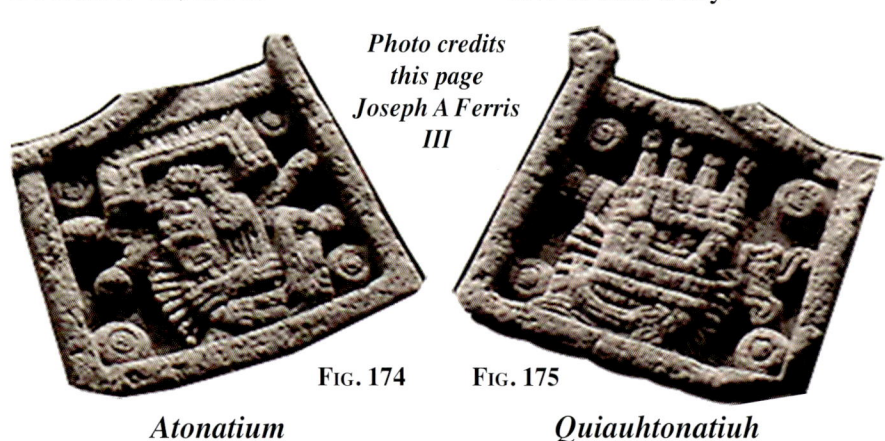

FIG. 174

Atonatium
Sun of Water

FIG. 175

Quiauhtonatiuh
Sun of Fire Rain

FIG. 177

Ocelotonatiuh
Sun of Jaguar

This is the glyph (Fig. 174) for the *Lord of the Fourth Epoch*. It was the *Flood Age* when all the *wooden people* died and even the tallest mountains were *covered in water*. This *Sun's* glyph is located to the bottom right of the current *Sun*, *Tonatiuh* in the *Center* of the *Aztec Calendar Stone* (Fig. 168).

This stone was known to the *Aztecs* as *Cuauhxicalli* or the *Eagle's Bowl*.

This is the glyph for the *Lord of the Third Epoch* (Fig. 175). According to tradition, at the end of this period everything was extinguished by the rain of *Lava* and *Fire*. People were saved from a molten death by being transformed into *Birds* by the Gods.

Using our same *Julian Day* count and adding onto the *Fourth Epoch*, the start of the *Fire Rain* rule was *December 20th, 13,365 BC*.

This final glyph was the *First Lord of the Epochs*. It was a time populated by *Great Giants* created by the Gods. They did not work the soil and lived in caves, like the *Cyclops* of *Odysseus* fame, eating wild fruits and roots, but were finally attacked and devoured by the *Jaguars* at the end of the Age.

Sun of Jaguar / Gold	Sun of Wind / Silver	Sun of Fire Rain /
23615 BC	18490 BC	13365 BC

This *Epoch* would have begun the entire *Five-series cycle*, and once again using our *Julian Day baseline*, we come up with *April 28th, 23,615 BC*.

If we want to know *how long ago* that was, we add on, say, 2012 years (to bring it up to the end of the calendar), which gives us a grand total of 25 thousand 627 years.

The current industry standard for the *Great Year* is 25,765 years.

There's a 138 year difference between what the Aztecs calculated as their *Great Year* and what contemporary science now says it will be. If we reduce that to a more recognizable figure, we come up with 1/187th. The Aztecs were within 1/187th of coming up with the same figure we use today.

FIG. 178

Photo credit Helen Sanders

Hesiod

If *we're* correct about the wobble.

In the four decades I have been researching the *Great Year*, I have seen the calculations for the length of the *Great Year* change, and I suspect it will be modified again.

Now we're going to switch gears and listen to another tale, this time a *Greek myth*. It is our oldest depiction of the **Ages of Man**, and it, too, describes periods of time that went before. The author is *Hesiod*, who, together with *Homer*, represents our earliest western writers.

In his '*Works and Days*' Hesiod tells us of the World's beginnings, in another *Creation myth*.

FIG. 179

Gold
Photo credit Oscar Anton

"Or, if you will, I will sum you up another tale well and skillfully—and do you lay it up in your heart—how the Gods and mortal men sprang from One Source.

First of all the deathless Gods who dwell on Olympus made a **Golden Race** of mortal men who lived in the time of Cronos when he was reigning in Heaven. And they lived like Gods without sorrow of heart, remote and free from toil and grief: miserable age rested not on them; but with legs and arms never failing they made merry with feasting beyond the reach of all evils. When they died, it was as though they were overcome with sleep, and they had all good things; for the fruitful Earth unforced bare them fruit abundantly and without stint. They dwelt in ease and peace upon their lands with many good things, rich in flocks and loved by the blessed Gods.

But after the Earth had covered this generation—they are called **pure spirits dwelling on the Earth**, and are kindly, delivering from harm, and guardians of mortal men; for they roam everywhere over the Earth, clothed in mist and keep watch on judgement and cruel deeds, givers of wealth; for this royal right also they received;

FIG. 180

Silver
Photo credit Martin Tremblay

—then they who dwell on Olympus made a **Second generation** which was of **Silver** and less noble by far.

It was like the Golden Race neither in body nor in spirit.

A **child** was brought up at his good **mother's side** an **hundred years**, an utter simpleton, playing childishly in his own home. But when they were grown and were come to the full measure of their prime, they lived only a little time and that in sorrow because of their foolishness, for they could not keep from sinning and from wronging one another, **nor would they serve the Immortals**, nor sacrifice on the holy altars of the blessed Ones as it is right for men to do wherever they dwell.

Then Zeus the son of Cronos was angry and put them away, because they would not give honour to the blessed Gods who live on Olympus.

Bronze Age	Sun of Water / Heroic Age	Tonatiuh / Race of Iron
8239 BC	3114 BC	2012 AD

Center and Circle

But when Earth had covered this generation also—they are called blessed Spirits of the Underworld by men, and, though they are of **Second order,** yet honour attends them also—

FIG. 181
Bronze
Photo credit Miriam Boy

Zeus the Father made a **Third generation** of mortal men, a brazen race, sprung from oaks and stones and ash-trees, and it was in no way equal to the Silver Age, but was terrible and strong. They loved the lamentable works of **Ares** (War) and deeds of violence; they ate no bread, but were **hard of heart** like adamant, **fearful men.** Great was their strength and unconquerable the arms which grew from their shoulders on their strong limbs. Their armour was of **bronze**, and their houses of **bronze**, and of **bronze** were their implements: there was no black iron. These were destroyed by their own hands and passed to the dank house of chill Hades, and left no name: terrible though they were, black Death seized them, and they left the bright light of the sun.

But when Earth had covered this **generation** also, Zeus the son of Cronos made yet another, the **Fourth**, upon the fruitful Earth, which was nobler and more righteous, **a god-like race of hero-men** who are called demi-Gods, the race before our own, throughout the boundless earth.

Heroic
FIG. 182
Photo credit Mike Fitzpatrick

Grim war and **dread battle** destroyed a part of them, some in the land of Cadmus at seven-gated Thebes when they fought for the flocks of Oedipus, and some, when it had brought them in ships over the great sea gulf to Troy for rich-haired Helen's sake: there death's end enshrouded a part of them. But to the others Father Zeus the son of Cronos gave a living and an abode apart from men, and made them dwell at the ends of the Earth. And they live untouched by sorrow in the **Islands of the Blessed** along the shore of deep swirling **Ocean**, happy heroes for whom the grain-giving Earth bears honey-sweet fruit flourishing thrice a year, far from the deathless Gods, and Cronos rules over them; for the Father of men and Gods released him from his bonds. And these last equally have honour and glory.

Iron
Photo credit Marco Franchino
FIG. 183

And again far-seeing Zeus made yet another generation, the Fifth, of men who are upon the bounteous Earth.

Thereafter, would that I were not among the men of the **Fifth Generation**, but either had died before or been born afterwards. For now truly is a **Race of Iron**, and men never rest from labour and sorrow by day, and from perishing by night; and the Gods lay sore trouble upon them. But, notwithstanding, even these shall have some good mingled with their evils.

FIG. 184
Photo credit Sandro Menzel
Father Zeus, Father Sky

And Zeus will destroy this race of mortal men also when they come to have grey hair on the temples at their birth. The father will not agree with his children, nor the children with their father, nor guest with his host, nor comrade with comrade; nor will brother be dear to brother as aforetime. Men will dishonour their parents as they grow quickly old, and will carp at them, chiding them with bitter words, hard-hearted they, not knowing the fear of the gods. They will not repay their aged parents the cost of their nurture, for might shall be their right; and one man will sack another's city.

There will be no favour for the man who keeps his oath or for the just or for the good; but rather men will praise the evil-doer and his violent dealing. Strength will be right and reverence will cease to be; and the wicked will hurt the worthy man, speaking false words against him, and will swear an oath upon them. Envy, foul-mouthed, delighting in evil, with scowling face, will go along with wretched men one and all. And then Aidos (reverence) *and Nemesis* (undeserved prosperity), *with their sweet forms wrapped in white robes, will go from the wide-pathed earth and forsake mankind to join the company of the deathless Gods: and bitter sorrows will be left for mortal men, and there will be no help against evil.* [95]

The comparisons between the *Five Ages* are interesting, especially the *First*. The *Aztecs* call them *Great Giants* who did not work, soil or toil and lived in caves eating wild fruits and roots. *Hesiod* says that they lived like Gods without sorrow of heart, remote and free from toil and grief. Both traditions speak of the *Divine Disappointment* with each of the previous *Ages*.

Including this one.

If these two myths derive from truly independent sources, there is absolutely *no logical reason* they should agree in this much detail.

Am I correct, Mr. Spock?

The prophecies *Hesiod* speaks of strike a particularly somber chord. We will come back to them later. Together with the *Norse mythologies*, these *End Times* predictions bear some investigation.

Ovid- detail from 1632 edition of Ovid's Metamorphoses

The *Aztecs* and *Hesiod* speak of *Five World Ages*, whereas *Ovid* and the *Hopi* speak of *Four*. *Hesiod* is 8th century BC *Greek*, while *Ovid* is 1st century AD *Roman*.

You're going to be quizzed on this, you know. You'd better start taking some notes.

As with the *Mayan* and *Old World* chronologies, one system isn't necessarily wrong because it's different from one we know. *Ovid's* system is close to *Hesiod's*; they both work from purer to more base metals. *Ovid's* four elements are the same as *Hesiod's*, *Gold*, *Silver*, *Bronze* and *Iron*. *Ovid* supplies us with additional detail, however.

Ovid — **Publius Ovidius Naso**

From his *Metamorphosis*,

"*Whereas other animals hang their heads and look at the ground, he made man stand erect, bidding him look up to Heaven, and lift his head to the stars. So the Earth, which had been rough and formless, was moulded into the shape of man, a creature till then unknown.*"

"*In the beginning was the* **Golden Age**, *when men of their own accord, without threat of punishment, without laws, maintained good faith and did what was right. There were no penalties to be afraid of, no bronze tablets were erected, carrying threats of legal action, no crowd of wrong-doers, anxious for mercy, trembled before the face of the judge: indeed, there were no judges, men lived securely without them.*

Never yet had any pine tree, cut down from its home on the mountains, been launched on ocean's waves, to visit foreign lands: men know only their own shores.

Center and Circle

Their cities were not yet surrounded by sheer moats, they had no straight brass trumpets, no coiling brass horns, no helmets and no swords. The peoples of the world, untroubled by any fears, enjoyed a leisurely and peaceful existence, and had no use for soldiers.

The Earth itself, without compulsion, untouched by the hoe, unfurrowed by any share, produced all things spontaneously, and men were content with foods that grew without cultivation. They gathered arbute berries and mountain strawberries, wild cherries and blackberries that cling to thorny bramble bushes: or acorns, fallen from Jupiter's spreading oak.

It was a season of everlasting spring, when peaceful zephyrs, with their warm breath, caressed the flowers that sprang up without having been planted.

In time the Earth, though untilled, produced corn too, and fields that never lay fallow whitened with heavy ears of grain. Then there flowed rivers of milk and rivers of nectar, and golden honey dripped from the green holm-oak." [96]

In both the Greek and the Mesoamerican myths, *corn* is identified as the first of *Mother Nature's* blessings.

In the *Silver Age*, the Earth apparently tipped on *Her Axis*, and the *Seasons* came into being.

Either that or humanity moved out of the Earth's equatorial regions, where temperate conditions helped ease the burden of existence. Either way, the myth is remembering a time long ago when Creation was something new.

We've been looking at a few stories that deal with larger tracks of Time, during eras that are remote at best, but let's return to our Bovine path in the Mediterranean.

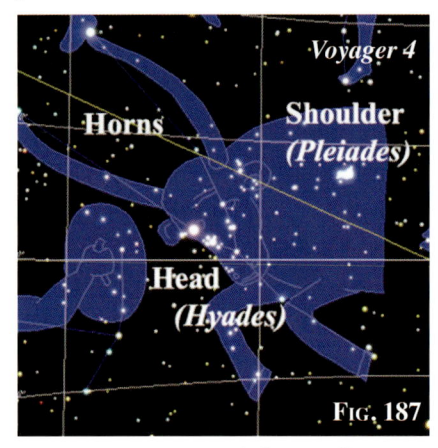

Anatomy of the Bull
Fig. 187

Thus far, we've observed as the *Vernal Equinox* has worked it's way down along the anatomical structure of what Gilgamesh refers to as 'the Bull of Heaven'. For a thousand years agricultural techniques were developed, tested and improved.

This was while the VE moved through the 'Horns of the Bull', from 4246 BC (Zeta) until 3247 BC (Aldebaran).

At the end of this pre-dynastic period, *Upper and Lower Egypt* are unified, somewhere around 3200–3100 BC.

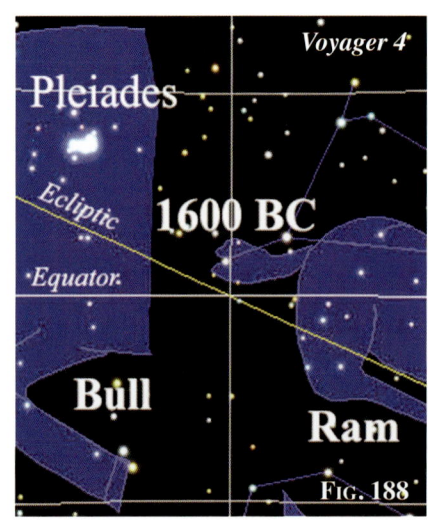

After the Shoulder... what?
Fig. 188

This is while the VE was in the '*Head of the Bull*', (Hyades) from 3247 BC (Aldebaran) until 2963 BC (gamma Tauri).

The *Epic of Gilgamesh* describes a period of time during which the VE (the astronomical cut) passed between the *Head* and the *Shoulder of the Bull*, from approximately 2963 BC (*Head*) until the last of the stars of the Pleiades (*Shoulder*) in 2170 BC (*Celaeno*).

And this is where the mythological world begins to run into a problem.

Some of Heaven's most brilliant stars come to a conclusion at this point, if you're using the *Vernal Equinox* as your marker.

The *stars of Gemini*, *Orion* and *Taurus* make up some of the *brightest* and most *dynamic* constellations of the night Sky. Two of these three groupings fall along the Ecliptic, while the other hugs the Earth's Equator and is therefore visible the world over. *Scorpio*, *Sagittarius* and the *Milky Way* all come together, but *Scorpio* is so low in the southern Sky as to not remain up long, leaving Sagittarius to fend for himself. Leo does well with Heart and Hair, but gets absolutely no help from the *Crab* or the *Virgin*. *Capricorn*, *Aquarius* and *Pisces* (together with *Cetus*, *Delphinus* and *Pisces Australius*) are all part of a vast area of the Sky collectively known as 'the Sea'.

The leader of this zodiacal territory (in a seasonal sense) is *Capricorn*, who was generally depicted with only the fore-quarters of a *goat*, together with the hind-quarters of a great *fish*. Like the centaur, it was considered a bi-corporeal being.

Last star of Gemini- *4634 BC	Tip of the Horns * 4246 BC	...Pre-Dynastic Period...
5000 BC	4500 BC	4000 BC 3500 BC

The *constellations* of the *Sea* that lie along the zodiac are a relative 'wash' as far as bright stars are concerned. Greek myth says the *Ram* left his brilliant fleece in a *Tree* in *Colchis* on the *Black Sea*. Except for the three 'medium-brightness' stars in the *Horns*, *Aries* really doesn't do a great deal to reverse this trend of 'lackluster' stars.

After the cluster of the *Pleiades*, some of *Heaven's brightest visual markers* fall behind, leaving mortal man with a heavenly dilemma to ponder.

How do you now mark the exact position of the *Vernal Equinox* in the Heavens without a bright star to guide you?

And what of our mythical theme? As you can see from the Sky map (Fig. 188), we're falling between two myth makers, old and new. The Greeks tell a tale that describes this celestial period beautifully. Both *Homer* and *Apollodorus* refer to it, but Ovid provides additional details.

"No sooner had he spoken than the bullocks, driven from their mountain pastures, were on their way to the beach, as Jove had directed...

Sky is telling us it's Spring, whether Season or Ingress. It is time to take the cattle to their summer grazing pastures.

"...they were making for the sands where the daughter of the great (Phoenician) king, used to play with the young girls of Tyre, who were her companions."

"Majesty and love go ill together, nor can they long share one abode. Abandoning the dignity of his scepter, the Father and Ruler of the Gods, whose Hand wields the flaming three-forked bolt, whose Nod shakes the universe, adopted the guise of a Bull; and, mingling with the other bullocks, joined in their lowing and ambled in the tender grass, a fair sight to see. His Hide was white as untrodden snow, snow not yet melted by the rainy South wind. The muscles stood out on His Neck, and deep folds of skin hung at his flanks. His Horns were small, it is true... (a curious statement constellationally) ...but so beautifully made that you would swear they were the work of an artist, more polished and shining than any jewel. There was no menace in the set of His Head or in His Eyes; he looked completely placid."

"Agenor's daughter was filled with admiration for one so handsome and so friendly. But, gentle though He seemed, she was afraid at first to touch Him; then she went closer, and held out flowers to His shining lips. The lover was delighted and, until He could achieve His hoped-for pleasure, kissed her hands. He could scarcely wait for the rest, only with great difficulty did He restrain Himself."

"Now He frolicked and played on the green turf, now lay down, all snowy white on the yellow sand. Gradually the princess lost her fear, and with her innocent hands she stroked His Breast when He offered it for her caress, and hung fresh garlands on His Horns: till finally she even ventured to mount the Bull, little knowing on whose back she was resting. Then the God drew away from the shore by easy stages, first planting the Hooves that were part of his disguise in the surf at the water's edge, and then proceeding farther out to sea, till He bore His booty away over the wide stretches of mid ocean. The girl was sorely frightened, and looked back at the sands behind her, from which she had been carried away. Her right hand grasped the Bull's Horn, the other rested on His Back, and her fluttering garments floated in the breeze." [97]

This Phoenician princess is being borne off to Crete to give birth to a new civilization.

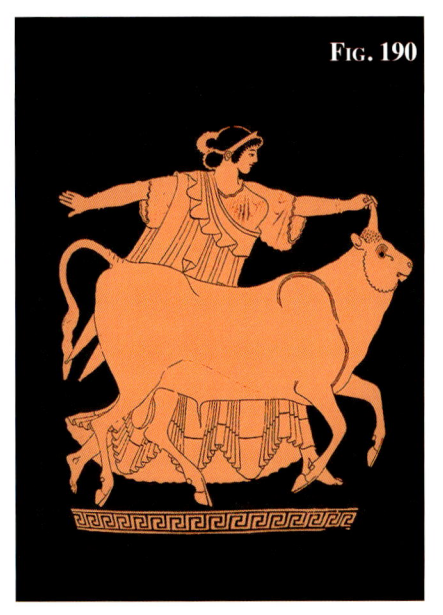

Photo credit Google Art Project

FIG. 189

Europa reconsidering?

FIG. 190

Europa in red figure pottery

Aldebaran * 3247 BC	* 2963 BC Last star of the Hyades	Last star of the Pleiades * 2170 BC
3500 BC	3000 BC	2500 BC	2000 BC

Center and Circle

While there, Zeus (the Skies) will sire three children; *Sarpedon*, *Rhadamanthys* and *Minos*. This last will become the founder of the most powerful dynasty on the Mediterranean, the *Minoan*.

Minoan civilization had its origins on Crete while the *Vernal Equinox* was still passing along the body of the *Bull*, beginning about 2700 BC. Judging by the architecture, it is not until 1900 BC that Minoa begins to rise in commercial prominence during a period known as the *Protopalatial*. [98]

Although it's of Greek origin, the myth suggests that Phoenician overlords may have come and helped organize the local agricultural community, crafting it into a trade-empire forged upon the back of the Sea.

What do the Minoans have to say about it?

Not much. By the time *Homer* put quill to papyrus, **Minoa** was a distant memory, centuries old. We have some archaeological evidence, the palaces of *Knossos* and *Phaistos*, for instance. There's pottery, and a few murals and that's about it.

No written records.

One other piece of evidence that does come down to us is a fresco of what has come to be called the *Bull-leaping ceremony*. Apparently, a Bull, whether trained or wild (we have no way of knowing), would charge at an athletic youth, who would, at the proper moment, grab the horns and do a flip up and over the head and onto the back of the Bull. From there, they would presumably jump off the rear of the animal, and scurry as quickly as possible to safety.

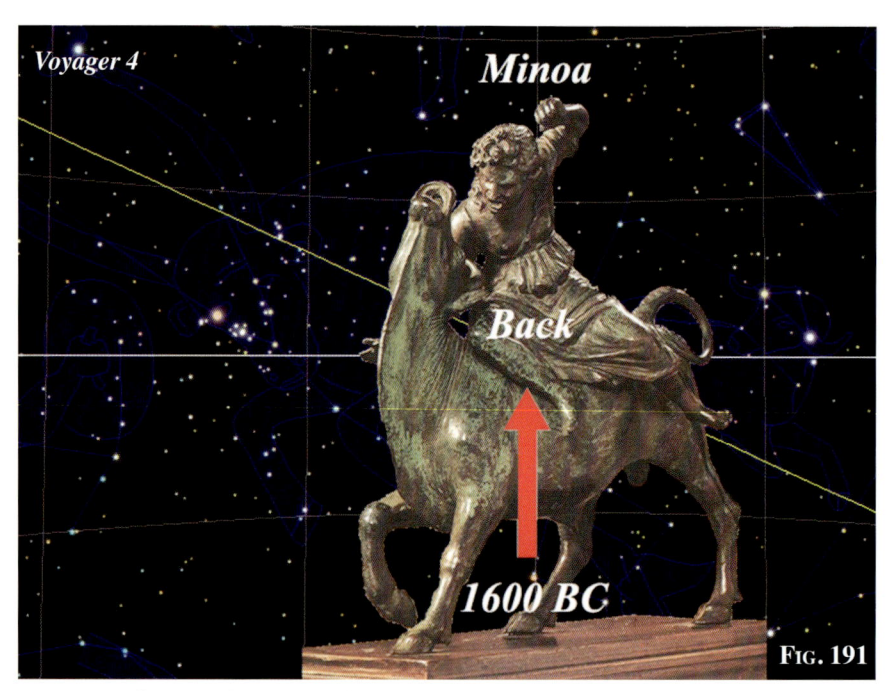

The emphasis in the myth is on the back of the Bull

Look at where the central youth is holding onto the *Bull* (Fig. 192). They are marking their spot in the *celestial parade*, their time in the Sun. For an agrarian community that had been working the land for eight centuries, *Fortune* finally fell their way. This rural ascendancy involved no signs of military defense or fortifications, yet rose to the top of the commercial world before the Mycenaeans later wrested this ripe olive from them (Fig. 1).

We have two sources we will take a look at, one *Greek*, the other *Minoan*. They're both describing the same event, a moment in Time during which *Mother Nature* was to rise supreme among those riding the wine-dark sea's broad back.

If this Minoan fresco represents the cut along the *Ecliptic*, our *Circle*, do we have any evidence of what our *Center*, the Dragon may be up to?

Would I ask?

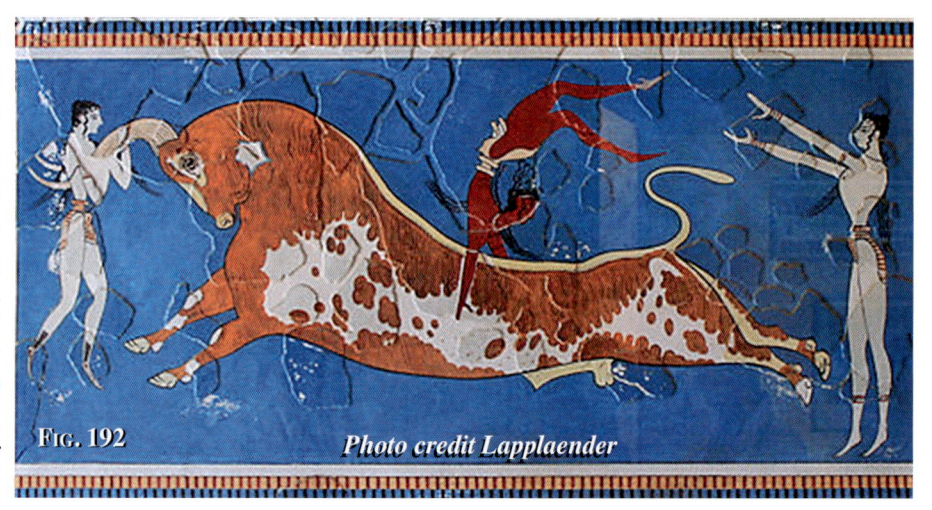

The emphasis in the fresco is on the back of the Bull

Center and Circle

The **Minoan Snake Goddess** is *Mother Earth* herself. She is the source of all that the *Good Earth* provides: timber for homes, food for the table, etc. Huge clay storage jars found at the Minoan palaces helped feed the population. Keeping food in these jars below-ground helped to keep them cool during the long, hot Mediterranean summers. Many of these jars featured the *Breasts of the Goddess*. Everyone nurses from the *Breasts of Mother Nature*. She feeds us all. Then, or now.

FIG. 193
Photo credit Chris 73

Minoan Snake Goddess

The *Twin Serpents* derive from our friend the *Caduceus*. They deal with the two calendar systems ruled over by the *Sun* and *Moon*. By heeding the *heavenly highlights*, agricultural hierarchies can maximize production and efficiency. This is the very same model we have seen being so deeply 'ingrained' in *Sumerian*, *Egyptian*, *Greek* and later *Mesoamerican* cultures.

Artwork courtesy Soa Lee   FIG. 194

In the prime of youth this Spring-crowned Europa looks forward to her destiny as the matriarch of a new dynasty.

You work with Heaven to prepare the Earth by harnessing the *Harmony of Nature* and her bridesmaids, the *Seasons*. In practical terms this end is achieved by coordinating civilization's efforts. In listening to the cadence of the calendar, by monitoring the measure of Time's metronome, they were able to harness the populace and with a *one, two, three, heave...*, everyone learned to pull together. Time tells you when. The obstinate roots of this system are still preserved in the *Farmer's Almanac* and a number of celestial gardening calenders designed to dance in tune with the Moon.

One advantage of working with the image of the *World Tree* is that it is obviously well connected to both *Heaven* above and the *Earth* beneath.

Here we are seeing (Fig. 193) the *Divine Feminine* at the end of an epoch which honored *Nature* and the natural world. With a bountiful harvest revered as one of life's true treasures, it was a time of general peace and contentment. Like the *World Tree*, the spinal column, back and body form the conduits which link *Heaven and Earth*. Like Her priestesses, themselves reflections of the Goddess, *She dances in rhythm to the Seasons*. But why does She hold the serpents away from Her Crown? And why is there a Cat on Her Head?

If we look back in time, to when the '*Heart of the Dragon*' was still approaching the Pole Position, our Egyptian *Uraeus* was worn as part of a *Northern Crown* which represented the *Dome of Heaven* (Fig. 199).

Center and Circle

Photo courtesy dominotic

FIG. 195

Wadjet

Meet **Wadjet**,[99] another Egyptian Dragon. The image of *Wadjet* with the Solar Disk is called our *Uraeus* (Fig. 195, 198, 199). It was the emblem on the crown of Upper (Southern) Egypt, as the Hedjet Crown (Fig. 196) was worn by rulers of Lower (Northern) Egypt.

FIG. 196

Photo courtesy Udimu

The Hedjet Crown worn by Narmer (above), and Deshret Crown (Fig. 199), our Uraeus

As a patron goddess, *Wadjet* was associated with the land and depicted as either a *Snake-headed Woman* or a *Snake*—usually an Egyptian cobra, a poisonous snake common to the region. Sometimes She was depicted as a *Woman with Two Snake Heads*, and this is where we begin to see the mythological roots and their ties to the Minoan Goddess.

At other times, she is depicted as a *Snake with a Woman's Head* (Fig. 198). They are all simply variations on a theme centered on Serpent.

Photo credit Paul Hessell FIG. 197

Hedjet Crown on the left, Deshret Crown on the right

Her oracle was in a well-known temple in *Per-Wadjet* that was not only dedicated to Her worship, but gave the entire city Her name. These Serpent themes stem from a common oracular source that later spread to Greece. We know when 'Truth' was sought, when 'astronomical accuracy' was 'conjured'. The *Going Forth of Wadjet*[100] was celebrated on December 25 with chants and songs.

Photo courtesy OnceAndFutureLaura

FIG. 198

Wadjet

It's the *Winter Solstice*, as the Sun begins to reclimb His *latitude ladder of declination*. It's Time to recalibrate the stars, beginning with a marker for the *Sky's Celestial Center*—the Dragon, Egypt's *Cobra*.

They're measuring ALL the seasonal ingresses as they keep pace with the *Stars of Heaven* and the *Center of the Circle*.

Of course, this is important only for cultures that want to know how to harness the Sky, mark its Center and arrange their calendar. Another annual festival of *Per-Wadjet* was held in Spring, on April 21st. Still another was held on the *Summer Solstice*. Are you surprised?

Center and Circle

Photo credit Armagnac-commons

As time moved on, the *Dragon Star* passed *Heaven's Pole*, and during the *Middle* and *New Kingdoms*, the *Serpent* was pushed out and away from Pharaoh's head with the diadem (Fig. 116). From the 19th through the 17th centuries BC, the *Celestial Serpent* appeared to move even further from the *spinal column* in the *Minoan Snake Goddess*. Draco still commanded the stellar 'territory' where the *North Celestial Pole* fell. It was no longer as precise as it once had been, since it's pole star was slowly slipping from the Central Axis of Heaven.

Wadjet wore the *Deshret Crown* (Fig. 199) identified with the *Uraeus* in early reliefs. The *Serpent* later gravitates 'down and out' into a diadem in Egyptian imagery. The Minoans translated this same notion into the hands of an out-stretched, dancing Goddess.

There's more than one way to slice a myth.

The mystique of the *Minoan Snake Goddess* is reminiscent of the *Eternal Feminine* in *Eurynome* dancing on the waters (Fig. 49) in an earlier incarnation. From *Menes* to *Minoa* is about 15 centuries. The *Celestial Serpent* is slowly sliding further from its pole position, while remaining closer than any other constellation during it's day.

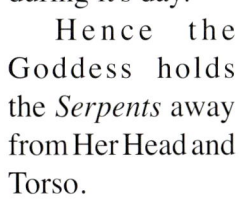

Hence the Goddess holds the *Serpents* away from Her Head and Torso.

Taking a giant Goddess step backwards now, what does the statuette represent?

FIG. 199
Deshret Crown

Marine life in the middle of the Minoan Mediterranean

A *Fertile* and *Bounteous Earth* dancing in tune with the *Seasons*.

C'mon everybody!
Let's eat, drink and be merry!

Note how the left arm balances for the flip, while the right marks the 'spot' on the back of the Bull

Athena's Web

Center and Circle

What about the figure of the cat, sitting on *Lady Minoa's Head*? We have been working with our drumbeat of *Center and Circle*. In *Minoa*, we had to draw upon two different images to complete our quest rather than finding a single story of two separate, yet combined images; one from the many votives, representative of the *Celestial Bull*, and one from the *Goddess*. Thus far, we have confined ourselves only to myths that first establish the *Center of Heaven* and then demonstrate where the *celestial cut* should be made along the *constellations of the Zodiac*.

Until now, the Dragon has served as our *Northern Celestial Marker*, but in *Minoan* times, the Dragon (*Serpent*) was beginning to move off-*Center*. If we listen carefully to our mythological trail tales, the essence of the *Lion* has been there all along.

Did you catch it?

The **Cat** is a feline, cousin to the proud Lion. The oldest astronomical writers speak of the bond between *Lions*, the *Sun* and *Kings*. To command the NCP is to command the point of power, authority, and in this case, Pharaohship. Just as the *Bull* was Heaven's anointed, so was *Pharaoh*.

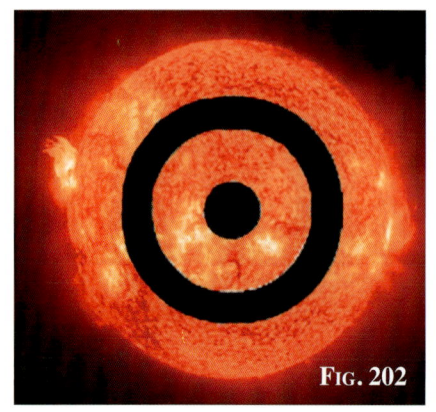

FIG. 203

African Lion by Jason Morgan

The Sun is identified not only with the *Vernal Equinox* and *North Celestial Pole*, our *Center and Circle*, but with all of the seasonal ingresses. It illuminates the *Equinoxes* on two days of the year, and the two *Solstices* on six other days. The Sun 'hangs' at the two Solstice extremes for three days each. The Sun provides the *Ecliptic*, the path of the Sun and central 'belt' around which all of the other 'wanderers' (planets) ride. Eight degrees either side of this 'belt' is the *Highway of Heaven* where the planets transit. It's therefore part of the '*Ring of Heaven*' we heard about in Celtic tales. The Circle with a dot in the Center (Fig. 202) is part of a design illuminated by the rising Sun at Newgrange on three days of the year, at the *Winter Solstice*.

We have watched the *Celestial Parade* follow the *Vernal Equinox* through the Sky, helping to define our *celestial shorthand* using both *Center and Circle*.

FIG. 202

The Circle with a dot in the Center is the symbol for the Sun

FIG. 204

The Sunlight striking this image marks the Solstice (see Fig. 70)

102 Athena's Web

Center and Circle

The Sun is a handy tool for *yearly* measurement. Using the *Vernal Equinox* 'illuminated' by the Sun (Spring), the constellations become a better measure of millennial motion. The Sun returns each year to the *Vernal Equinox*, but the constellations show a better path of where the *Vernal Equinox* has been, when it was there, and where it is going.

Beginning with these Minoan images, the *Serpent* has moved off-Center, so the Dragon no longer marks the exact Center (as the Cat does with the *Minoan Snake Goddess*). In earlier myths, this image of command and power was combined in the Dragon as a part of his formidable resume.

We have just seen this combination of motifs in *Wadjet* (Fig. 195). In this depiction, the *power*, *authority* and *command* are made clear in the form of the *Lion*, while the *Serpent* on top of his Head marks the then current position of Draco at the *Pole of Heaven*. With the *Minoan Snake Goddess*, we find the images reversed. Instead of a *Lion* with a *Snake* on the top of His Head, it's a *Goddess* with a *Cat* on Her Head.

That's one example. A second came from Enkidu, in his warning to Gilgamesh.

"*O my lord, you do not know this monster, and that is the reason you are not afraid.*

"*I who know him, I am terrified.*

"*His teeth are like dragon's fangs, his countenance is like a lion, his charge is the rushing of the flood.*

"*With his look he crushes alike the trees of the forest and reeds of the swamp.*" [101]

In this instance the Dragon is doubling up, both claiming his own mythological attributes, as well as establishing the visual marker for the *North Celestial Pole*. As Draco begins to slide from the Pole, he is stripped of some of his *central authority*, the power to mark the exact *Center of Heaven*.

Photo credit Nick Kaye

FIG. 206

Minoa's Bull of Heaven

Another example of '*Sun*' and '*Serpent*' is the *Royal Cat* of Re making the cut on *Apophis* next to the *Tree* (Figs. 152, 160, 161).

Like the Greek Sun God *Apollo's* taking the *Oracle at Delphi* from Draco, so the *Center of Heaven* has again been established in various mythological manners. Even Thor had difficulty picking up the 'paw' of the 'cat' *Jormungand*.

FIG. 205

Bast
Photo credit Richard Klug

"O my lord, <u>you</u> do not know this monster, and that is the reason <u>you</u> are not afraid..."

Athena's Web

Center and Circle

We've watched as the *Old Kingdom of Egypt* blossomed and reached full fruit on the banks of the Nile, but every Spring must come to an end.

United about 3200 BC, Egypt managed her *Agricultural Empire* with an efficiency that has since become legendary.

In *Genesis*, Abraham and his family are forced to travel to Egypt because there's a famine in the land. Only Egypt has the storehouses with enough grain not only to feed her own people, but also to those too who come in supplication.

They are given food.

As the *Vernal Equinox* pulls away from the *Pleiades*, the epoch of the *Celestial Bull* is drawing to a close, and with it passes the *Old Kingdom of Egypt*. Other traditions continue to honor the *Bull* with a great reverence, under a host of names, but the *Celestial Signature* will have passed.

They are no longer receiving the nod of *Zeus* in its *Bovine* form, which is to say the endorsement of *Sky*, the *Will of Heaven*.

This is what so many of the Biblical books rail against, following the traditions, habits and rituals of a by-gone era.

Looked at from a different perspective, *Yahweh* (Jehovah) has withdrawn his endorsement of the Age of Taurus.

It's not what the Skies say anymore. It's over. The Minoans somersault to a point on the *back* of the Bull, *behind* the constellational image. Much later, Mithraism will also rise in Roman form, with their Solar Hero (born on Christmas Day) actively sacrificing the Bull as their chief icon.

Photos courtesy Tour Egypt

FIG. 207

Ipuwer's report

The period had passed.

With every day, Time marches on. This image grows fainter as the Springtime point continues to move. You can't stop it.

After the ninety-year reign of **Pepy II** (2246–2152 BC), Egypt is plunged into a period of social darkness borne of famine.

Recent evidence has shown there was an abrupt global cooling circa 2200 BC which caused a drastic drop in the annual Nile flood levels.

Water is essential to an agrian community. High-resolution palaeo-climatic data for the period confirms an extended famine together with reports of squalor, pestilence, want, poverty, plunder, violence, murder, arson, hatred, treachery, enmity and fighting in the land. Egypt was a nation utterly dependent upon the *river*. They had learned to harness her so well that when she collapsed, she collapsed completely. It was to bring a kingdom down.

The following is an eyewitness account from an Egyptian sage named *Ipuwer* [102] (Fig. 207).

"*Lo, the desert claims the land. The towns are ravaged. Upper Egypt became a wasteland!*

"*Lo, everyone's hair [has fallen out].*

"*Lo, great and small say, 'I wish I were dead'.*

"*Lo, children of nobles are dashed against walls. Infants are put on high ground. Food is lacking. Wearers of fine linen are beaten with [sticks]. Ladies suffer like maidservants.*

"*Lo, those who were entombed are cast on high grounds. Men stir up strife unopposed. Groaning is throughout the land, mingled with laments. See now the land deprived of kingship. What the pyramid hid is empty.*

"*[The] People are diminished.*"

Within twenty years, fragmentary records suggest no fewer than 18 pharaohs and one queen ascended to the throne. With the de-stabilization of government, the Temple system which observed, compiled and recorded Heavenly data collapsed. Power passed on to local, de-centralized, independent warlords.

---------------- Sumerians --------- Akkad -- * ---* Babylonian culture --------- --- Assyrian*Chaldean Empire

3000 BC   2000 BC   1000 BC   0 BC

Center and Circle

Our investigation into the *Bull* quickly moved into the *Tigris-Euphrates River Valley* and we watched as many of the developments there mirrored developments in Egypt. The *End of the Epoch* was no exception.

In about 2300 BC, *Sargon of Akkad* was able to unite the central Mesopotamian River Valley, overthrowing his *Sumerian predecessors* in an explosive rise to power. Unfortunately for the *Akkadians*, the agrarian cycle was drawing to a close.

As we saw in Egypt, *farming communities* are dependent upon the soil and the water needed to nourish it. In the irrigated farmlands of southern Iraq, the traditional yield had been 30 grains returned for each grain sown, making it more productive than contemporary farming methods. What was termed '*the Upper Country*' of the newly conquered *Akkadian territory* was also dependent upon rain as a water source.

During the *famine* which swept the end of the 22nd century BC, rainfall approached its modern levels of less than 20 mm (1 in) per year in Iraq. Agriculture there became *totally* dependent upon irrigation. Fighting and militarism increased as agrarian communities collapsed. Nomadic herders sought reliable water supplies, crowding and competing with the needs of hard-pressed farmers. The *high level of agricultural productivity* of both Egypt, the Tigris-Euphrates and Indus cultures had enabled *high population densities*, providing agricultural muscle for the harvests, as well as military muscle for the army.

With the *famine*, those advantages began to evaporate quickly. Populations dwindled rapidly.

Within a hundred years, *the Akkadian Empire collapsed*, ushering in what has come to be known as a '*Dark Age*'. We don't know much about this period which spanned from the fall of Akkad (2083 BC) until the *Sumerian renaissance* (2050 BC).

Evidence shows that skeleton-thin sheep and cattle died as a result of the drought. Trade collapsed. One settlement, *Tell Brak*, shrank in size by 75%.

This climate-induced catastrophe impacted the entire Middle East, *coinciding with the collapse of the Egyptian Old Kingdom*. The *Great Bull* had been brought to His Knees.

The Curse of Akkad

For the first time since cities were built and founded,
The great agricultural tracts produced no grain,
The inundated tracts produced no fish,
The irrigated orchards produced neither wine nor syrup,
The gathered clouds did not rain,
the masgurum did not grow.
At that time,
one shekel's worth of oil was only one-half quart,
One shekel's worth of grain was only one-half quart....
These sold at such prices in the markets of all the cities!
He who slept on the roof, died on the roof,
He who slept in the house, had no burial,
People were flailing at themselves from hunger.[103]

Victory Stele of Naram-Sin
Akkadian, c. 2200 BC,
6' 7" tall

Naram-Sin's stele depicts the king's defeat of the Lullubi in present-day Iran.

Photo credit Thomas Schuman

FIG. 208

The Akkadians under Sargon dominated the Sumerians about 2300 BC. His grandson Naram-Sin pays homage to the Gods after a victory

Athena's Web

A pastoral post card from the Garden of Eden

"But in the 'Fertile Crescent' and in Egypt, on the other hand, cultured and highly developed civilizations jostled each other in colourful and bewildering array. For 1,000 years 'the Pharaohs had sat upon the throne. About 2000 B.C. it was occupied by the founder of the XII Dynasty, Amenemhet I. His sphere of influence ranged from Nubia, south of the second cataract of the Nile, beyond the Sinai peninsula to Canaan and Syria, a stretch of territory as big as Norway. Along the Mediterranean coast lay the wealthy seaports of the Phoenicians. In Asia Minor, in the heart of present day Turkey, the powerful kingdom of the ancient Hittites stood on the threshold of its history. In Mesopotamia, between Tigris and Euphrates, reigned the kings of Sumer and Akkad, who held in tribute all the smaller kingdoms from the Persian Gulf to the sources of the Euphrates.

"Egypt's mighty pyramids and Mesopotamia's massive temples had for centuries watched the busy life around them. For 2,000 years farms and plantations, as big as any large modern concern, had been exporting corn, vegetables and choice fruits from the artificially irrigated valleys of the Nile, the Euphrates and the Tigris. Everywhere throughout the "Fertile Crescent" and in the empire of the Pharaohs the art of cuneiform and hieroglyphic writing was commonly known. Poets, court officials and civil servants practised it. For commerce it had long been a necessity.

"The endless traffic in commodities of all sorts which the great import and export firms of Mesopotamia and Egypt despatched by caravan routes or by sea from the Persian Gulf to Syria and Asia Minor, from the Nile to Cyprus and Crete and as far as the Black Sea, is reflected in their business correspondence, which they conducted on clay tablets or papyrus. Out of all the rich variety of costly wares the most keenly sought after were copper from the Egyptian mines in the mountains of Sinai, silver from the Taurus mines in Asia Minor, gold and ivory from Somaliland in East Africa and from Nubia on the Nile, purple dyes from the Phoenician cities on the coast of Canaan, incense and rare spices from South Arabia, the magnificent linens which came from the Egyptian looms and the wonderful vases from the island of Crete.

"Literature and learning were flourishing. In Egypt the first novels and secular poetry were making their appearance. Mesopotamia was experiencing a Renaissance. Philologists in Akkad, the great kingdom on the lower Euphrates, were compiling the first grammar and the first bilingual dictionary. The story of Gilgamesh, and the old Sumerian legends of Creation and Flood were being woven into epics of dramatic power in the Akkadian tongue which was the language of the world. Egyptian doctors were producing their medicines in accordance with text-book methods from herbal compounds which had proved their worth. Their surgeons were no strangers to anatomical science. The mathematicians of the Nile by empirical means reached the conclusion about the sides of a triangle which 1,500 years later Pythagoras in Greece embodied in the theorem which bears his name. Mesopotamian engineers were solving the problem of square measurement by trial and error. Astronomers, admittedly with an eye solely on astrological prediction, were making their calculations based on accurate observations of the course of the planets.

"Peace and prosperity must have reigned in this world of Nile, Euphrates and Tigris, for we have never yet discovered an inscription dating from this period which records any large-scale warlike activities.

"Then suddenly from the heart of this great 'Fertile Crescent', from the sandy sterile wastes of the Arabian desert whose shores are lashed by the waters of the Indian Ocean, there burst in violent assaults on the north, on the north-west, on Mesopotamia, Syria and Palestine a horde of nomadic tribes of Semitic stock. In endless waves these Amorites, 'Westerners' as their name implies, surged against the kingdoms of the 'Fertile Crescent'." [104]

Aries

Fig. 209

Maguire Gallery

Mythological Ramifications

As Time *marches* on, the *Vernal Equinox* slips into the constellation **Aries**, and the need for a *new mythological motif* became apparent to all wise in the ways of Heaven.

It's a new vibration.

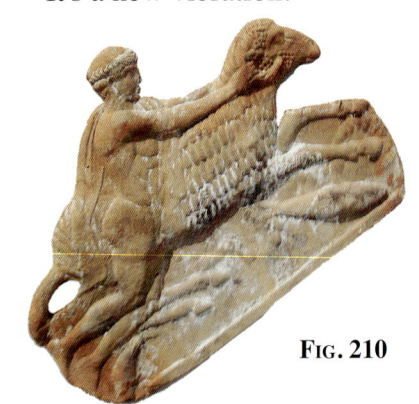

FIG. 210

Flying high through the Sky
Photo credit Peter Roan

After the collapse of the last of the Semitic dynasties, there begins a slow shift in society's focus; from that of social collective working with the Seasons and tilling the Earth together, to an emphasis on the *development of individual excellence* and *personal mastery* over the environment, eventually emerging with the Hebrew and Hellenistic cultures among others. As this new Age unfolds, each pioneered a new attitude toward life. Rather than collective societies compiling information and services under a centralized bureaucracy, *individual names* and *achievements* start to stand out. We learn of the daring of *Abraham* and *Moses*, admire the poetry and grace of *Homer*, wonder at the philosophy of *Aristotle* and *Plato* and remember the origins of the scientific method in the works of *Thales*, *Meton*, *Hipparchus*, *Ptolemy*, and others. In religion, philosophy, democracy, medicine and science, contemporary western society identifies with the reflections of its own cultural youth in these sprouting seeds.

One example of an attempt to bridge the mythological gap between these two may be found in the *Creation Myth* of the Bible.

In the story of *Adam and Eve*, all the important *celestial clues* are there. The choreography is still *Center and Circle*, but this time in a manner that's wrapped around a *Tree of Knowledge* in the middle of the *Garden of Eden*. It looks back to the period before the fall, depicted as a '*Garden*'.

It's not so much *where* was the Garden as *when* was the Garden?

The Garden was a period when the *Cornucopia of Life* was, in and of itself, a spiritual goal.

After *Adam and Eve* are expelled, their return is blocked by a *Weapon of War*. If we once again examine the oldest writers on the topic we find that the constellation *Aries* is linked to the *God of War*.

"So Yahweh God expelled him from the garden of Eden, to till the soil from which he had been taken. He banished the man, and in front of the garden of Eden he posted the cherubs, and the flame of a flashing sword, to guard the way to the tree of life." [105]

To *guard the way* to the *Tree*.

Not only is *Aries*, the *constellation* of the *Ram*, appropriately represented by a *Flashing Sword*, it is also one of the 'FIRE *signs*' of the Zodiac, adding still another clue to its *Celestial Identity*.

The '*flame of a flashing sword*' is a particularly appropriate symbol for our new *Vernal Equinox* indicator. Any imagery associated with the *Ram* may be used. *Aries* has now taken over from Taurus as the *Leader of the Twelve*, and so, the "*flame of the flashing sword*" (*Aries*) has been placed in front of the '*Garden*' (Taurus) (Fig. 147).

Aries has been placed in front of Taurus. It has replaced the old model, together with its themes.

Listen as God curses Adam.

"Accursed be the soil because of you. With suffering shall you get your food from it every day of your life... with sweat on your brow shall you eat your bread, until you return to the soil." [106]

Now compare this to an earlier myth which marked NCP and VE, for when *Gilgamesh* tells God what he has done, he is also met with a curse.

"Why did you do this thing? From henceforth may the fire be on your faces, may it eat the bread that you eat, may it drink where you drink..." [107]

The parallels are striking if we allow enough room for the mythic links and the mirroring of Dragon to *Snake*.

Last Pleiad- *2170 BC	--Phrixus*Cadmus*Jason---	--Mycenaeans--------------	---Troy *
2400 BC	2000 BC	1600 BC	1200 BC

Fig. 211

**The Ram
as art in ancient Greece**
Photo credit Luke Penrod

"*Of the sons of Aeolus, Athamas ruled over Boeotia and begat a son Phrixus and a daughter Helle by Nephele. And he married a second wife, Ino, by whom he had Learchus and Melicertes. But Ino plotted against the children of Nephele and persuaded the women to parch the wheat; and having got the wheat they did so without the knowledge of the men. But the earth, being sown with parched wheat, did not yield its annual crops; so that Athamas sent to Delphi to inquire how he might be delivered from the dearth. Now Ino persuaded the messengers to say it was foretold that the infertility would cease if Phrixus were sacrificed to Zeus. When Athamas heard that, he was forced by the inhabitants of the land to bring Phrixus to the altar. But Nephele caught him and her daughter up and gave them a ram with a golden fleece, which she had received from Hermes, and borne through the sky by the ram they crossed land and sea. But when they were over the sea which lies betwixt Sigeum and the Chersonese, Helle slipped into the deep and was drowned, and the sea was called Hellespont after her. But Phrixus came to the Colchians, whose king was Aeetes, son of the Sun and of Perseis, and brother of Circe and Pasiphae, whom Minos married. He received Phrixus and gave him one of his daughters, Chalciope. And Phrixus sacrificed the ram with the golden fleece to Zeus the god of Escape, and the fleece he gave to Aeetes, who nailed it to an oak in a grove of Ares.*" [108]

In this myth, the *World Tree* is an *Oak*.

And what is hung in its branches? A *Golden Fleece*.

The *Ram* has surged to a place of *celestial preeminence*. The new *Leader of the Pack* is looking for a *Good Shepherd*. In yet another *celestial signature*, the *Vernal Equinox*, the *first day of Spring*, now begins to slip into the *Stars of the Ram* from the rear.

The metal and color of the *Sun* has always been *gold*, and here we see that mythological theme combining with the *Celestial Ram* to become a *Golden Fleece*.

We hear the new vibration in the mythic outline, but if we listen carefully, it's roots go further back as these stories morphed from one to the other as the Sky-image changed. Tales of shape-shifters are commonplace in Greek mythology. You have to hold on tight.

Phrixus and *Helle* are Twins. The roots of this story extend further back than is at first apparent here, but we'll confine ourselves to the transition into *Aries*. What is happening through this period of cultural development?

The entire Mediterranean has just gone through a terrible drought. In desert regions this loss of rainfall would be most dramatically felt. Constellationally, we are moving off the *Pleiades* and into the nether-zone between Taurus and Aries (Fig. 188). Our mythic metaphors are preparing to put on a whole New Fleece of meaning.

In this story line, the King has married a new Queen, while the old Queen is still hanging around, but without her former power and influence.

Fig. 212

Phrixus reaches for Helle, who has fallen off the back of the Ram

---
| ---Dark Ages--- | ---Archaic--- | Greek Golden Age-- | -------- | Battle of *Corinth 146 BC |
| 1200 BC | 800 BC | | 400 BC | 0 BC/AD |

Center and Circle

The *Land* (Kingdom) is now married to a *New Heaven*, and they are producing children of a *New Dynasty*.

And what happened after the King married this new Queen?

The wheat was parched and the crops did not grow.

What were people trained to do during times of great calamity?

Turn to the Gods for advice.

The myth is telling us, in part, how they coped with this tragedy. While Greece (Boeotia) did not suffer as much as its peer desert communities, the rocky terrain of the region has never been conducive to vast agricultural development, and therefore the mountainous homeland would have also been hit fairly hard.

Is the myth remembering this set of events, mixing in the character names of those who actually lived through it, combining 'mythology' and history?

Of course it is.

The myth is describing the same meteorological shift, but as the *Hellenes* (Greeks) saw it.

The Old Queen (the old Agricultural Epoch) had been fine. People liked *Her* and *Her Children* (the years). The *Milk* (Fig. 122) from her *Udder* had sustained them and kept them happy. But the *New Queen* (the stellar shift since the great famine), is more belligerent. Times have been tough and She's angry, jealous of the children of the Old Queen. Even though she carries the official Seal of State, it's not enough.

She's quite testy and wants to do away with any vestige of the old ways. We will see similar examples of new habits cutting ties with tradition in other cultures.

Photo courtesy of Moyan Brenn flickr.com/aigle_dore"

FIG. **213**

"With sweat on your brow you shall eat your bread."

As they approach the Altar (VE) *Phrixus* and *Helle* escape on the back of the *Ram* as He rides through the Sky, not too difficult an imaginative hurtle if we understand that we're dealing with the stars as they transit overhead.

The problem is that the *Vernal Equinox* is just beginning to 'mark' the back of the *Ram*, and we haven't gotten a real handle on it yet.

There's no room at the back of the bus for both of them to sit on. As a result *Helle* 'slides off' the *Ram's Rear* and falls into the Hellespont, '*the Sea of Helle*'.

We know that later in Greece, overpopulation becomes a problem (not enough food/rocky environment). Different city-states would send out expeditions with a sizeable portion of the population in an effort to 'colonize' outlying areas, as they did in southern Italy and the Black Sea. It's my guess that the results of this famine pressured people to reach out to new environments in an effort to ease the burdens of the local population.

What do you bet these Quests set out in the Spring? There's a lot to be done before winter after you conquer someone else's territory.

It was a period of increased migrations and militarism. Supply and demand. The fewer the resources, the greater the demand.

Give it to me.

The arrival of *Phrixus* in Colchis seems to have gone well. *Helle's* drowning in the Hellespont may or may not have been a failed colonial enterprise.

Myths are not merely the fanciful wanderings of overactive imaginations. They are the collective memories of the people who lived through their reality, their life, their world.

Is Hollywood so different?

In a tradition that lands squarely in Abraham's lap, here's the Loeb footnote [109] to this myth.

"*These traditions point to the conclusion that in the royal line of Athamas the eldest son was regularly liable to be sacrificed either to prevent or to remedy a failure of the crops, and that in later times a ram was commonly accepted as a substitute for the human victim.*" There's a pattern here.

Oh, the *Tree* in the *Grove*? *Center and Circle*.

Isn't all of this getting a little old?

110   *Athena's Web*

Center and Circle

Jason must first tame the Bulls before taking the Golden Fleece

Now our *trail of the Dragon* leads us back into more familiar waters, on a voyage across the *Black Sea* to *Colchis* with a truly legendary crew. The rhythmic rise and fall of our craft lulls us into a false sense of security as the *Argo* speeds to her destination. The waters are parted into seafoam and spray by a prophetic oak cut from *Dodona*. Lost in their thoughts as they gaze across the prow are *Heracles*, *Jason*, *Theseus*, *Atalanta*, *Castor and Pollux* and other names echoed by the epic bards. The best heroes Greece has to offer are here with us today.

As in our previous myths, one of the essential clues is the link of *Center to Circle*, *Celestial Pole* to *Vernal Equinox*, Draco to *Aries*.

Apollonius of Rhodes rose to the occasion in his **Argonautica**,[109] the tale from which we draw regarding these *Heroic Exploits* with their legendary heroes.

Jason's quest is for the *Golden Fleece*, but to win it he is first required to perform another labor. He must harness two FIRE-*breathing Bulls* and plow the field known as *the Plain of Ares*, sowing *Dragons teeth* into the furrows, and then defeat the fully *armed warriors* that grow from the Earth.

No problem.

At this time in culture, observationally deprived of bright stars, it makes sense to determine the *Vernal Equinox* by using the brighter and still nearby stars of Taurus in mythic metaphor. This Jason does, as...

"*...tracking the countless traces of the bulls*" (by observing the rotations of the *daily Bull*), "*and they from some unseen den beneath the ground, where were their strong stalls*" (the entire Earth forms the walls of these stalls; this is the *constellation rising* from beneath the horizon),

"*all wrapt in smoke and flame, rushed forth together, breathing flaming fire.*"[110]

Here we have an interesting *mythic juxtaposition*. The new stars marking the Vernal Equinox belong to the *Ram*. *Aries* is linked to the element FIRE. The old constellation on the Vernal Equinox, and the one with the brighter stars, is Taurus. It's written that Taurus is EARTH. Here *Apollonius* is mixing his elemental metaphors by giving us FIRE-*breathing Bulls*, re-animating the Bovine archetype, but with the *element* of the incoming constellation, FIRE. Like two ends of a frayed rope, the traditions are being tied together. In this manner, the old themes could still be used to determine the latest (annual) *celestial geography*.

The old stars could still be used to determine where the *new Springtime point* had precessed to, but this is only the second of Jason's greatest labors. The first had just been getting there. No one had ever constructed a craft like the *Argo* before, nor taken it on seas such as these.

Jason vs. fully armed warriors arising from the Earth

Center and Circle

FIG. 216

Photo credit Peter Roan

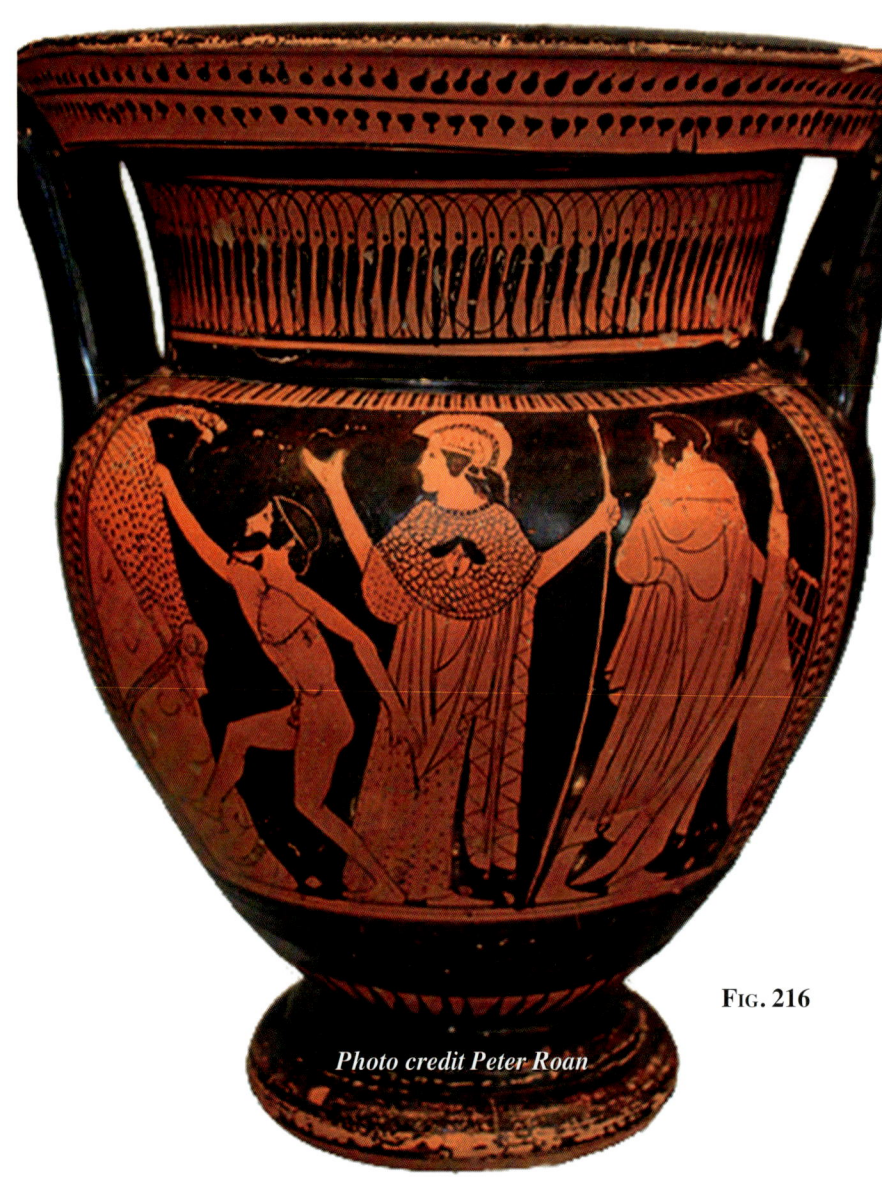

Jason seizing the Fleece as Athena looks on

After Jason ploughs the 'field', he releases the Bulls and "*scared them in flight over the plain.*"[111]

He has visually observed Spring, and the Bulls continue on their path until they set in the West. He then pauses to relieve his thirst, bending down to drink from the River.

While *Aea*, the *capital of Colchis* was built near the mouth of the river *Phasis*, there's another river that *Apollordorus* could also be referring to, and that's the *Eridanus*.

We've been tracking the Vernal Equinox across the stars of *Taurus* and *Aries*. Beneath these two constellations is the *Nile of Heaven*, a river that runs from its origins in Orion to some non-descript constellations in the southern hemisphere. This *celestial stream* is known as the *River Eridanus*.

After successfully completing his labor, Jason journeys to the *Dragon's lair*, where our *sleepless Serpent* is wrapped around the *Tree* guarding the *Golden Fleece*.

Center and Circle. Again.
Medea lulls the Dragon to sleep, and our hero runs for the *Argo* with their *golden prize* and escapes.

At the outset of this new mythic overture, we said that the development of *individual excellence* and *personal mastery* were earmarks of this Time. The Greeks looking back see it as a wave of '*humanism*'.

Coming to full blossom later in the *5th and 4th centuries BC*, a collection of city states led by Athens believed that intelligence and reason were what set man apart from the animals. This personal pride in one's own humanity led to the *anthropomorphication* of the *Divine Pantheon*.

Man made God in his own image, both in Heaven and out of Earth.

The strength of *Ares*, the beauty of *Aphrodite* and the majesty of *Zeus* were preserved on Earth in marble, while their personalities fully embraced the realities and spirit of day-to-day Greek life.

The Hellenes were responsible for the development of the concept of the *Individual Self*, which in turn led to the creation of *democratic ideals*. The successful commercial enterprises of this maritime power nurtured opportunities for *science*, *mathematics*, the *arts* and *literature* to be explored and developed.

Using *celestial geography*, *verse*, *myth* and *metaphor* together with their place in Time as threads of various textures and hues, *Athena's* nimble fingers tied off the knots of this story in *poetry* and *song*, displaying *Her* work for generations to follow.

*As once She guided this crew,
so now She guides you.*

Athena's Web

Center and Circle

Athena

Photo credit Sebastià Giralt

By hanging on tight to the *Tail of our Dragon*, we have prevented ourselves from being thrown by *Her shape-shifting abilities*. We have seen Her emerge as the *Guardian of the Forest*, wrapped around a *Tree* in a sacred grove, pierced by an arrow, and cut by a *Cat*. She also appeared as the *Serpent* that *encircles the Earth (World-wide Serpent)*, *bites its own tail (Ouroboros)*, *spirals* about a *Caduceus*, or wraps another Tree, this time in the *Garden of Eden*. In yet one more feat of magical mastery, we'll now gaze in fascination as our *Tree-Serpent* reappears as a *Sea-Serpent*.

In a variation on the familiar story of *Jason and the Argonauts* (Fig. 217), the Dragon which guarded the *Golden Fleece* was a *Sea-monster*. In this twist on the usual story line, Jason not only doesn't fare as well as in the more familiar version, but he comes out the loser; he gets swallowed whole by the Dragon. It's not until after the intercession of *Athena* that the *Dragon* is forced to *disgorge Jason*, and our dubious hero, somewhat the worse for wear, and in definite need of a shower, finally manages to obtain the *Fleece*.

Take note of the spear *Athena* holds, which, like the *Celestial Pole* (and *Equinox*) the story illustrates, divides the circle of the vase neatly in half.

We will see the Dragon in His more familiar *Serpentine* form again, as he battles *Cadmus* and his band of not-so-merry men.

and the Dragon

FIG. 217

As we move through the cultures found in Africa, Asia, Europe and even the Americas, we've noted many subtle shifts in the images used, but once the essential theme is understood, these differences become not perplexing, but fascinating.

We have observed the mythic dilemma by the lack of prominent stars in the '*Fleece*' of the *Ram*. Greek pride responded that the power and brilliance of those stars had been left in the Tree in Colchis with the *Golden Fleece*. As the *Ram* had led *Phrixus* to safety, he sacrificed the *Ram* to *Zeus*, the *God of Escape*.

Not only do we have the *Fleece*, but the *Tree* is in a *Grove sacred to Ares*. Because the *Fleece* was hung on the Tree and left there, the stars (or *Fleece*) of the constellation are said to be comparatively dim.

Center and Circle.
Center and Circle.

For this Age, a military drum beats out the cadence. Our Creation Myth is learning to digest a new archetype. It is discovering what Life under a Sky-god with *Ram's horns* is like, but it will take awhile before they move past the '*backside*' of the *Ram*.

We've now examined two new story lines, together with all the Art History which follows them 'on-down' the line. We have learned of *Phrixus* riding the *Ram*, and we have heard two versions of *Jason* in his *Quest* for the *Ram*. Let's take some time to reestablish our rein on the Dragon.

Athena's Web

Center and Circle

An even earlier *Greek myth* associates Dragon and *Ram*. A *Tree* is thrown in for good measure.

It's the story of the founding of **Thebes** in **Greece** by **Cadmus**, son of a Phoenician king, and one who apparently brought *stellar wisdom* with him from the East.

Having been told by the *Oracle at Delphi* to found a city wherever the wanderings of a *Cow* should *stop* (a mythological clue to the Vernal Equinox slipping out of Taurus), Cadmus fulfills the Oracle and wishes to offer a sacrifice to Zeus for this new endeavor. He sends his men out to fetch water for the offering, but they are killed by a horrid *Serpent* with scales that glitter like gold.

This *Serpent* protects the sacred waters needed for the blessing. Guess who?

"*The serpent, twisting his scaly body in a huge coil, raised his head so as to overtop the tallest trees, and while the Tyrians from terror could neither fight nor fly, slew some with its fangs, others in its folds, and others with its poisonous breath.*" [112] (Fig. 218)

Coming to look for his friends later in the day, Cadmus saw the lifeless bodies of his men and the *bloody jaws* of the *Serpent* and swore to avenge them, or die trying.

He threw a boulder at it, but it made no impression. He then threw his javelin, penetrating the *Serpents' scales*, piercing him through the entrails. The *monster* bent back (Fig. 219), trying to pluck out the lance with *His jaws*.

Then, coming down to the ground, he advanced on *Cadmus*, attempting to bite his spear.

The Center of the Web marks the NCP of Heaven about 1400 BC. "The serpent, twisting his scaly body in a huge coil, raised his head so as to overtop the tallest trees..." Head and jaws are high overhead.

FIG. 218

Stellar Wisdom from the East...

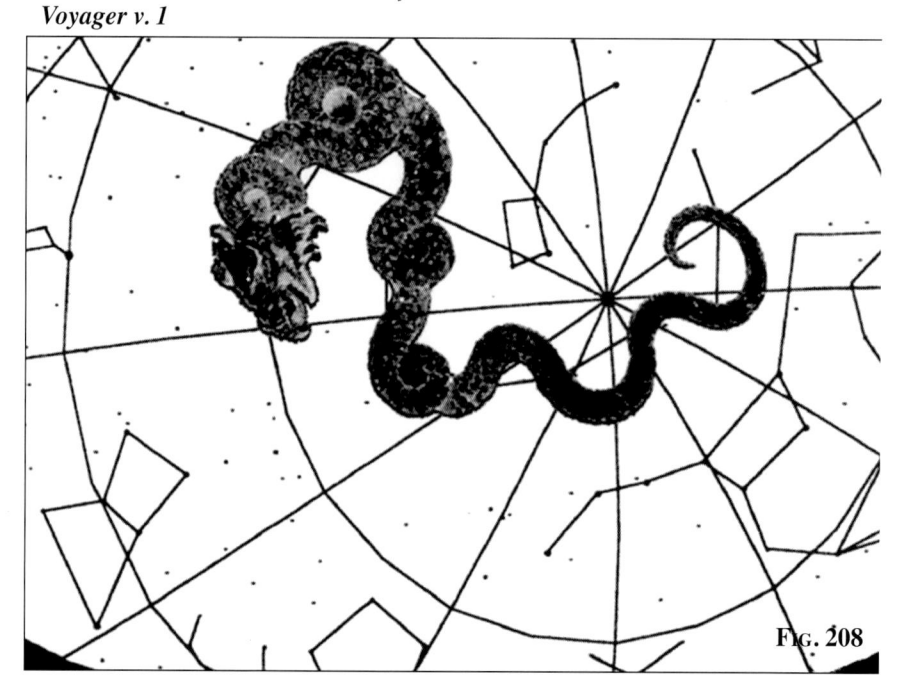

Voyager v. 1

FIG. 208

Here the Dragon is bent back trying to pluck the lance with his jaws. Notice the approaching horizon, seen in the corners. Many myths of antiquity were designed with a celestial focus in mind.

"At last Cadmus, watching his chance, thrust the spear at the moment when the animal's head, thrown back, came against the trunk of a tree, and so succeeded in pinning him to its side." [113] (Fig. 220)

As before, many of the images represent a precise awareness of the *motions of Heaven*. Of course the *scales that glitter* are the *stars*. Draco overtopping the Trees and then coming down to the ground is how this constellation visually rotates in Heaven as we saw in the seasonal Chinese version with the *Imperial Dragon* (Fig. 28). In the Greek the pinning of his head to the side of the *Tree* illustrates the constellation's *Nadir*, much like a huge hour hand rotating to the six o'clock position on the face of a *giant star clock*, one which lies just above the *northern horizon*.

From Ovid's *Metamorphoses* we obtain further insights into the Latin version of this tale.

"Hidden in this cave dwelt the serpent of Mars, a creature with a wonderful golden crest; fire flashed from its eyes, its body was all puffed up with poison, and from its mouth, set with a triple row of teeth, flickered a three-forked tongue...

"It was as huge as the serpent that twines between the two Bears in the sky, if its full length were seen uncoiled." [114]

No kidding.

Besides making a definitive reference of Dragon to *constellational source* in the Latin version, the Dragon is transformed back into a *Serpent*. These two kindred spirits, of Dragon and *Serpent*, seem to be relatively interchangeable in many traditions, as does the gender of the creature.

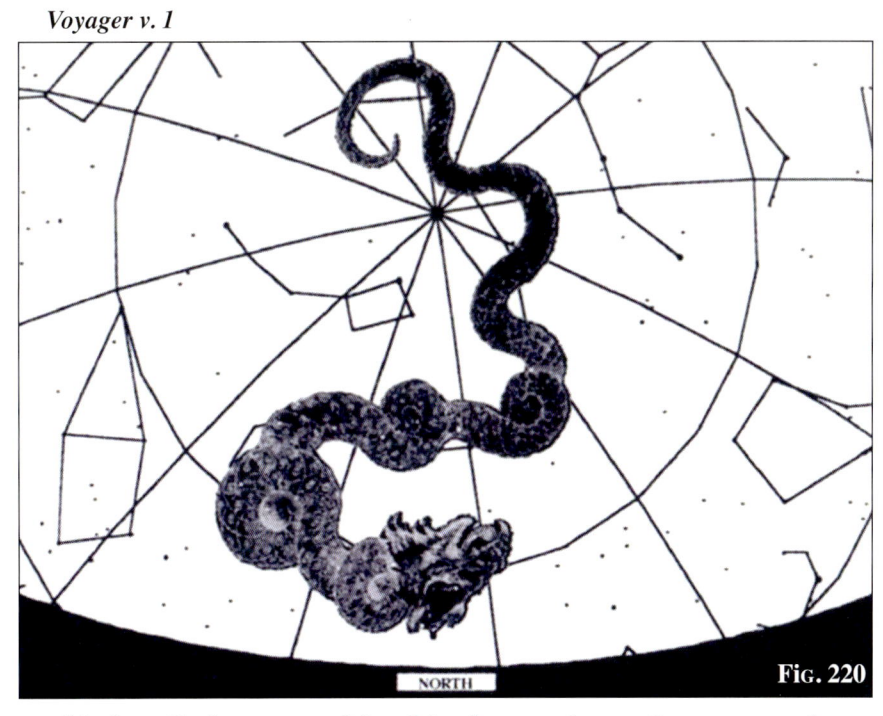

Voyager v. 1

Center and Circle

FIG. 220

"At last Cadmus, watching his chance, thrust the spear at the moment when the animal's head, thrown back, came against the trunk of a tree, and so succeeded in pinning him to its side."

It further substantiates the claims for a *Serpent* in the Tree in the *middle* of the *Garden of Eden* being derived from a common source. *One Earth-One Sky*.

FIG. 221

Cadmus slaying the Serpent of Mars, referred to as a Dragon in other versions of the story Later in life, Cadmus and his wife, Harmonia, reigned over Illyria. They were changed into Dragons and transported to the Island of the Blessed.

Both mythological and astrological traditions link the *Sword* to the *constellation Aries*, the *Serpent* to *Mars*, or the Dragon protecting the Tree with the *Ram's Golden Fleece* in a *Grove sacred* to *Ares*. These *woolen images* are interwoven into a rich tapestry that, once you see them are hard to miss. Yet these are not the only myths that demonstrate an awareness of the *circumpolar constellations* and link them to the Vernal Equinox. Like other myths, *Genesis* is neither first nor last in the series. It draws on older tales from *Gilgamesh*, and yet has similar theological themes to those found under the *Mesoamerican Tree*.

These images spin around and around so much it can make you dizzy. Like one of Agatha Christie's best tales, this is a real murder mystery. We have the body, method, means, and the *motif*.

Athena's Web

Center and Circle

Fruit from the Tree

**The Bush (Tree) and Ram
Center and Circle**

"It happened some time later that God put Abraham to the test. "Abraham, Abraham," he called. "Here I am," he replied. "Take your son," God said, "your only child Isaac, whom you love, and go to the land of Moriah. There you shall offer him as a burnt offering, on a mountain I will point out to you.

"Rising early next morning Abraham saddled his ass and took with him two of his servants and his son Isaac. He chopped wood for the burnt offering and started on his journey to the place God had pointed out to him. On the third day Abraham looked up and saw the place in the distance. Then Abraham said to his servants, "Stay here with the donkey. The boy and I will go over there; we will worship and come back to you.

"Abraham took the wood for the burnt offering, loaded it on Isaac, and carried in his own hands the fire and the knife. Then the two of them set out together. Isaac spoke to his father Abraham, "Father," he said. "Yes, my son," he replied. "Look," he said, "here are the fire and the wood, but where is the lamb for the burnt offering?" Abraham answered, "My son, God himself will provide the lamb for the burnt offering." Then the two of them went on together.

"When they arrived at the place God had pointed out to him, Abraham built an altar there, and arranged the wood. Then he bound his son Isaac and put him on top of the wood. Abraham stretched out his hand and seized the knife to kill his son.

"But the angel of Yahweh called to him from heaven. "Abraham, Abraham," he said. "I am here," he replied. "Do not raise your hand against the boy," the angel said. "Do not harm him, for now I know that you fear God. You have not refused me your son, your only son." Then looking up, Abraham saw a ram caught by its horns in a bush. Abraham took the ram and offered it as a burnt offering in place of his son. Abraham called this place, "Yahweh provides," and hence the saying today: On the mountain Yahweh provides.

"The angel of Yahweh called Abraham a second time from heaven. "I swear by my own self- it is Yahweh who speaks- because you have done this, because you have not refused me your son, your only son, I will shower blessings on you, I will make your descendants as many as the stars of heaven and the grains of sand on the seashore. Your descendants shall gain possession of the gates of their enemies. All the nations of the earth shall bless themselves by your descendants, as a reward for your obedience."

—Genesis 22:1-18

When the *stars of the Ram* were inflamed by the Vernal Equinox, themes of *self-sacrifice* would not be that uncommon among various cultures.

All *Athamas* got for his willingness to sacrifice both his son and daughter was a *curse*. It was decreed that if any of the eldest males of *Phrixus* line entered the Achaean Town hall, he (they) would be sacrificed.

Why is *Isaac* to become a *burnt offering*? Because the new *spiritual element* is FIRE. The *Ram* is offered up as a *burnt offering* instead. As with many other *Spiritual Mountains*, herein is our frame of reference for a *map of the Sky*. We have our *Celestial Ram* (they expected a *Lamb*), and we have a cut that must be made.

On the *Mountain*, Heaven shows the way.

116 *Athena's Web*

While we have paved a broad *celestial highway* in a case for the *Ram* from the myths, we've not yet touched upon the Dragon in this *new vibration*. We have seen the *Serpent in the Tree* in the *Garden of Eden*, but there are other examples of our *Center*, having now *fleeced the Ram* with our Circle. Besides the *Tempter of Eve*, are there other examples of the *Serpent and Shaft* (Tree) to be found in the Bible?

Most of the myths we've looked at so far, with the exception of these last few examples, have derived from indigenous peoples. Traditionally we've considered tribal cultures to be savages, incapable of understanding higher codes of moral conduct. To some, it's obvious that images of *flame-spitting Serpents* (*Uraeus*) or Dragons mounted to the *top of poles* (Indian/Sumerian/Babylonian myths) were primitive prototypes taking a crude shot at spirituality. After all, these folks were, according to the Greeks, raving barbarians!

This sort of thinking tends to place spirituality in the *'us vs. them'* category.

Well, what if it's all just us? What if we were all *One People* on *One Planet* sharing *One Sky* at a Time? The ultimate focus of spiritualtity is *Unity*.

On that note, it's Time to weave the thread of the *Serpent* more snugly into the *Judeo-Christian tapestry*, together with the same images, themes and underlying *Spirit* we've witnessed everywhere else we've looked. The only difference this time is, instead of Pharaoh, Brahmin, or pagan Hero, our central figure is a chap named *Moses*.

FIG. 223
A 'Brazen Serpent' upon the Staff

From Numbers 21:4-9;

"*They left Mount Hor by the road to the Sea of Suph, to skirt the land of Edom. On the way the people lost patience. They spoke against God and Moses, "Why did you bring us out of Egypt to die in this wilderness? For there is neither bread nor water here: we are sick of this unsatisfying food.*

"*At this God sent fiery serpents among the people: their bite brought death to many in Israel. The people came and said to Moses, "We have sinned by speaking against Yahweh and against you. Intercede for us with Yahweh to save us from these serpents."*

Moses interceded for the people, and Yahweh answered him, "Make a fiery serpent and put it on a standard. If anyone is bitten and looks at it, he shall live."

So Moses fashioned a bronze serpent which he put on a standard, and if anyone was bitten by a serpent, he looked at the bronze serpent and lived."

The image described is linked to a theme already centuries old when Moses led the Hebrews out of Egypt in the 13th century BC.

An *Iranian bronze* (Fig. 128) depicting a Dragon with a socket beneath the body, designed to be mounted upon a shaft representing the *Axis of Heaven*, dates from the 19th century BC, some six centuries prior to the *Biblical Exodus*.

To translate this tale, one must dip into the mythology. The venom of being bitten is, in part, passion. Passion may fuel many engines, but it is a corrosive mixture. Those who were bitten by their passions, let's say—fear of dying alone in the desert—could freak out and really lose it.

They could hurt somebody.

If they look at the Sky and determine what to do with their passion in a constructive way (astrologers do this all the time for their clientele), they lived.

Whether the *flame-spitting Serpent* atop the Pharaoh's *Uraeus* heads bear any relationship to the '*fiery Serpents*' on a *staff* in the Bible's Book of *Numbers* is open to conjecture, but the fact that the Hebrews have just spent several centuries in Egypt prior to the *Exodus* suggests more than simply a passing acquaintance.

Center and Circle

Here in *Numbers*, we can now count on Moses and our *Serpent* to open a real can of worms.

This is not the only time in the *Old Testament* that *Heaven's chosen* has been observed wielding a *Serpent-staff*. If we return to *Exodus*, we will find another reference to precisely this image.

From Exodus 7:10,

"Aaron threw down his staff in front of Pharaoh and his court, and it turned into a serpent."

Catching it by the tail reverts it to its former appearance.

The *Serpent-staff* was not simply an ornamental *accoutrement* of ritual, nor was it simply an aid for walking across vast desert wastelands. This *staff* served, in the same manner as the *Maypoles*, *standing stones*, *ziggurats* and even *pyramids*, as a *measuring rod for Heaven*.

It was an instrument used to read and interpret *Divine dictum*. Obviously, more permanent markers (Temples) could be better used for greater sighting accuracy, which is why Aaron holds the *staff*. Moses can stand back and make his observations using a point on the Horizon, just as the *Hopi* (Fig. 158) and *Dogon* did (Fig. 42). Unlike more permanent *celestial stations*, a *staff* can be carried on trips. Think of it as a portable laptop for quick calls to check in with the boss at the office. It's not as efficient, but it works in a pinch.

But, if we're on the right track and this *staff* is a *celestial weather vane* for determining the *Will of God*, don't you think the Egyptians would have known about it? Haven't we already examined this same *motif* in their mythology?

Before Pharaoh Moses displays his Serpent-staff wisdom

Indeed we have.
From Exodus 7:11-13,

"Then Pharaoh in his turn called for the sages and sorcerers, and with their witchcraft the magicians of Egypt did the same. Each threw his staff down and these turned into serpents. But Aaron's staff swallowed up the staffs of the magicians."

It's been fourteen centuries since Thuban visually marked the *Pole of Heaven*. The exact location of the *Sky's Center,* and its correlation to the Vernal Equinox was apparently better known to Moses at this time than to the Egyptians. His wisdom is stronger. Remember *Utanka the Brahman* (p.62) probed the hole the *Serpent* went down with his *staff*, but without luck? Even if you're a Brahman, this is not an easy feat to master. Our mythic imagery, here outfitted in *Biblical garb*, changes from *staff*, to *Serpent*, and back to *staff* again!

I want one!

And what happens in Exodus after the *Serpent staff* makes its appearance? Aaron takes the *staff* and stretches out his hand toward Heaven, for Yahweh 'fixed the hour'. He's taking his sighting. The river turns to blood; there are plagues of boils, frogs, mosquitoes, locusts and gadflies; death to the livestock, hail, darkness, and finally, *death of the firstborn*.

Moses has cast his Sky Map and realizes that there are *dark days on the Horizon*, and he attempts to leverage this information in a confrontational power play with Pharaoh. Moses sees, predicts and successfully warns Pharaoh to '*Let my people go'*, and he finally does, but only after all Egypt's firstborn sons are swept away.

It's *Passover* that is being remembered. That Full Moon evening when the *Angel of Death* passed over Egypt and marked a dividing line between Hebrew and Nile native. Our story line all along has been *Center and Circle*.

It continues here.

If this is *Passover*, and the *Serpent-staff* has been warning us of its arrival, which Full Moon of the year are we talking about?

The Full Moon following the Vernal Equinox.

The Full Moon of Spring.
Center and Circle.

With *Gilgamesh* there had been death. So too, with *Apophis*, *Ti'amat*, *Cadmus*, *Phrixus* and *Isaac*, there's death and sacrifice.

Our *Celestial drumbeat* has taken on a decidedly more *klezmer* tone. We'll just sit for a while and listen beneath the cool shade of the *World Tree*.

Center and Circle

Spring throws back the heavy, woolen blanket of winter and unthinkingly rubs the sleep from her eyes. Tossing her long auburn hair over her shoulder, she stands up, looks about, and slowly stretches her youthful body caressed by the warming, golden rays of the newly risen Sun. Taking hold of her waiting dress, vibrant in white iris, yellow daffodils and red tulips, she races barefoot out across the meadow, still wet with the morning dew, running eagerly to greet the new dawn.

Spring has a personality which stands out from the other Seasons in her youthfulness, enthusiasm, and promise of things to come. During Spring, the *Earth's body softens* and gives birth, her fingers reach to the Sky for Sunlight while she digs her toes into the soil below for nourishment. This *Season* initiates a race against time, the elements, and everything else thirsting for *Life*. After the hard, cold grip of Winter has released its spell, all eyes turn to watch expectantly for *Her Joyous Coming*.

The vibratory shift between the two constellational periods was both deep and pervasive. As the *Celestial Bovine* morphed into a *Golden Ram*, it brought with it a moral dilemma.

How do you best interpret this new Celestial Sign to God's (singular or plural) satisfaction? As we have already seen, a single symbol can have a multitude of meanings.

There was little doubt that the first cut, the 'prime' cut, should be given to God. You give It (He/She/Them) your best. Both Hebrew and Hellene sought cattle without blemish to offer at the altar. Pretty much everybody agreed with this practice. But beyond that, being first in line can be interpreted in any number of different ways.

As we have seen, *Minoan culture* (Figs. 192, 201, 206 *et al.*) continued to employ Bovine imagery to help find their Celestial Mark, while the Hebrews found a *Ram* with his *Horns* caught in a *Bush*, in this case, our desert Tree. *Yahweh* (Jehovah) demands the *first child*. Jacob tricks Esau (his older brother) out of his *primogeniture* birthright.

These two distinct paths bring with them ritualistic choices. Those who continued to honor and use the brighter stars of the Bull were more likely to embrace tradition and preserve the old ways. Those who forged ahead with dimmer *mythic* (astronomical) *materials* were looking at Life in a new way, and were attempting to discover just what that way meant.

The metaphor calls for a *Good Shepherd* to come and tend *His Flock*, to show them the way.

FIG. 225

Spring awakens to the New Year

Center and Circle

Photo credit St Paul Lutheran Church

FIG. 226

Moses on the Mountain

When the people saw that Moses was a long time before coming down the mountain, they gathered around Aaron and said to him,

"Come, make us a god to go at the head of us; this Moses, the man who brought us up from Egypt, we do not know what has become of him."

Aaron answered them, "Take the gold rings out of the ears of your wives and your sons and daughters, and bring them to me." He took them from their hands and, in a mold, melted the metal down and cast an effigy of a calf.

"Here is your God, Israel." they cried, "who brought you out of the land of Egypt! Observing this, Aaron built an altar before the effigy. "Tomorrow," he said, "will be a feast in honor of Yahweh."

And so, early the next day they offered holocausts and brought communion sacrifices; then all the people sat down to eat and drink, and afterward got up to amuse themselves.

Then Yahweh spoke to Moses,

"Go down now, because your people whom you brought out of Egypt have apostatized. They have been quick to leave the way I marked out for them; they have made themselves a calf of molten metal and have worshiped it and offered it sacrifice. 'Here is your God, Israel,' they have cried, 'who brought you up from the land of Egypt!'"

Yahweh said to Moses,

"I can see how headstrong these people are! Leave me, now, my wrath shall blaze out against them and devour them: of you, however, I will make a great nation."

But Moses pleaded with Yahweh his God. "Yahweh," he said, "why should your wrath blaze out against this people of yours whom you brought out of the land of Egypt with arm outstretched and mighty hand? Why let the Egyptians say, 'Ah, it was in treachery that he brought them out, to do them to death in the mountains and wipe them off the face of the earth?' Leave your burning wrath; relent and do not bring this disaster on your people. Remember Abraham, Isaac and Jacob, your servants to whom by your own self you swore and made this promise: I will make your offspring as many as the stars of heaven, and all this land which I promised I will give to your descendants, and it shall be their heritage forever."

So Yahweh relented and did not bring on his people the disaster he had threatened. Moses made his way back down the mountain with the two tablets of the Testimony in his hands, tablets inscribed on both sides, inscribed on the front and on the back. These tablets were the work of God, and the writing on them was God's writing engraved on the tablets.

Joshua heard the noise of the people shouting. "There is sound of battle in the camp," he told Moses. Moses answered him:

"No song of victory is this sound, no wailing for defeat this sound; it is the sound of chanting that I hear."

As he approached the camp and saw the calf and the groups dancing, Moses' anger blazed. He threw down the tablets he was holding and broke them at the foot of the mountain. He seized the calf they had made and burned it, grinding it into powder which he scattered on the water; and he made the sons of Israel drink it. To Aaron Moses said,

"What has this people done to you, for you to bring such a great sin on them?"

"Let not my lord's anger blaze like this," Aaron answered. *"You know yourself how prone this people is to evil. They said to me, 'Make us a god to go at our head; this Moses, the man who brought us up from Egypt, we do not know what has become of him.' So I said to them, 'Who has gold?', and they took it off and brought it to me. I threw it into the fire and out came this calf."*

**The Golden Calf
Notice the Crescent Horns**

Photo credit
Stephanie and Joshua McFall

FIG. 227

When Moses saw the people so out of hand- for Aaron had allowed them to lapse into idolatry with enemies all around them- he stood at the gate of the camp and shouted, "Who is for Yahweh? To me! And all the sons of Levi rallied to him. And he said to them, "This is the message of Yahweh, the God of Israel, 'Gird on your sword, every man of you, and quarter the camp from gate to gate killing one his brother, another his friend, another his neighbor.'

The sons of Levi carried out the command of Moses, and of the people about three thousand men perished that day.

"Today," Moses said, *"you have won yourselves investiture as priests of Yahweh at the cost, one of his son, another of his brother; and so he grants you a blessing today."* On the following day Moses said to the people, *"You have committed a grave sin. But now I shall go up to Yahweh: perhaps I can make atonement for your sin."*

And Moses returned to Yahweh. "I am grieved," he cried, *'this people has committed a grave sin, making themselves a god of gold. And yet, if it pleased you to forgive this sin of theirs..! But if not, then blot me out from the book that you had written."*

Yahweh answered Moses, "It is the man who has sinned against me that I shall blot out from my book. Go now, lead the people to the place of which I told you. My angel shall go before you but, on the day of my visitation, I shall punish them for their sin."

And Yahweh punished the people for molding the calf that Aaron had made. —Exodus 32

Center and Circle

We have, in these lines, a collision of cultures. Celestially speaking, it's *Bull* vs. *Ram*.

First of all, we have Moses spending a long time on the mountain. That's where you go to get the best look at the Sky. This is the same place (not the same mountain) Abraham went to sacrifice his son. This is where you confer with Heaven, to better understand the *Will of God*.

The twelve tribes are frightened, and with good reason. They've been jettisoned into the wilderness, are uncertain about where they will wind up, are unappreciated by the people already living in the lands they enter, have ticked off those they left behind and, to top it all off, they haven't seen nor heard from their captain in over a month.

"*Come, make us a god to go at the head of us...*"

Like the *staff* (Fig. 128) inserted into the *Iranian Dragon* socket, the *'Lord of Note'* would also lead the way in Egyptian processions. The Hebrews are familiar with the stellar traditions, having observed them in Egypt. Frightened, hungry and confused, they want to return to the security of the past, but there is no going back, and it would renounce the God of their Fathers, the God of *Abraham*, *Isaac*, and *Jacob*. Their forefathers had left the Bull Gods behind centuries earlier. Moses understands the distinction between *Bull* and *Ram* astronomically, mythologically and spiritually. Aaron does not. Moses had been brought up in the Egyptian court and given a royal education. We know the Egyptians were heavily invested in stellar observation. Could this be the source of the wisdom of Moses?

When *Athamas* marries a new Queen, it stirs up trouble with the old Queen and her offspring. Are we seeing the same celestial rivalry here between Pagan and Jew?

Aries is a FIRE *Sign*. God's wrath will *blaze out* against them, but Moses convinces God to be lenient. Then Moses goes down the mountain. What happens? *Moses's anger blazes out* against the people. What does he do? He summons the sons of Levi and they *butcher three thousand* of their kinsmen, killing brothers, friends and neighbors alike. All this for a theological error that not even Aaron fully understood.

It's a good thing Moses talked the Lord out of being *angry*, huh?

This is what it meant to live during a FIRE *Sign* under a *Lord of War*.

FIG. 228

Photo credit hampel-auctions.com

Raising the Maypole while the Sun (gold) is passing through the sign of the celestial Bull.

Center and Circle

Hooves
Lock
Horns

FIG. 229

FIG. 230

In many books of the Bible, we witness a tug-of-war between *Jehovah* (Yahweh) and the *Ba'als*. It's a see-saw celebration, with the people of Israel first adhering to the counsel of *Jehovah*, but then falling back on the traditions and practices of the peoples all about them, namely the *Canaanites*.[115]

The following lines from the Bible are typical of this struggle. Age-old rituals are embraced by the Israelites, in violation of the commandments of Moses. These paragraphs become a moral explanation of why the Hebrew tribes were taken into captivity by the Assyrians in 587 BC.

The term 'Lord' is generally used for one who holds power over us. As a title, it may address those on Earth or in Heaven. We will witness the psychological shift that occurs as the *celestial emphasis morphs* from one to the other.

The terms '**Ba'al**,' '**Baal**' and '**Bel**' permeate the Old Testament. *Ba'al* meant '*Lord and Master.*' It was added to names as veneration, whether of persons or places.

Ba'al was revered among the *Canaanites*, the *Western Semitic peoples* which generally included the *Tyrians*, *Moabites*, *Philistines* and the *Phoenicians* among others. The Hebrews of the Old Testament were distancing themselves from this association with the Bull of Heaven—both the *Deity* and the *fertility rituals* associated with Him.

We see this break quite clearly in Hosea 2:18-19, 23 where it is expressly stated,

'When that day comes—
it is Yahweh who speaks—
she will call me "My husband,"
no longer will she call me
"My Baal." ('My Lord')
I will take the names
of the Baals off her lips,
their names shall never more
be uttered again...
When that day comes—
it is Yahweh who speaks—
the heavens will have
their answer from me,
the earth its answer from them...'

Sacred poles of wood and stone were being set up in *high places* and on *hill tops*, steeped in traditions that celebrated the *fertility of the Earth* together with some triple-X rated fare.

Both *Abraham* and *Moses* climb the Mountain to pray and meditate. It's not 'high places' *per se* that are at fault, but the sacraments and customs associated with well-established shrines. The network of priests who once gathered and managed stellar intel collapsed as a result of the *famine*. After the 21st century BC, star records began to fall out of alignment as precession fails to be observationally or mathematically accounted for.

Athena's Web

Center and Circle

The level of astronomical sophistication was compromised. Those closely monitoring the *celestial choreography* oriented on the *distant horizons* had long ago passed away. The open air, *natural observatories,* once carefully positioned to help calibrate precessional passage, lapsed into the lap of a local populace out to have a good time, without regard to the specific *stellar information* which could be passed on in a systemized, orchestrated and co-ordinated fashion.

The Temples fell into darkness. Embedded by habits of centuries-old ritual, the people continued to celebrate what they knew, conducted on *hill tops*, by *standing poles*, with the sacred FIRE-*ritual rites*, without really understanding what they were doing astronomically. They went through the motions as well as they knew how, but without the precision of a guiding hand by an educated priesthood.

FIG. 231

Semitic Ba'al
Photo credit Jastrow

The original 'guiding star' which inspired these rituals has lost its *light*, has lost its *power*.

They have come to pursue 'emptiness.' They missed the mark.

From *Kings II*, 17:7-18,

'This happened because the Israelites had sinned against Yahweh their God who had brought them out of the land of Egypt, out of the grip of Pharaoh king of Egypt. They worshiped other gods, they followed the practices of the nations that Yahweh had dispossessed for them. The Israelites, and the kings they had made for themselves, plotted wicked schemes against their God. They built high places for themselves wherever they lived, from watchtower to fortified town. They set up pillars and sacred poles for themselves on every high hill and under every spreading tree. They sacrificed there after the manner of the nations that Yahweh had expelled before them, and did wicked things there, provoking the anger of Yahweh. They served idols, although Yahweh had told them, "This you must not do."'

'And yet through all the prophets and all the seers, Yahweh had given Israel and Judah this warning, "Turn from your wicked ways and keep my commandments and my laws in accordance with the entire Law I laid down for your fathers and delivered to them through my servants the prophets." But they would not listen, they were more stubborn than their ancestors had been who had no faith in Yahweh their God. They despised his laws and the covenant he had made with their ancestors, and the warning he had given them.

FIG. 232

El, Phoenician Supreme God
Photo credit Michael Martin

They pursued emptiness, and themselves became empty through copying the nations around them although Yahweh had ordered them not to act as they did. They rejected all the commandments of Yahweh their God and made idols of cast metal for themselves, two calves; they made themselves sacred poles, they worshiped the whole array of heaven, and they served Baal. They made their sons and daughters pass through fire, they practiced divination and sorcery, they sold themselves to evildoing in the sight of Yahweh, provoking his anger.

For this, Yahweh was enraged with Israel and thrust them away from him. There was none left but the tribe of Judah only.'

The '*pillars*', '*sacred poles*' and '*high hills*' are the essential ingredients of stellar observation. '*Every spreading tree*' is our *World Tree*. The package is complete.

---

* Trade with Egypt and Cyprus.........   ....Phoenician merchants........................

3000 BC   2500 BC   2000 BC   1500 BC

Fig. 233

Photo credit Rafael Gomez

A Baal Temple at Night

Center and Circle.

The local populace didn't know *celestial mechanics*, any more than the general populace today knows about *precessional motion*, the *ecliptic*, or *Right Ascension*, never mind how to track them. In this time of scarcity, during and following the *Famine*, Life was given a new lesson and a *New Theology*. They were to learn how to 'Lord it Over' the present.

Fig. 234

Photo credit petrus.agricola

Ba'al in his Earthly temple

They wanted to *Master* their moment in time. IF they managed to rise above all the competing conditions about them and triumph over adversity and foes, then fame and legend would take care of themselves.

While the common populace may not have known about the mechanical specifics, what they DID know were the oral traditions, stories and images associated with the holidays. Many of these continue to this day.

Myths were the historical record of the relationship of Heaven to Earth, merged with the *Divine*. It doesn't matter whether *we believe* the stars had a powerful and dramatic influence over the events here on Earth; what matters is that is how *they* saw it. It'd be historically dishonest to represent it in any other way. It is how *they* viewed God in Heaven, and they honored that relationship the best they knew how.

The terms *Ba'al* and *Bel* represent both the God [116] and his pal, *Babe the Sky-Blue Ox*. The forms were interchangeable and found everywhere. The name of Carthaginian general Hanni*bal* is one example. *Belteshazzar* means Prince of the Bull-god, *Bel. Baalgad* (Lord of Good Fortune) and *Baal-ze'bub* (Lord of the heavenly habitation) are others. *Ba'al* becomes associated with Satanic rituals and is sometimes called '*Lord of the Flies*.' Anyone who's spent any time with cattle understands precisely where this association comes from. This 'Holy Cow' is indeed '*Lord of the Flies*.'

They're *fly factories*.

No occult mysteries here.

One Egyptian incarnation of the Bull was as *Mnewer*.

Sound it out.

Ba'al Hadad is a term translated as "*Lord of Thunder*." On a cylinder seal in the Boston Museum his image is depicted as a *Man with a Lance*; while behind Him is represented his *animal attribute*, his *familiar*, the *Bull*.

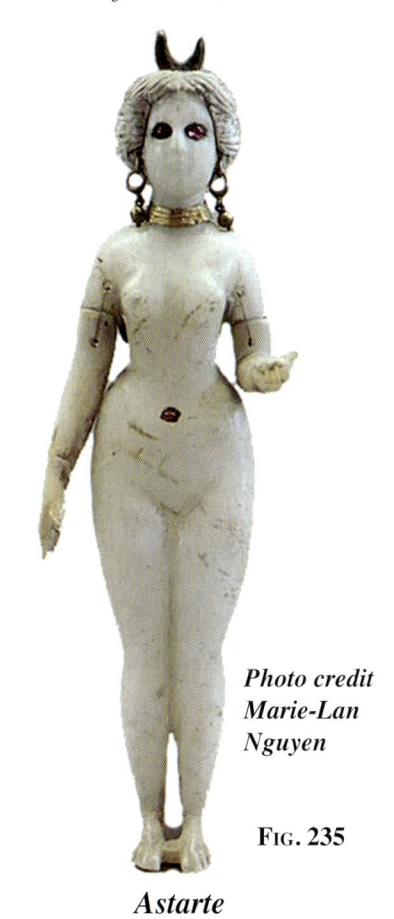

Photo credit Marie-Lan Nguyen

Fig. 235

Astarte

At the beginning of the 2nd millennium, along the eastern shores of the Mediterranean Sea, *Hadad*, armed with a thunderbolt and standing on a *Bull*, faces a God who points his spear towards a man stretched at his feet. According to *Canaanite* [117] tradition, *Ba'al* was the first and most powerful son of *El* (Fig. 232).

.... 1150 BC .... Golden Age to 850 BC .... 333 BC ...... Carthage 146 BC

1500 BC   1000 BC   500 BC   0 BC

Center and Circle

One of *El's* epithets was 'Tor,' which also means 'Bull'.[118]

The font of Spring, the source of *Nature's Creation*, was born of the stars of the Bull between 4300 and 2150 BC, but the rituals continued for centuries thereafter.

Unlike contemporary *astronomy* born of the *mathematics* of *spherical trigonometry*, this was an *observational astronomy*, a child of *standing stone* and *Temple*, born of FIRE and Earth.

While the Vernal Equinox moved through *Gemini* and *Taurus*, *observational astronomy* focused on bright *stellar benchmarks* for both the North and East *Cardinal Points*. But once the VE moved beyond the *Pleiades*, there were no Eastern 'stellar' markers. The dimmer stars along the back of the *Ram* offered a poor visual substitute compared to those of the *Bull*. As a result, many cultures 'remember' the use of the *cluster of Taurus* as a final lingering visual reference to Spring.

This is one reason why myths about various *Bulls* perpetuate well beyond 2150 BC among Mediterranean cultures, and why *Judaism* was so eager to disassociate itself from these neighboring peoples. Moses was raised and *educated in the Egyptian court*, in a secret tradition long-steeped in their abilities of *stellar observation*, which Moses apparently learned, and learned well.

Like our Megalithic culture in Ireland, the *Canaanites* conducted their rituals in Temples and on hill top "*high places*." From these vantage points they could use the *horizon* as part of an *observational network* to monitor the motions of Heaven, using the natural contours of both hills and islands, or they could *light fires* in previously selected locations on pre-designated days we now think of as 'holidays' (holy days).

By the time events in the Old Testament unfold, the pasage of the Vernal Equinox through the *constellation of the Bull* was overlong over. Yet tradition continued to revere the rituals rooted in over two thousand years of repetition.

In Judaic folk memory, *Abraham* is the *Patriarch* who sires a new culture. It's symbolism is the (*self-*) *sacrifice* of the *Ram*. Moses is later. The 'Law' (Torah) that Judaism guards (Aries) is the *Theology of the Ram*, a treatise on *Self-Mastery*.

We're in trouble.

During the Exodus (12th c. BC) and later period of Assyrian captivity (6th c. BC), the Vernal Equinox had precessed deep into the *constellation of the Ram*, and Moses knew it.

Astronomically, the stars of Taurus, with all its *mythological*, *spiritual*, *religious* and *ritual associations*, were now incorrect. They had had their day, but that day was over. The out-dated Age and its themes had, in the eyes of the *Old Testament authors*, become false gods. The *Bull* was no longer leader of the '*celestial parade*.'

The religion of the *Ram* had taken over, throwing a new emphasis upon the words,

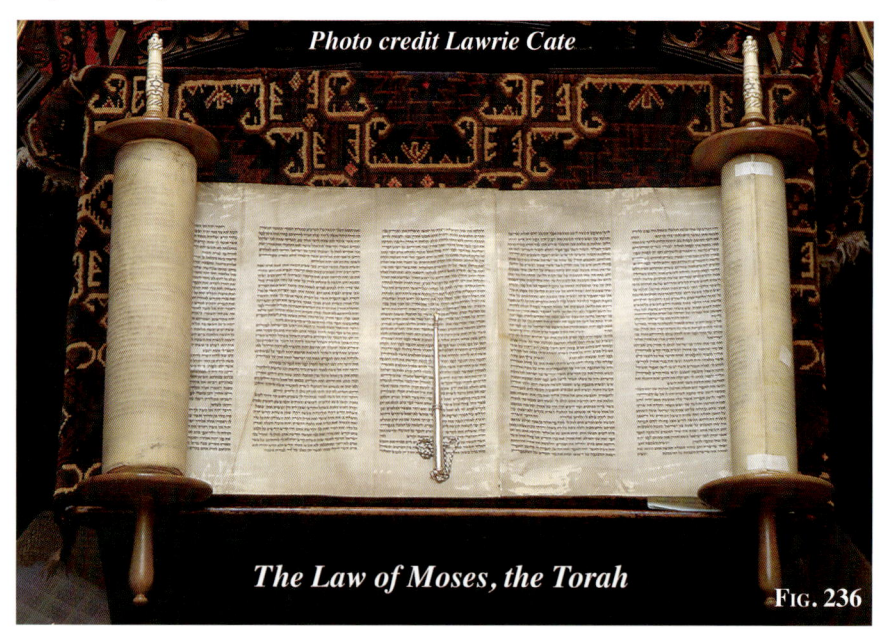

Photo credit Lawrie Cate

The Law of Moses, the Torah

FIG. 236

"God said, "Let us make man in our own image, in the likeness of ourselves, and let them be masters of the fish of the sea, the birds of heaven, the cattle, all the wild beasts and all the reptiles that crawl upon the earth". Gen. 1:26

Man is in charge.

Pause.

> "**Thou shalt have no other gods <u>before</u> me**."

Yahweh, or in more contemporary terms, *Jehovah*, was now *Number One* in Heaven; having overtaken the *Divine Bovine*. This leads us to speculate on a celestial assumption.

Center and Circle

Photo credit Johan Anglemark

FIG. 237

Phoenician writing

Jehovah is the *Ram*, or more precisely, the *Good Shepherd* of the *Flock*. The angel that shall go before them is *Mars* (and also it's astrological ruler the *Ram*).

After all of this Biblical background, imagine our surprise to find both *Bel* and *Ba'al* dancing in the *Fire-rituals* of Ireland.

Before we pack the bags for our journey to visit the *wee folk*, let's do a quick review of some **Phoenician roots**, starting with the geography. These are sacred lands.

The eastern Mediterranean was the cultural and trading crossroads of the world where East met West; and Africa, Asia, and Europe all merge. Given the barren terrain and lack of fertile river valleys for agricultural expansion and sustenance, the *Phoenicians*, one branch of the *Canaanite* family, turned to the timber of the forests and the sea for transportation. They would eventually develop trade routes with *Cyprus* for *copper ingots*, the *Egyptians* for *papyrus* and *bales of linen*, and *herbs* and *spices* from the caravans of *Arabia*.

Besides import/export, they also developed extensive home industries of purple dye and glass.

But in addition to these, they traded ideas and opinions as well. The Phoenicians had been part of the Canaanite world, whose beginnings lie in the dim mists of early history. They were part of the Semitic migrations that had settled in Lebanon and Palestine.

The mythological tradition of the Phoenicians was thus exposed, as their trading partners might suggest, to a background common to a widely extended ethnic group. Some of the earliest contacts we are familiar with extend back to the Old Kingdom of Egypt, about the beginning of the third millennia BC, shortly after the period of unification.

We also know the Phoenicians had extensive contacts with the Hebrews, both as friend and foe. During Israel's ascendancy to Empire, they turned to the Phoenicians for help in the construction of Solomon's most famous building, the *Temple in Jerusalem*. The Bible tells us it was constructed by Phoenician workmen from King Hiram of Tyre.[119] The two pillars *Jachin* and *Boaz* have Phoenician precedents, while the interior *accoutrement* was either Phoenician, or influenced by the Phoenicians. These skilled craftsmen had done the same for the Egyptians.

Photo credit Elite plus FIG. 238

Phoenician ship

Athena's Web

Center and Circle

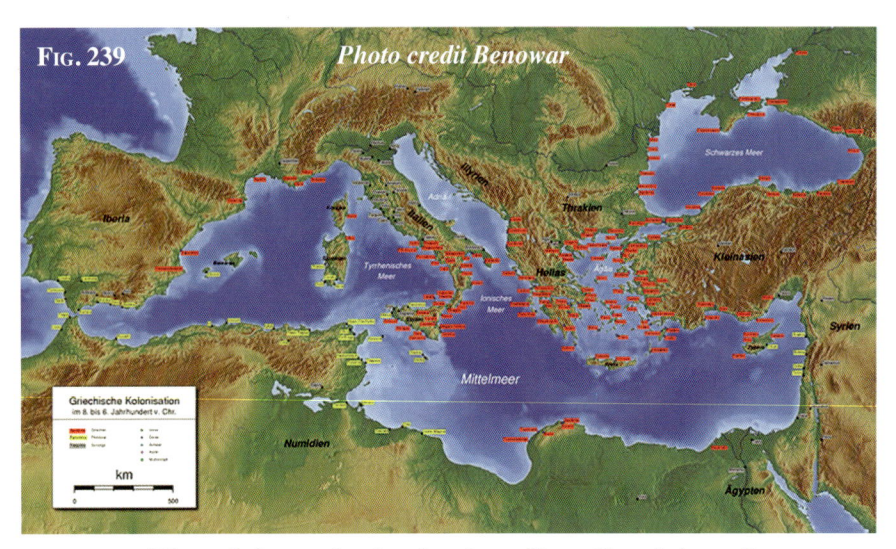

Phoenician colonization in yellow, Greek in red

"*In the search for models of Solomon's Temple in Jerusalem it seems that all the indications point to Canaan and Phoenicia.*"

So says Werner Keller, author of '*The Bible As History.*'

The Phoenicians also gave us the alphabet. They passed it on to the Greeks, and from there into Europe, sometime around 1400 to 1000 BC. *Philo of Byblos* wrote a treatise on the Phoenicians, claiming they were the ones who taught the Greeks their myths. It's from a Phoenician myth that we know their Temples contained 'Skylights' to monitor the Sky. Like the other cultures with whom they traded, the Phoenicians focused on the *Bull*, known to them as '*Ba'al*'.

The Bull of Heaven was used to visually fix the position of the VE from about 4200 to approximately 1200 BC. The Phoenicians were observing the same Sky as everyone else. These images are reflected in their myths.

Along with their wares and dyes, these traders passed along mythological and astronomical tips to the Greeks. Since the earliest recorded times, they had been active in trade. First, trading insect resistant cedar wood with the Egyptians, and more than a thousand years later, with most of the eastern end of the Mediterranean including the Hebrews. They were neighbors to Egypt, the Babylonians, Damascus, and Israel. They were connected.

We only know about the Phoenicians through their enemies, even though they invented writing. The Bible is one of the best continuous sources of Phoenician reference. They and their extended family lived right next door. Although they left behind no great work by any native historians, they made it possible for everyone else to leave the written records we have today, as they also invented the script, pencils, pens and ink.

Located in *Lebanon* and *Palestine*, the *Canaanite* tradition is a long way from the *Emerald Isle*.

We know the Phoenicians later developed colonies in *Corsica*, *Sardinia*, *Sicily*, *Spain* and along the *northern coast of Africa*. We also know they sailed through the *Pillars of Hercules* (Gibraltar) and 'beyond.' But what does that mean? How far did they go and when did these voyages commence? In *Ireland*, *Britain* and *northwestern Europe* there's an abundance of place names dedicated to *Ba'al*. [120] Baal-y-gowan, Baal-y-Nahinsh, Baal-y-Castell, Baal-y-Moni, Baal-y-ner and Baal-y-nah are a few found in *Ireland*. *Signal-Fires* swayed in time to the *rhythms* of the *distant Fires* of Heaven.

The Phoenician Tree of Life

Pagan Fire Ceremony

In many parts of the world, children dance and play around the *Maypole* on *May Day* (Fig. 228). As a general rule, people like to get together, eat, drink and have fun. From ancient times this sort of activity was associated with *Ba'al worship*.

The word '*Belus*' or '*Ba'al*' is the same in *Babylonian*, *Phoenician*, *Gaelic*, and other languages. The old *May Day* ritual was called *Beltina* in the *British Isles* and *Ireland*. Such place names as Ball Hill, Val Hill, and Baalbeg are examples of *Ba'al* names in *England*.

Baalbeg is a deserted village above Loch Ness in *Scotland*.[121]

Whether the *Phoenicians* had a much earlier *maritime connection* with the peoples of *Ireland* than has commonly been supposed (they were known to have had maritime trade contact with the Egyptians as early as 3200 BC), or whether the *Irish* had earlier overland *contacts* with the *Middle East* should be also be considered. We know that at their peak the *Celts* sacked *Rome* and *Delphi*. They extended from *Ireland to Asia Minor* (Turkey).

It's possible their predecessors made the journey from one end of Europe to the other, either as *traders* or on what we might consider today '*diplomatic*' missions, to share information in the form of celestial navigation, agricultural tips and stories about the myths and wisdom of their own lands.

If we look seaward (Fig. 239), we know the *Phoenicians* followed the coastline to *Egypt*, *Crete*, founded *Thebes* in *Greece* (note how the map makes no reference to this) and on through the *Straits of Gibraltar* at the mouth of the Mediterranean.

What then prevented these seafaring adventurers from following the coast north along the *Iberian Peninsula*, *Gaul* and on to modern day *Pas-de-Calais*, from whence the *White Cliffs of Dover* are easily observable and a mere 20 miles across the *English Channel*?

Suddenly, *Ireland* does not seem so far away.

The *Phoenicians* were thought by some to have carried **Ba'al worship** to the *western* and *northern coasts* of *Europe*. *Ba'al* gave his name to the *Baltic Sea*, the Great Belt and Little Belt Channels of *Denmark*, to towns such as Baleshaugen, Balestranden, and to many localities in the *British Isles*, such as Belan, and the Baal Hills in *Yorkshire*. Even today there are over fifty *Irish towns* which begin with Bal-, Ball- or Bel-, including *Baltimore*.

'*Peor*,' is another word associated with this tradition. It appears in *Numbers*, *Deuteronomy*, *Joshua*, *Psalms* and *Hosea*. *Peor* was the name of a *Mountain* located somewhere in the vicinity of Jericho from which *Baalam* last blessed Israel. This word combined with *Ba'al* to become *Baalpeor*.[122]

Peor is a *Hittite word* for FIRE,' and is related to the Greek '*Pyr*', found in many English words.

We have *Fire on the Mountain*. '*Baalpeor*' can be translated as the '*Lord of Fire*,' '*Fire of the Lord*,' or '*Ba'al's Fire*.' Fortunately, *Irish tradition* remembers these rites and preserves them in folk traditions from centuries ago. The following is from *Ancient Legends, Mystic Charms, and Superstitions of Ireland*, Lady "Speranza" Wilde, London, 1888.

Center and Circle

'This season is still made memorable in Ireland by lighting fires on every hill, according to the ancient pagan usage, when the Baal fires were kindled as part of the ritual of sun-worship, though now they are lit in honour of St. John. The great bonfire of the year is still made on St. John's Eve, (April 30th) when all the people dance round it, and every young man takes a lighted brand from the pile to bring home with him for good luck to the house.'

In ancient times the sacred fire was lighted with great ceremony on Midsummer Eve (Summer Solstice); and on that night all the people of the adjacent country kept fixed watch on the western promontory of Howth, and the moment the first flash was seen from that spot the fact of ignition was announced with wild cries and cheers repeated from village to village, when all the local fires began to blaze, and Ireland was circled by a cordon of flame rising up from every hill. Then the dance and song began around every fire, and the wild hurrahs filled the air with the most frantic revelry.'

'Many of these ancient customs are still continued, and the fires are still lighted on St. John's Eve on every hill in Ireland. When the fire has burned down to a red glow the leap is done backwards and forwards several times, and he who braves the greatest blaze is considered the victor over the powers of evil, and is greeted with tremendous applause. When the fire burns still lower, the young girls leap the flame, and those who leap clean over three times back and forward will be certain of a speedy marriage and good luck in after life, with many children.'

The married women then walk through the lines of the burning embers; and when the fire is nearly burnt and trampled down, the yearling cattle are driven through the hot ashes, and their back is singed with a lighted hazel twig. These hazel rods are kept safely afterwards, being considered of immense power to drive the cattle to and from the watering places.'

'Their back is singed', marking Europa's seat and Minoa's flip.

'As the fire diminishes the shouting grows fainter, and the song and the dance commence; while professional story-tellers narrate tales of fairyland, or of the good old times long ago, when the kings and princes of Ireland dwelt amongst their own people, and there was food to eat and wine to drink for all the comers to the feast at the king's house.'

When the crowd at length separate, every one carries home a brand from the fire, and great virtue is attached to the lighted brone which is safely carried to the house without breaking or falling to the ground.'

F<small>IG</small>. 242

Bonfires dancing...
Photo credit Graham Green

Photo credit Ayaz Asif

F<small>IG</small>. 243

***Mystique of the Highlands—
Loch Ness, Scotland***

Many contests also arise amongst the young men; for whoever enters his house first with the sacred fire brings the good luck of the year with him.'[123]

Embodied in these rituals are the *constellation's blessing*. The *Great Bull* was the source of all Life, power, and fertility.

Agricultural tips, some as simple as using a hazel twig to drive cattle, or of when to plant and harvest crops, were part of the *Cornucopia* this *Divine Creature* was thought to bring and are found embedded in myths from various lands. The precise moment of the *Winter Solstice*, *Equinox* or *Cross-quarter holiday*, carefully calibrated by the monuments at *Newgrange, Dowth, Knowth* and others, was passed on as quickly as possible, as part of a *Divine dance* between *Heaven and Earth*.

It was a leaping regional *Heartbeat of Living Flame*, a *Pulse of Life*, drummed out by the *Seasons*. It carried the *Good News* of the celebration in the *Return of Spring*, rotating yet one more cog in an eight-spoked *Wheel of Time* pivoting about the *World Tree*.

Taurus had been a gathering of the *Herd*. One of the attributes of the out-dated Goddess was *fun*. It's not surprising Judaic leaders had a hard time keeping their followers from joining these infectious *celebrations of Life*.

As the Vernal Equinox slipped into the stars of the *Ram*, *Fire* took on an added spiritual dimension. The *agrarian themes* of *fertility* pass easily from *Bull* to *Ram*, but we have moved from EARTH to FIRE as the new *elemental medium* of choice. It assumed an even greater role in the ever-evolving pagan Sky-dance as the Stars, these tiny '*Fires*' in the Sky, were harnessed, addressed and undressed.

The country is where the roots of this ancient agrarian tradition was most deeply nourished, in the indigenous farming community.

In the rhythms of the Moon and Seasons there was considerable wisdom to be conveyed, and it maintained its hold, through both tradition and memory over the country-folk. Because both fun and fertility were part of the celestial endorsement of the Age of Taurus, it was popular with the people.

The Age of Aries was to bring in a much more somber tone in its celebrations with the seriousness of the manly 'science' of War.

What we are witnessing in the Bible is Judaism's attempt to stomp out the embers of this ancient ritual because the time of the Bull is over, dead and buried by 578 BC.

Photo credit David Barlow

Fig. 244

Megalithic Sunset—Union of Heaven and Earth

Center and Circle

Fourteen centuries have passed, '*As the fires diminish and the shouting grows fainter.*'

The folk traditions of *Ireland* preserve the memory of these rituals thousands of years later, right up to the 19th century AD. The mortar that secures these tales to their *Phoenician connections* is the place names under which they honored the Divine in both Heaven and on Earth.

These *Fire-rituals* transport us from one end of a darkened Megalithic tunnel right through to today. From construction in the 4th millennium to the Biblical protests with Moses in the 12th century, twenty-four centuries have passed. From Moses's day through to the Celtic Fire-ceremonies continued to this day, another thirty-two centuries have elapsed.

These *Fires* and illuminated *Standing Stones* were carefully selected *Sacred Sites* where *Sky* and *Earth* joined to make *Love*. This had been the Taurean ('ruled' by *Inanna* Fig. 129, *Astarte* Fig. 235, *Aphrodite* Fig. 245 and *Venus*- Fig. 246) metaphor, their myth. Led by the Druids in Ireland, the common people became part of a *Living Flame Natural Observatory* emphasizing the *Cardinal Points* of the compass and more.

The Hebrews followed a later mythological tradition, one that began twenty-one centuries after the Bull was tamed by the *mythological Yoke*.

In the centuries following the *Great Famine*, the *agricultural empires collapsed*. It's probable that hill top festivities degenerated into state-endorsed orgies, which, like the stars they represented, offered little light by which to guide their ever-changing *Star of the East*.

Baalpeor, the '*Lord of Fire*' on the Mountain wasn't limited to a hill outside Jericho, but was part of a *World-Wide System* of communicating the *Rhythms of the Seasons*, remembering a Time when an enthusiastic adolescent princess gathered Spring flowers by the Sea and Creation was still young indeed.

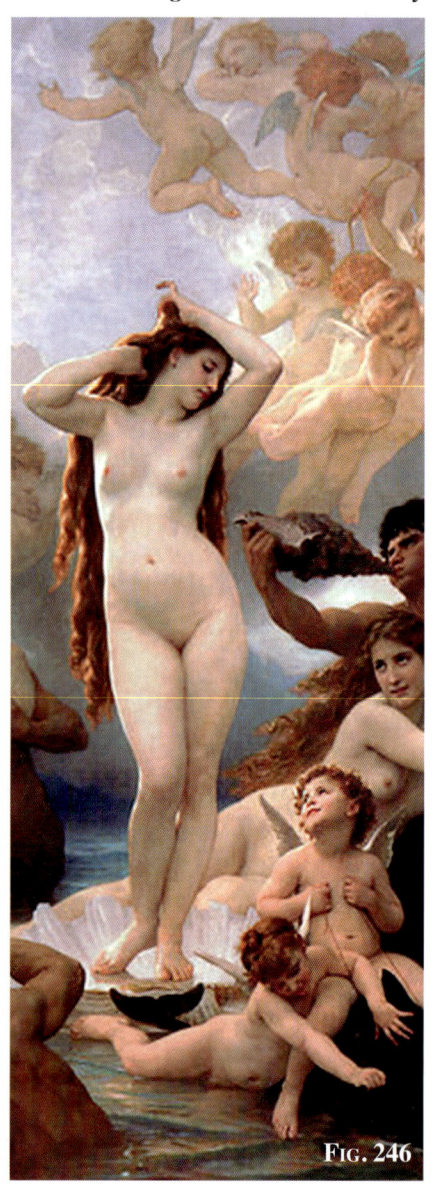

Image credit musée d'Orsay

FIG. 246

**Bouguereau's Venus
... a Goddess of Love**

Both the *Greek heroes* and tales of *Abraham* stem from the early days of the passage of the Vernal Equinox through the *stars of the Ram*. After *Cadmus* challenges and defeats the Dragon guarding a *Grove sacred to Ares*, he settles down and marries *Harmonia*. Together they have a daughter, *Ino*, who becomes the second wife of *Athamas*, the next generation. Presumably, the *Cadmus* myth is older than the *Phrixus* myth.

Photo credit Shawn Lipowski

FIG. 245

**Aphrodite of Milos
In any language...**

Phrixus is the hero who flies to *Colchis* with the *Ram of the Golden Fleece* prior to the arrival of the Argonauts, which ranks Jason's tale third in the sequence. Yet all these *heroes* and their stories predate the Trojan War, generally thought to have occurred circa 1200 BC. *Moses* and the *Exodus* are usually dated to this same century. *Abraham* is a number of generations earlier than *Moses* (Fig. 268).

These are all the older myths of the Vernal Equinox, while it is still passing through the pioneering stars of the *Ram*, learning and discovering new sets of laws by which to live *Life*.

But if we look later in our mythological history, do we still see that the *Ram* is our *central figure*, the *guardian* of the *start of the Sacred Circle*?

At the *Babylonian New Year festival* of 600 BC, during the reign of *Chaldean King Nebuchadnezzar II*, there was a solemn procession that wound its way through the *Gate of Ishtar* (Fig. 249). This procession coincided with the Vernal Equinox in a ceremony lasting for eleven days.

Basically, they were recreating *Heaven on Earth*, and getting ready for the *New Year*. This moment of the *Birth of the New Year* contained in it all the wisdom and prophecy of what the Gods had to say about the coming year at its inception. The planets and skies might be read, and the accumulated wisdom of the Ages applied to the '*signs and portents*' that *Destiny decreed*. This is why they studied history and compared notes.

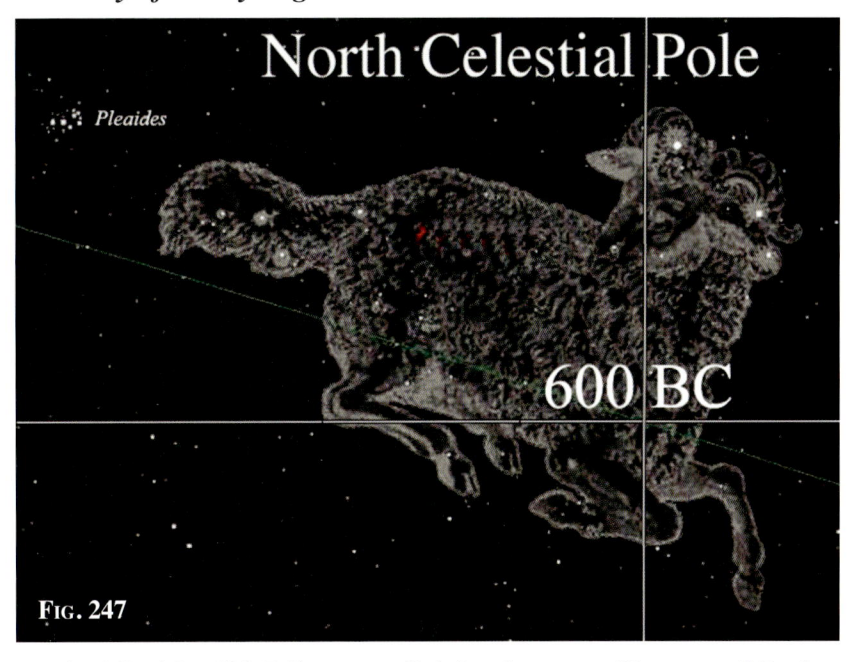

Courtesy of Starry Night Education

FIG. 247

The 'Cut' in 600 BC — note division between Horns and Body

So they're checking their charts. They're reading a map of the Sky. They're studying the horoscope for the moment the year is born. They're consulting with the Sky God (Marduk) in the days leading into the ceremony to make sure everything is as it should be. The correct *Center* must be established, both for Heaven and Earth.

The precise mark on the *Circle* must be predicted, observationally confirmed, and executed.

On the fourth day [124] the priest invokes *the constellation of the Ram* (Aries), rising three and a half hours before Sunrise, so the stars might still be seen. Last minute calculations and observational checks are being put into place.

Photo credit Josep Renalias

FIG. 248

The Gate of Ishtar was rebuilt by Nebuchadnezzar II

Athena's Web

133

Center and Circle

On the fifth day the royal sword bearer cuts off the *Head of the Ram* (Fig. 247), *marking the start of the New Year*.[125] After the beheading, another priest, reciting spells, wipes the sanctuary with the carcasss. He then carries the body to the river and throws it into the water, facing west (direction of Sunset, Death and the Underworld), while the executioner does likewise with the head. As both of them are now in a 'taboo condition' after the performance of this rite (i.e. considered 'unclean' because they had come in contact with a corpse), they are required to retire to the country until *Nabu* (the Divine Scribe) leaves the city on the twelfth day of Nisan.

The priests were forbidden to see any part of the sacrifice or purification of the Temple as they too would be considered unclean.

Later, the king enters the shrine of Marduk (under the dome of Heaven, the Sky that Marduk represents) and allows the high priest to remove his crown, ring, scepter and harp. These are then taken to the inner altar and laid there, back under the auspices of the Divine. This Earthly mortal, now stripped of his symbols of power and naked before Heaven is being reminded of the real chain-of-command.[126] Power is not supposed to go to the King's head since he is the *Divine Representative of Heaven's Will*, not his own. What's interesting is that *Nebuchadnezzar II* is king in 600 BC. He's the one the Bible remembers as being humbled by Heaven precisely because he forgets him/Self and feels all his accomplishments are due to his own superlative efforts.

FIG. 249

The Gate of Ishtar

Originally built by Hammurabi, re-dedicated and used by Nebuchadnezzar II centuries later

Image credit Ingrid Ramsay

FIG. 250

Nebuchadnezzar II receiving gifts on New Year's Day

Before we examine this myth in greater detail, let's finish the day's ceremonies.

In a clear demonstration of this chain-of-command, the priest then strikes the king on the cheek and forces him to his knees before the statue. It is in this posture that the king must truly humble himself before God, presumably by expressing deep regret about his shortcomings to the present.

At sunset on the fifth day, the priest digs a hole in the courtyard and plants honey, cream and oil of the best quality. The riches of the Earth are being seeded back to the Earth, so the Divine Cycle may be repeated for another year. A white bull is then led to the spot where the hole was dug and the king kindles a fire in the middle of this trench. Both King and priest recite a prayer, the contents of which are lost apart from the opening lines addressed to the *Bull of Anu* as

'*the shivering light illuminating the darkness…*' [127]

Starlight.

Here we have an excellent glimpse of the problem of the 'dim stars of Aries' being worked with and overcome. The '*shivering light*' are the stars of the Bull, in this case *Anu*, who points the way to the correct Vernal Equinox in the constellation Aries.

Which means that as late as 600 BC the brighter stars of the Bull are still being used to 'illuminate' the way to the *head and horns of the Ram*. We've only looked at the 4th and 5th days of the ceremony, but the rituals of the other days are also revealed.

The full ceremony is called the **Rites of Bit Akitu.** [128]

"There were festivals recognized on the seventeenth and thirtieth days each month. Apart from the lamentations for Tammuz in the summer, the largest festival is that of the new year, which falls in the Babylonian month of Nisan and coincides with the Spring Equinox."
The First Great Civilizations,
—Jacquetta Hawkes [129]

Nebuchadnezzar the king, to men of all people; nations and languages, throughout the world:

Photo credit W. A. Spicer **Fig. 251**
Daniel and Nebuchadnezzar II

"May peace be always with you. It is my pleasure to make known the signs and wonders with which the Most High God has favored me.

*"How great are his signs,
how mighty his wonders!
His sovereignty
is an eternal sovereignty,
his empire lasts
from age to age…"*

The signs are the *Zodiac Signs*. The precessional motion we are studying takes us from Age to Age, from sign to sign. Neb is talking. Are we listening?

I, Nebuchadnezzar, was living at ease at home, prosperous in my palace. I had a dream, it appalled me. Dread assailed me as I lay in bed; the visions that passed through my head tormented me. So I decreed that all the sages of Babylon be summoned to explain to me what the dream meant. Magicians, enchanters, Chaldeans and wizards came, and I told them what I had dreamed, but they could not interpret it for me. Daniel, renamed Belteshazzar after my own god, and in whom the spirit of God Most Holy resides, then came into my presence.

I told him my dream:

Belteshazzar, I said, chief of magicians, I know that the spirit of God Most Holy resides in you and that no mystery puts you at a loss. This is the dream I have had; tell me what it means.

"The visions that passed through my head as I lay in bed were these:

Everybody put their *World Tree* hats on. We're opening with stock footage from the Universal Myth. Ready? Think *World Tree*.

*"I saw a tree
in the middle of the world:
it was very tall.
The tree grew taller and stronger,
until its top reached the sky,
and it could be seen
from the ends of the earth.
Its foliage was beautiful,
its fruit abundant,
in it was food for all.
For the wild animals
it provided shade,
the birds of heaven* (Fig. 137)
*nested in its branches,
all living creatures
found their food on it.*

Center and Circle

This image of the *World Tree* is familiar as part of our Universal theme. Dreams select images from the archetypal library, and in this case, our opening framework is an ancient one. Everything to this point is fairly generic.

The *World Tree* is a metaphor for Life itself, a combination of the textured Earth tones, deep, moist and rich, placed in the *middle of the world*. Its branches reach as *high as the Sky*, and it can be seen to the *ends of the Earth*.

Nebuchadnezzar continues:
*I watched the visions
passing through my head
as I lay in bed.
Next a watcher, a holy one
came down from heaven.
At the top of his voice
he shouted,
'Cut the tree down,
lop off its branches,
strip off its leaves,
throw away its fruit;
let the animals
flee from its shelter
and the birds
from its branches.*

The generic opening serves two purposes. First, it grounds us in familiar mythic territory. For those who have an ear for the pattern, this pulse strikes a resonant cord. The *Tree* is the *Almighty*, but it's also *Nebuchadnezzar*, in the same manner that the *strength of the Bull* (Fig. 96) can be seen standing behind *Pharaoh*, 'backing him up' with the agricultural power and resources of the kingdom. Just as *Pharaoh* was identified with the *Bull of Heaven*, so *Nebuchadnezzar* may be identified with the *World Tree*. Rulers are *Heaven sent*, and the power of God (Marduk, Hathor, Jehovah, etc.) stands behind them.

The narrative continues:

*But leave stump and roots
in the ground
bound with hoops
of iron and bronze
in the grass of the field.
Let him be drenched
with the dew of heaven,
let him share the grass of
the earth with the animals.
Let his heart
turn from mankind,
let a beast's heart
be given him
and seven times
pass over him!
Such is the sentence
proclaimed by the watchers,
the verdict
announced by the holy ones,
that every living thing
may learn
that the Most High rules
over the kingship of men,
he confers it
on whom he pleases,
and raises
the lowest of mankind.'*
—Daniel 3:31 to 4:14

Photo credit Prof. Richard T. Mortel, Riyadh

Like Hammurabi befote him, Nebuchadnezzar II's Dragon Fig. 252

Daniel successfully interprets the dream and is given great power and recognition. One year later *Nebuchadnezzer* is afflicted with some unknown malady that causes him to lose his wits. He grovels, feeds with and sleeps among the wild animals for days on end. When his reason is finally restored seven years later, he 'gets it' that all things come from God, and returns to run a righteous Empire.

Daniel 4:31-34 continues,

"When the time was over, I, Nebuchadnezzar lifted up my eyes to heaven: my reason returned. And I blessed the Most High,

"*praising and extolling him
who lives for ever,
for his sovereignty
is an eternal sovereignty,
his empire lasts
from age to age.
The inhabitants of the earth
count for nothing:
he does as he pleases
with the array of heaven,
and with the inhabitants
of the earth.
No one can arrest his hand
or ask him,
'What are you doing?'*

"At that moment my reason returned, and, to the glory of my royal state, my majesty and splendor returned too. My counselors and noblemen acclaimed me; I was restored to my throne, and to my past greatness even more was added. And now I, Nebuchadnezzar,

"*praise and extol and glorify
the King of heaven,
his promises are always
faithfully fulfilled,
his ways are always just,
and he has the power to humble
those who walk in pride."*

We are seeing the same concerns being expressed in Babylonian ritual and in the lines of the Bible. Personal ego must not get in the way of God's Will. In this they are in complete agreement.

Photo credit Web Gallery of Art

Fig. 253

Nebuchadnezzar's dream

While we're here in the Book of Daniel, let's take a quick look at this theme of *Hooves Locking Horns* once again, as it reoccurrs throughout the Hebrew Bible. In this instance, the old, honored symbols of *Center and Circle* are being worshiped, but they are no longer the exact astronomical orientation. The time of *Bull* and *Dragon* is winding down, and these archetypes, while still used, are losing their power and prestige.

They don't get no respect.

There was a big dragon in Babylon, and this was worshipped too. The king said to Daniel,

"*You are not going to tell me that this is no more than bronze? Look, it is alive; it eats and drinks; you cannot deny that this is a living god; worship it, then." Daniel replied,*

"*I worship the Lord my God; he is the Living God. With your permission, O king, without using either sword or club I will kill this serpent."*

"*You have my permission," said the king. Whereupon Daniel took some pitch, some fat and some hair and boiled them up together, rolled the mixture into balls and fed them to the dragon; the dragon swallowed them up and burst. Daniel said,*

"*Now look at the sort of thing you worship!" The Babylonians were furious when they heard about this and began intriguing against the king.*

"The king has turned Jew," they said, *"he has allowed Bel to be overthrown, and the dragon to be killed, and he has put the priests to death."* So they went to the king and said,

"*Hand Daniel over to us or else we will kill you and your family." They pressed him so hard the king found himself forced to hand Daniel over to them."*

Of course, from here Daniel is thrown into the lion's pit where he stayed for six days. The lions don't eat him, but they do eat his accusers as soon as they are thrown into the pit in his place.

The question is, have the Babylonians under *Cyrus the Persian* (time has passed and a new king has taken charge) truly *lost celestial perspective*, or are we misinterpreting the way Daniel perceives it?

Center and Circle

In most Western Myths, you're *supposed* to kill the Dragon. How did *Marduk* do it with *Ti'amat*?

Between the *Jaws*, down the *Throat*, and into the *Stomach* and *Heart*. The only element missing in Daniel is the *Heart*, but that's OK because by this point the *Dragon's Heart*, *Thuban*, hasn't marked the *Center of Heaven* since its peak some twenty-one centuries earlier. The myth winds up being decisive in the *Stomach* instead of the *Heart*. The digestive system shuts down and the Dragon 'dies'. As we witnessed with the *Minoan Goddess*, the *Heart*, *Sun*, *Lion*, *King* and *cat* are all part of a stream of theme that keeps popping up in relation to Dragon myths.

Once again, the *Center* is segregated from the '*body*' of the Dragon. The Times they are a-changin'.

Like the *Minoan Goddess* of a thousand years earlier, 'Draco' is slowly moving off the *North Celestial Pole* of the Earth, just as the 'Lion's pit' is now separate from the Dragon. It is no longer the constellational 'heart' that is being pierced (burst), but the 'stomach' (*eta Draconis?*). The tarballs are being used to determine the Center of Creation. The myth still 'needs' Draco to find Heaven's Center.

Daniel is comfortable with his Center (the Lion's pit). The Lions are content with his geometric accuracy. When the priests are thrown into the Lion's den, *their* Center is off, *their* astronomy is off, and therefore any predictions they may make about the future (part of their job) will be off.

This is the same old 'Finding the Center of Heaven' myth, but this time, instead of the usual instruments of severance such as lance, arrow, club or common pin (Chinese), Daniel used tarballs. Interestingly, there are tales of *Alexander the Great* later killing a Dragon with tarballs, so Daniel is not alone. The mythic thread is consistent and continuous.

Another myth that parallels many of the metaphorical elements found in *Daniel* is a *Mesoamerican* tale. Instead of being thrown into a pit, we have a volunteer willing to get his hands dirty and dig his own. This section is appropriately called, '**The Pits**'.

From the *Popol Vuh*, Chapter 7:

Here now are the deeds of Zipacná, the elder, son of Vucub-Caquix.

"I am the creator of the mountains," said Zipacná.

"Zipacná was bathing at the edge of a river when four hundred youths passed dragging a log to support their house. The four hundred were walking, after having cut down a large tree to make the ridge-pole of their house."

This ridge poll is the *Celestial Axis*. Their 'house' is the world.

"Then Zipacná came up, and going toward the four hundred youths, said to them:

"What are you doing, boys?"

"It is only this log," they answered, "which we cannot lift and carry on our shoulders."

"I will carry it. Where does it have to go? What do you want it for?"

"For a ridge-pole for our house."

"All right," he answered, and lifting it up, he put it on his shoulders and carried it to the entrance of the house of the four hundred boys.

"Now stay with us, boy," they said.

"Have you a mother or father?"

"I have neither," he answered.

"Then we shall hire you tomorrow to prepare another log to support our house."

"Good," he answered.

The four hundred boys talked together then, and said:

"How shall we kill this boy?"

FIG. 254

Carrying the ridge-pole

FIG. 255

Zipacná and the 400 boys
Photo credit Bruce Rimell

What's happening here? Why do they want to kill him, someone who has done nothing but help with their project? This does not seem to be a just return for his efforts.

"Because it is not good what he has done lifting the log alone. Let us make a big hole and push him so that he will fall into it."

Daniel didn't have to dig his own grave. The lions were already waiting.

"Go down and take out the earth and carry it from the pit, we shall tell him, and when he stoops down, to go down into the pit, we shall let the large log fall on him and he will die there in the pit."

So said the four hundred boys, and then they dug a large, very deep pit. Then they called Zipacná.

"We like you very much. Go, go and dig dirt, for we cannot reach," they said.

"All right," he answered. He went at once into the pit. And calling to him as he was digging the dirt, they said:

"Have you gone down very deep yet?"

"Yes," he answered, beginning to dig the pit. But the pit that he was making was to save him from danger. He knew that they wanted to kill him; so when he dug the pit, he made a second hole at one side in order to free himself.

"How far?" the four hundred boys called down.

"I am still digging; I will call up to you when I have finished the digging," said Zipacná from the bottom of the pit. But he was not digging his grave; instead he was opening another pit in order to save himself."

Zipacná knows the correct *celestial orientation* and his pit accommodates it. The *four hundred boys* are using an older mark. The ridge-pole is the *Tree of Life*, without its branches. It's the foundation of their world, whether *spiritual*, as a reflection of their *Sky World*, or *terrestrial* (material), as a roof over their heads.

They are one and they are two.

At last Zipacná called to them. But when he called, he was already safe in the second pit.

He has the correct mark, the *four hundred boys* do not.

"Come and take out and carry away the dirt which I have dug and which is in the bottom of the pit," he said, *"because in truth I have made it very deep. Do you not hear my call? Nevertheless, your calls, your words repeat themselves like an echo once, twice, and so I hear well where you are."*

So Zipacná called from the pit where he was hidden, shouting from the depths.

Then the boys hurled the great log violently, and it fell quickly with a thud to the bottom of the pit.

"Let no one speak! Let us wait until we hear his dying screams," they said to each other, whispering, and each one covered his face as the log fell noisily. He [Zipacná] spoke then, crying out, but he called only once when the log fell to the bottom.

"How well we have succeeded in this! Now he is dead," said the boys. *"If, unfortunately, he had continued what he had begun to do, we would have been lost, because he already had interfered with us, the four hundred boys."*

FIG. 256

Classic Chicha

Photo credit Raquel Gil Fotografía

Zipacná represents a new *celestial orientation*. The boys are the old *celestial orientation* on its way out. They no longer have the 'strength' to lift the pole or reach the depths. The sentiment expressed is the same as between *Athamas* and his wives, or the attitude of *Daniel* to *the priests of Bel*. It's the essence of our *Hooves Locking Horns*. Eventually, it must end in death. Whether '*Epoch, Age or Season*' each must come to its end.

Center and Circle

And filled with joy they said: "Now we must make our chicha within the next three days. When the three days are passed, we shall drink to the construction of our new house, we, the four hundred boys."

The new Epoch, the new house, is ready to begin. They are looking forward to the celebration.

Then they said:

"Tomorrow we shall look, and day after tomorrow, we shall also look to see if the ants do not come out of the earth when the body smells and begins to rot. Presently we shall become calm and drink our chicha," they said.

But from his pit Zipacná listened to everything the boys said. And later, on the second day, multitudes of ants came, going and coming and gathering under the log. Some carried Zipacná's hair in their mouths, and others carried his fingernails.

When the boys saw this, they said, "That devil has now perished. Look how the ants have gathered, how they have come by hordes, some bringing his hair and others his fingernails. Look what we have done!" So they spoke to each other.

Nevertheless, Zipacná was very much alive. He had cut his hair and gnawed off his fingernails to give them to the ants.

And so the four hundred boys believed that he was dead, and on the third day they began the orgy and all of the boys got drunk. And the four hundred being drunk knew nothing any more. And then Zipacná let the house fall on their heads and killed all of them.

The old Epoch is over.

Not even one or two among the four hundred were saved; they were killed by Zipacná, son of Vucub-Caquix.

In this way the four hundred boys died, and it is said that they became the group of stars which because of them are called Motz, but it may not be true. [130]

Seven Little Sisters, the Pleiades

Photo credit Rob Gendler

Do Daniel, the priest and *Zipacná* really dig pits, or is there more at work? The ridge-pole is our *Tree of Life*, the *Celestial Axis*. If we had a reference to the stars, we'd have *Center and Circle*. Is there a stellar link to the 400 boys in Mesoamerican culture?

According to the *Aztecs, Motz* is the *Seven Little Sisters*. We think of them as the *Pleiades*.

Omuch qaholah is another name for the *four hundred young men* who perished in an orgy, the same as those who were worshipped in Mexico under another name, *Centzon-Totochtin*, the *four hundred rabbits* who protected the pulque and drunkards.

Apparently the *Pleiades* are a hedonistic lot and like to party to excess.

Like *Zipacná*, Daniel's in a pit, but it's not a problem because they have the correct orientation. Besides, the Dragon has to live before you can kill it, and *Daniel 14* brings the Dragon to life in the same manner as other mythologies. Does the Bible believe in Dragons?

Apparently, Martin Luther didn't think so because *Daniel 14* doesn't make the jump from Catholic to Protestant Bible except in the King James version. As far as myths go, the astronomical reflection is a good one in that it still uses the *stars of the* Dragon and *Bull* (old *Center and Circle*) to determine the *updated* Vernal Equinox. They're making sure the myth downgrades the message and traditions associated with them. The nature-oriented themes of the Bull have been replaced by a new message: personal mastery amongst intense competition.

From Indo-European invasions to the standing army of Rome, the field of battle becomes one of the spiritual 'testing grounds' of this period; literally, a trail by FIRE.

Israel's most favored memories are for the Kingdoms of David and Solomon, when Judaism through the Hebrew Bible is at its cultural peak.

That's what the new *celestial orientation* has to say. This is what the Laws of Moses are all about. A new code of conduct for those bathed beneath this new stellar vibration.

There are actually two themes at work here, but each is important to the *celestial mechanics* needed to make the mechanism run right. The myths have supplied all the clues.

The *four hundred boys* have led us to an orientation that has long been a common *Celestial Focus* the world over, and that's the *Pleiades*.

Some have called them the 'Little Dipper' because the 'seven' stars that form the group look like the larger and brighter circumpolar constellation. 'Seven' is in quotes because there's a story, found from Indonesia through China to the Mediterranean, of a missing 'sister' that's been thought to be a missing 'star'. Most people are only able to see six stars, while those who are able to see seven are as likely to see nine or ten. The *Pleiades* is a cluster of literally thousands of stars depending upon the sharpness of your eyesight or the magnitude of your telescope, so the question naturally arises, why 'seven' if it's not the obvious visual answer?

We don't know, but the repetition of the story line makes one wonder. Did one of the sister stars nova, creating the icy-blue haze that these heavenly bodies smile at us through? They're very pretty.

The *Pleiades* are among the oldest stars to be mentioned in astronomical literature, possibly first appearing in Chinese annals of 2357 BC. They were once described as clusters of *golden bees* or *flying pigeons*.

Photo credit Allison Lucas   FIG. 258

The Pleiades once opened and closed the nautical season

Later in China they become the *Seven Sisters of Industry*, and were worshipped by girls and young women to whom they were believed to relate.

The Greeks also associated *flocks of Doves* with this cluster. It's been suggested that this forms part of the etymological root of the word, *peleiades*, 'a flock of doves', from the Indo-European, *pel-*, '*pale, dark-colored gray*', the grey bird. We are witnessing a very old tradition of augury here.

A number of the Greek poets, among them *Athenaeus*, *Hesiod*, and *Simonides* described the *Pleiades* as *Rock-pigeons* flying from the Hunter *Orion*, and this suggests an ancient inquiry.

We noted that during the time the Vernal Equinox was moving through the stars of the constellation Gemini, there was a marked increase in the amount of art depicted on statuary, vases and house walls that represented *Birds of the* AIR. Naturally, other themes abound, such as the *Twins* and *Serpent*, but, nevertheless; during this time there's a decided increase in bird imagery. The artistic focus is a reflection of how strongly they felt the connection at the Time.

Because 'AIR' was the elemental medium of the Time, this was the preferred method of most directly accessing the *Will of the Almighty*. *Bird augury* and *flight* were widely used as one of the best methods for determining *Divine approval*.

Center and Circle

FIG. 259 — Photo credit Tiago C Lima

Was the Parthenon aligned to the Pleiades?

Random appearances of birds as well as seasonal migrations were closely scrutinized, not only as a self-delivered food source (from Heaven!), but for the affirmation or disapproval of future enterprises.

Pigeons were released at the start of each sailing season with the resultant auspices carefully interpreted. Checking with the Gods first goes back a long way. Building with mortar and stone was frowned upon unless the proper entreaties had been sought, and the appropriate Deities honored.

Birds were released at the start of the season before the first ships set out, as both *Celestial Sign* and *Terrestrial Emissary* took flight together.

Those wise in such matters were held in high favor as having 'insider information' about God's Plan.

We have focused on the *Celestial Markers*, and their relationship to the *East Point*, or *Vernal Equinox*, but this was obviously not the only way to measure Time. As we have seen, both Moon and Sun were regular tabulators of the calendar, but so were *Helical Risings* or *Settings*. The *zenith transit*, if there was one, might also be used.

In contemporary usage the *Autumnal Equinox* is preferred to the Vernal Equinox in the Hindu tradition. As we have seen though, the trail of NCP and VE seem to take storytelling precedent in many Mediterranean models.

The *Pleiades*, with their distinct visual appearance, were used by various cultures on both sides of the Atlantic in various ways.

Lockyer felt the *Parthenon* and several other Greek Temples (the archaic *Temple of Athena* and *Bacchus* in Athens, the *Asclepieon* in *Epidaurus*, as well as the *Temple of Athena* in *Sunium*), were aligned to the *Pleiades*, although some have questioned these findings.

The whole point here is that different people used the stars in a variety of methods to tell Time. Rather than dilute our premise, it reinforces it. Some used Sun, others the Moon or Venus. The 18.6 year Nodal cycle, orbits of the outer planets, all were monitored.

Which brings us to a final note. The *Popol Vuh myth* depicts a great social gathering with everybody getting drunk in an orgy. Here's what Manilius, writing in Latin during the 1st century AD had to say about this cluster:

"The Pleiades, sisters who vie with each other's radiance. Beneath their influence devotees of Bacchus (god of wine and ecstasy) and Venus (goddess of love) are born into the kindly light, and people whose insouciance runs free at feasts and banquets and who strive to provoke sweet mirth with biting wit." [131]

Translation? They like to eat, drink and be merry. Do you suppose that's what the four hundred boys were doing when they brought the house down? There seems to be a whole lot of *Pleiades partying* going on. Is it a coincidence that these two ethnically distinct cultures, separated by both Time and Space, would share the same interpretation for the same stellar cluster? Or is it more than that?

In our opening section on the *Ram*, we first examined a few Greek myths involving the *Ram*, and from there turned to a little of the Hebrew view, and on to Babylon. What about the *Egyptians*, those champions of *Nationalized Farming* we had left behind on the famished banks of the *Nile*?

They, too, moved on, celestially speaking, although the symbol of the *Bovine* with the *Solar Disk between the horns*, whether male or female, would continue throughout the length of Egypt's history. They were remembering the time of their *Messiah's birth*, the *Genesis* of *their* culture.

We have mentioned that, from the *Hyades* to the *Pleiades*, a thousand years had passed, another thousand as Spring moved through the *Horns of the Bull*. But with the Great Famine, the *Old Kingdom* collapsed, and two later Kingdoms arose, the *Middle* followed by the *New*. There were extended '*Intermediate Periods*' between each of these.

The continuity of the *Old Kingdom* was broken. The later empires had their day, but never reached the level of stability, longevity and power of the earlier epoch. They were now a people racing in an attempt to catch up. They never again regained the full strength of the Bull, which had at last buckled.

There was one later line of pharaohs, an Egyptian lineage, that recaptured some of her former splendor, if not its duration. They were the *Eighteenth Dynasty*, the founders and driving force of the *New Kingdom*.

What was happening during these Times? What are the '*Second Intermediate*' and '*New Kingdom*' periods all about?

The *Second Intermediate Period* (circa 1720–1570 BC) had meant domination by outsiders, by foreigners known as the **Hyksos.** [132]

Here's some background.

After the last Pharaohs of the 13th dynasty, Egypt entered a period of social decline: with anarchy, civil war and a continual rivalry for the throne among the populace. The *North East Delta* was occupied by Palestinians who launched successive waves of attacks, but were held back for a time by the Hittites and Mitannians. However, the last Egyptian sovereigns finally succumbed to the infiltration of nomadic invaders.

These *Semitic nomads* of the *Eastern desert* were allied with the *Phoenicians* and *Canaanites*. The *Hyksos* were *nomadic shepherds* and among the first to develop new tactics with both iron and arms that led them to victory.

FIG. 260

A Bacchanal

Center and Circle

Photo credit Miltiade

FIG. 261

Amenophis III, peak of the New Kingdom

They weave together powers of horse, bow and arrows.

The *Hyksos* initiated an engagement by harassing Egyptian foot soldiers from horseback; using both bows and compound-bows. The Egyptian infantry, after poundings from these aerial assaults, were unable to mount a successful counterstrike and were worn down under successive onslaughts. The *Hyksos* arrows had bronze points for added penetration. By riding close to the enemy with a concentrated nearly horizonal bombardment, the *Hyksos* were able to shoot their arrows in a heated exchange, withering the native line.

The Delta and lower Nile eventually succumbed to these attacks, forming a **Hyksos Kingdom** about 1720 BC, with its capital in *Avaris* (Tanis) in the Eastern Delta. *Avaris* was the new Mediterranean doorway to the Nile, no longer in Egyptian hands.[133]

The Earth-based Egyptians began to take lessons at their local *Martial Arts Academy*. Eventually, they learned from these beatings.

Through contact with the *Hyksos*, the Egyptians started to incorporate the use of horses, war chariots, and new tactics with the bow (Fig. 261), as well as better battle axes and daggers made of iron. Once the *Hyksos* were overthrown in 1550 BC, native Egyptians continue to open to commercial ties with the outside world, becoming less insular as a nation. While Egypt yet remained in the hands of these outsiders, it built a strong sense of personal patriotism for being an Egyptian, an identity hammered home on the anvil of personel loss.

They had lost land, river and heart. They were a disenfranchised people, forlorn and outcast, feeling like second-class citizens in their own backyards. They'd been kicked out of their own homes.

They were not happy campers.

The mountainous Southern regions of Upper Egypt were able to hold out against the *Hyksos* and eventually turn the tide: reclaiming the river, their throne, and their sense of personal pride. Resistance was centered around the city of Thebes.

Photo credit Tour Egypt

FIG. 262

Amun-Re

Center and Circle

Photo credit Tour Egypt

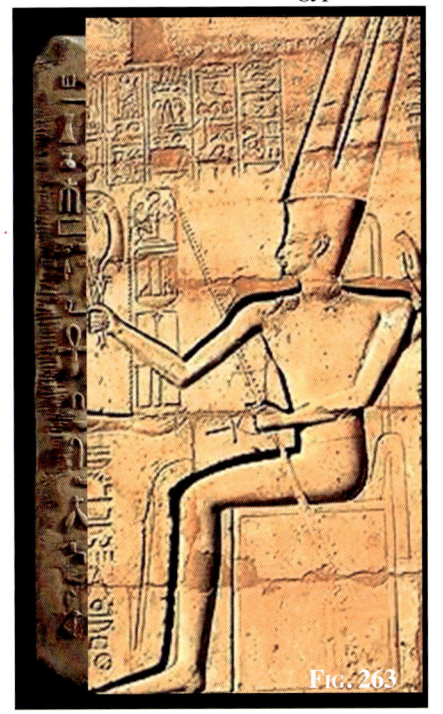

Fig. 263

Amun-Re

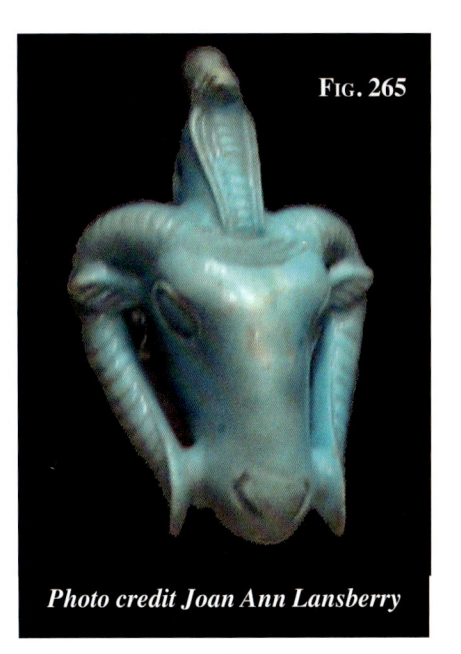

Fig. 265

Photo credit Joan Ann Lansberry

Uraeus-Ram / Center & Circle

While it lasted, the *Empire of the Hyksos* extended from middle Egypt through the Delta, Palestine and into Syria.

During this Time, a *new champion* steped to the *Celestial Forefront*, and this was our hero *Amun, the Lord of Thebes*. [134]

Say hello to the *God of War* in Egyptian garb.

When *King Ahmose I* expelled the *Hyksos* from Egypt, *Thebes* became Egypt's most important city and the dynastic capital of the *New Kingdom of Egypt*.

The patron deity of *Thebes* had been *Amun*, now elevated to a new national importance. The *Eighteenth Dynasty Pharaohs* attributed their success to this God and lavished much of their spoils on his Temples. These military advances and the resulting security thrust Egypt into a cultural renaissance, restoring trade and advancing architectural design to a level unmatched for another thousand years. The *Hyksos* had begun this practice during their Time in Lower Egypt, opening the Fatherland (Geb) up to trade contacts with the outside world. Once forced ajar, this Mediterranean door was never again to close to the same degree.

Egypt of the past was gone.

Egyptians considered themselves cruelly oppressed during the period of *Hyksos* rule. As a result, their final victory over their enemies carried with it the suppressed chorus of centuries, bursting forth in songs of praise for their new champion, their new Egyptian Heavenly Savior, *Amun*.

And this is where our *God of War* dons his cultural robes.

Amun was a self-made Man (God). He had no father and no mother and was to become the most famous of the Egyptian Gods after *Osiris*. Part of the vibration of this *Celestial Design* was *self-reliance* and *personal determination*. Drawing their strength from deep inside the southern highlands, Upper Egypt slowly pushed the *Hyksos* back, learning to fight under these new rules of engagement, and were eventually able to turn the tide.

Amun was seen as a champion of the less fortunate, a hero of the rights of the poor and the oppressed. Feeling that their *Divine Equilibrium* had been upset, *Amun* was the avenger of *Maat- Justice*. He re-set the scales and allowed the lands' natural rulers a return to their once majestic ways.

Egypt had their hero.

Photo courtesy Frederic Chanal

Fig. 264

Criosphinx (Ram-Sphinx) backing the power of Pharaoh

Athena's Web

145

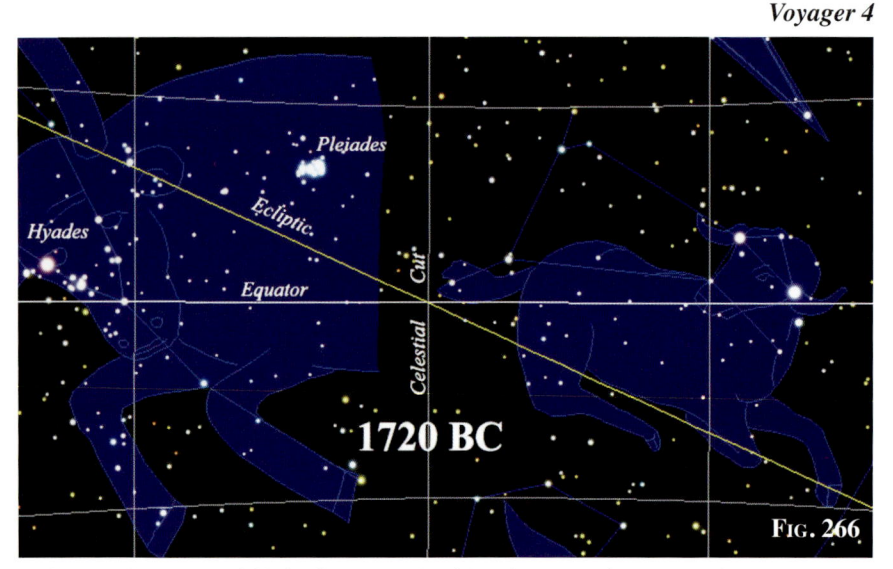

Center and Circle

Voyager 4

FIG. 266

The Hyksos establish their capital in the Northeast Delta at Avaris

"[Amun] who comes at the voice of the poor in distress, who gives breath to him who is wretched...

"You are Amun, Lord of the silent, who comes at the voice of the poor, when I call to you in my distress You come and rescue me...

"Though the servant was disposed to do evil, the Lord is disposed to forgive. The Lord of Thebes spends not a whole day in anger, His wrath passes in a moment, none remains. His breath comes back to us in mercy...

"May your ka be kind, may you forgive. It shall not happen again."
— *Deir el-Medina* stela. [135]

It's clear the Egyptians feel as though they're somehow at fault. They promise not to stray again.

After absorbing Kush, *Amun* identifies with their chief deity (Fig. 267). He was depicted as *Ram-headed*, or as a *woolly ram with curved horns*. So *Amun* becomes associated with our *Ram*. Those in Kush had been watching the Skies, but so had the Egyptians.

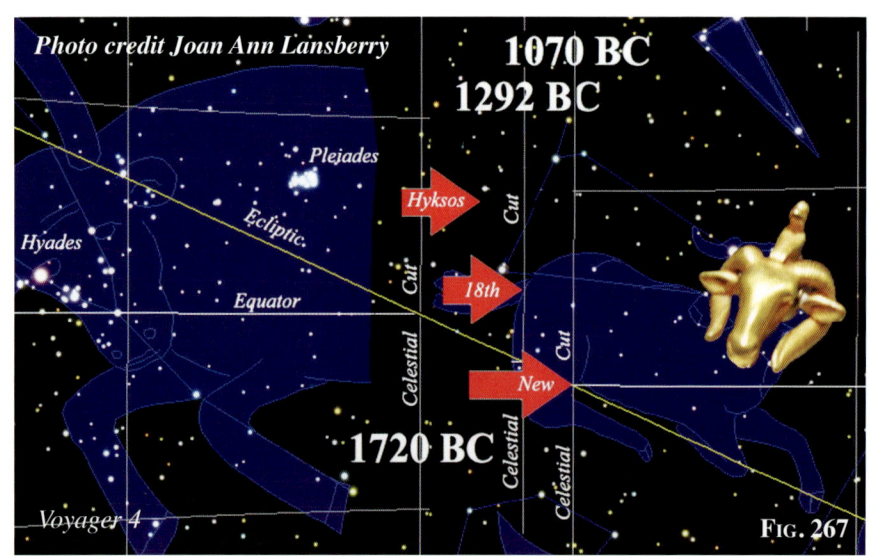

It wasn't a hard sell. Neither was unaware of the larger *Celestial Correspondence*.

Let's observe the Vernal Equinox during this period. All we have to do is date the cultures. The *Hyksos*, as mentioned, came in a series of waves but were finally able to establish a capitol in *Avaris* (part of Tanis) around 1720 BC. They are entrenched in their power a century later as *Botein (delta Arietis)*, the Tail of Aries, becomes the *Eastern Star* in 1624 BC. Another century and they're driven out as *Pharaoh Ahmose I* breaks their stronghold and expels them from Egypt, overthrowing their capital in the *Battle of Tanis* in 1550 BC. [136]

After centuries of schism, Egypt is reunified.

From 1550 to 1292 BC the *Eighteenth Dynasty* reestablishes its hold on power and the start of the *New Kingdom*, bringing back a renewed vitality to Egypt not witnessed in centuries.

What did the 'Eastern Frontier' look like? Precessional motion does not move very fast.

The *Second Intermediate Period* begins while the Vernal Equinox is travelling through the nether zone of the two constellations; following the *Pleiades* (Fig. 266), but prior to the first stars in the *Tail of the Ram*. We now think of the *Eighteenth and Nineteenth Dynasties* as the *New Kingdom of Egypt*.

In a lifestyle not entirely unknown as a precedent, the Age opens with a civilization that starts out as *nomadic shepherds* but then 'evolves' into an effective military force capable of taking over towns and establishing a kingdom.

FIG. 267

Photo credit Joan Ann Lansberry

Begun by the 18th Dynasty, the New Kingdom extends to 1070 BC

| Old Kingdom to 2152 BC | 1st Inter-- | -- Middle K. ---- to 1720 BC | Hyksos - | ---- to 1550 BC | New Kingdom ------- |

2400 BC — 2000 BC — 1600 BC — 1200 BC

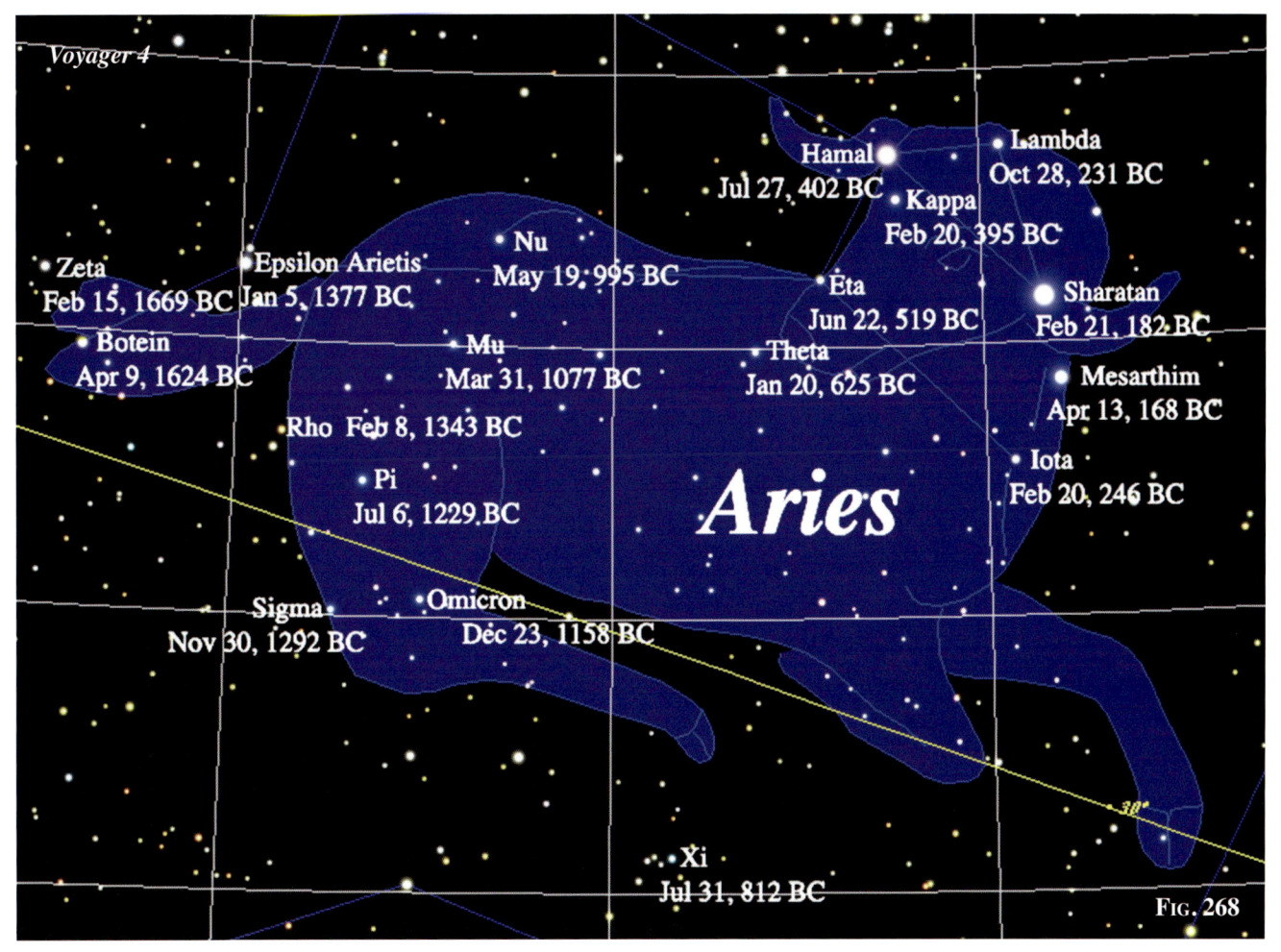

Dim though they may be, the stars of the Golden Fleece were carefully monitored

History has seen this pattern before. The *Ram* turns on Egypt, until Egypt learns to incorporate military strength into her program and change with the Times.

Life was being raised at the *breast of War*. FIRE is a powerful medium, but its true mettle is tested in battle, or so the story-line from this period ran.

The *Theological Themes* are *mythologically born of the Ram*. Different cultures emphasize varying facets of these themes, but history remembers the common *Ascendancy of Battle* with its advanced weapons and martial philosophy involving courage, bravery, valor and honor.

Egypt's contribution to this vision of a *God of War* was to make him a *Martial Hero*, like the *Fireman* who rushes in to rescue your only child from the *Raging Conflagration* called Life.

The Greeks saw this archetype somewhat differently. They had *Zeus* calling *Ares* the '*most odious of all the Gods*,' loudly crying foul whenever wounded, but thundering His euphoria upon firmly seizeing the laurel wreath in victory. He's a testy God.

Rome came to see *Mars* as the strength of the nation via the *Army*, where he held a position of high rank. *Aeneas, Rome's Trojan Founder*, was seen in this light.

Homer opens his epic work with a dirge to *Ares*, the *Lord of Man-wasting War*. It is an apt description for those attempting to survive while living admidst a time of belligerent, hostile neighbors during the *Age of Aries*.

And that's on a good day.

The *Fires of Heaven* were carefully monitored, meditated on and revered. Cultures all around the Mediterranean were attempting to honor the *Celestial Contract* as well as they knew how. The following is an excellent example of how this new bond between Heaven and Earth was being worked out, as represented by God, man and the Times.

| ------------ to 1070 BC -- 3rd Inter ------ | ---------- to 657 BC -- Late --------------- | -- to 332 BC -- Alex to Cleo ----- 30 BC |
| 1200 BC | 800 BC | 400 BC | 0 BC |

Center and Circle

"I am Yahweh," he said to him, "who brought you out of Ur of the Chaldeans to make you heir to his land.

"My Lord Yahweh," Abram replied, "how am I to know that I shall inherit it?" He said to him,

"Get me a three-year-old heifer, a three-year-old goat, a three-year-old ram, a turtledove and a young pigeon."

Photos credit
Rex Harris

*"To your descendants
I give this land,
from the wadi of Egypt
to the Great River,
the river Euphrates, the Kenites, the Kenizzites, the Kadmonites, the Hittites, the Perizzites, the Rephaim, the Amorites, the Canaanites, the Girgashites, and the Jebusites."*

—Genesis 15:17–21

FIG. 269

The Abrahamic Covenant

He brought him all these, cut them in half and put half on one side and half facing it on the other; but the birds he did not cut in half.
—Genesis 15:7-10

When the sun had set and darkness had fallen, there appeared a smoking furnace and a firebrand that went between the halves. That day Yahweh made a Covenant with Abram in these terms:

In the days before folks signed names on paper, the Covenant was seen as an agreement between two parties. Opening with a prayer for good wishes on the proposed outcome, the animals would be sacrificed, and the parties involved would walk between the animals, calling down their fate (death) on themselves if they failed to live up to their end of the bargain.

Since this Covenant is between God and Abram (before he became 'Abraham'), God manifests Himself in the form of FIRE, and is 'seen' as a spiritual witness to the contract.

Nowadays we use lawyers.

The contract states *Yahweh* will give Abraham (the name change occurs with the covenant) the whole land of Canaan to own forever. In return for which...

God said to Abraham, "You on your part shall maintain my Covenant, yourself and your descendants after you, generation after generation. Now this is my Covenant which you are to maintain between myself and you, and your descendants after you; all your males must be circumcised. You will circumcise your foreskin, and this shall be the sign of the Covenant between myself and you."

—Genesis 17:9-11

Remember how the *Solar Disk* between the *Horns of the Bull* was pervasive throughout the length of Egyptian history, even though it only marked the correct astronomical position at their unification? Like Gilgamesh, this was where the '*Cut of Heaven*' is being made. The Hebrews are doing the same thing. The *Ram* is Man. For them, one Hebrew male. The *Celestial Cut* is moving in from the rear, or as my wife said, from the 'ass-end' of the *Ram* first.

Say hello to *celestial circumcision*. The anatomy of the *Ram* is correct for the stellar Time indicated. Which star do you suppose they thought of as the foreskin; *epsilon*, *Botein* or *zeta*?

148

Athena's Web

Center and Circle

Welcome to Judaism 101.

This shall be the sign of the Covenant, the genitals and essence of the *Ram* (*Ovis aries*).

It's the Jewish snapshot of Heaven, *their* moment in Time and Space. It was their *Genesis* as a people, distinct from all others.

Well, that sure explains a lot of things. Like *Helle* falling off the back of the *Ram* (a contemporary myth to Abram's intended sacrifice of his son, they both represent the start of the new constellational pattern in Heaven), we now can understand why Abram's offspring will number more than the '*stars of Heaven*,' why Abram can still have a child at the age of 100, and why their initial *Martial Goals* (all the land between Egypt and the Euphrates?) fall so far short of the mark. According to astrological tradition, once an *Aries* gets past the start of a project, he or she tends to lose interest. As long as it's fresh, new and exciting, the enterprise has strength and holds their focus. Once it loses that 'cutting edge,' newness or sense of being 'special,' their heated *Fiery Enthusiasm* tends to go out.

The Egyptians expressed this same sentiment when they spoke of *Amun*, their *Aries Archetype*.

"*The Lord of Thebes spends not a whole day in anger, His Wrath passes in a moment, none remains...*"

They're short studies.

Aries often tend to start looking around for other '*Fiery Enthusiasms*.' They want to know where the *action is*.

Under this newly tapped reservoir of *celestial energy* for the '*Chosen People*' the potential is there, the promise is there, but a contract is a contract, and the people continually fall back on the older, more established rituals of tradition (which had been both a fun and festive blessing of Venus), forgetting these raw, new and unfamiliar *woolly rules*. It comes as no surprise that the people do not live up to their end of the contract, and therefore God is released from His end of the bargain.

Without the lawyers.

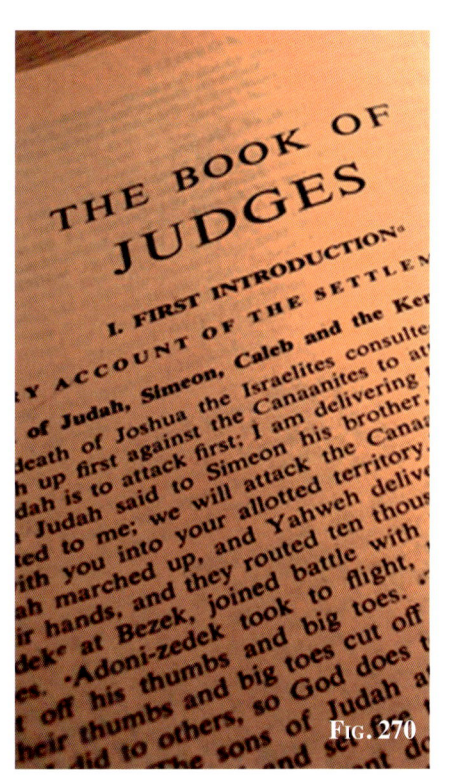

Fig. 270

7th Book of the Christian Bible

Or the fees.

The only way Judaism collectively got to the Euphrates was as slaves, and this only after they had spent four hundred years as slaves under the Egyptians. The blessing, the Covenant that had been made, had been turned into a curse.

Was God wrong?

No. The contract was simply broken. Much of the *Hebrew Bible* is a list of kings; those who honored Yahweh's laws, and those who failed to do so and succumbed to the pagan traditions everywhere around them.

If the kings and how they scored were part of a system with a letter grade given at the end, or even on a pass/fail basis, they wouldn't earn a passing mark.

Life is a constant challenge. In a curious placement of books, the verdict is found in Judges 2:1-3.

The angel of Yahweh went up from Filgal to Bethel and came to the House of Israel; and he said, "...and I brought you out of Egypt and led you into this land which I swore to give your fathers. I said: I shall never break my covenant with you. You for your part must make no covenant with the inhabitants of this country; you must destroy their altars. But you have not obeyed my orders. What is it that you have done? Very well, I now say this: I am not going to drive out these nations before you. They shall become your oppressors, and their gods shall be a snare for you."

According to *Judges*, this is why no Jewish king ever sat on a Babylonian throne. The terms are clear.

The deal's off.

Switch to Plan B.

This is a God born of FIRE.

From Deuteronomy 4:11,

"*...the mountain flamed to the very sky, a sky darkened by cloud, murky and thunderous.*

Then Yahweh spoke to you from the midst of the fire..."

And from Deuteronomy 4:24,

Center and Circle

FIG. 271

Agni- Persian/ Hindu God of Fire

"... for Yahweh your God is a consuming fire, a jealous God."

We're not here to do an exhaustive study on worldwide mythic folk culture. I'm exhausted already! But if one stirs the coals and digs in the embers, you'll find more than meets the eye.

We've just been looking at the *Ram* in the *Judaic tradition*.

From Exodus 19:13,

When the ram's horn sounds a long blast, they are to go up to the mountain.

We see these same themes all around the Mediterranean.

FIG. 272

Agni's mount is the Ram— his color Red

Alexander the Great (Fig. 274) was often depicted with *Ram's horns* coming out of his head on various coins. By Alexander's time, we're beginning to close in on the end of stellar passage through the constellation of the *Ram*.

Agni (Fig. 271) is the *Persian* and *Hindu God of Fire*, whose mount is the *Ram*. His color is *Red*.

FIG. 273

Meet Mars, Mr. Macho

Mars is the *Red Planet*, traditional ruler of *Aries*. As *Agni*, He is the acceptor of sacrifices. Having a good *Fire* is important if you're going to cook your food after you surrender its soul. He is the Chief Officer of religious duties, and is thought to be the *messenger* between Gods and mortals.

FIG. 274

Alexander with Ram's horns
Photo credit Wayne Sayles

Prayer and sacrifice are how you best reach the Gods, and *praying by the Fire at the sacrifice* was a common *Zoroastrian* practice, as it was with many other religions. Both the *sword* and *ax* are common attributes of *Agni*.

Zoroastrianism is a religion which pre-dates Christianity, arising in Iran about 1000 BC, right in the middle of the Vernal Equinox's passage through the constellation of the *Ram*. *Zoroastrianism* is a monotheistic religion and some of its original tenets are believed to have influenced both Judaism and Christianity, particularly with regard to their fundamental notions of *good* and *evil*. They believe they can fight evil through *good deeds*, *words* and *acts of charity*.

Fire plays a central role in their *worship of Truth* and the *Spirit of God*, and consecrated *Fires* are kept perpetually burning in the Temples. Rome did the same with the *Vestal's flame*.

FIG. 275

Celtic Ram-headed Snake

Center and Circle

Photo credit Bloodofox

Fig. 276

Cernunnos, Torc and Serpent

FIRE *first*.

What each of these cultures is doing is expressing these same over arching *celestial themes* in varying ways, while sticking to the same stellar principles.

Nobody owns the Sky. It was common to and interpreted by everyone, but there was a *'celestial vocabulary'* being spoken between various peoples. If you look to the *symbolism* rather than the name, it *springs* to life. Of course, each culture expressed it in their own slightly different fashion. For one thing, Gaia, *Mother Nature*, the land itself, is different wherever you go. What do river, forest, mountain or ocean mean to your culture, to your people?

All of these very personal philosophies are reflected by the people who must live them. It doesn't matter to the Sky; He marries the Earth, no matter what Her geographic landscape, watery seascape or seedy past.

Zeus sees beauty everywhere.

He's a lusty Lord, but then, during *Taurus* and *Aries*, they were all lusty Lords.

Greek traditions tell us *Mars rules Aries*, while the *Sun is exalted in Aries*. *Red* and *Gold* are the spectrum of the visual wavelength that we're working with here; they're not that dissimilar. They're a fiery, brilliant lot. *Mars* supplies *enthusiasm* and your basic *martial drives*, while the Sun represents *Life* itself, a spirit for living, what the *Zoroastrians* called the *Spirit of God*.

It was a period *inflamed*, a *Time animated*. Thus, the Lord speaks to Moses from the '*burning bush*' (*Burning* desert *Tree*, *Fire* and *Tree*, *Circle* and *Center*) in the desert. It's also why Moses uses the mountain as his observatory, 'trimming it' so that he can better monitor the rotations of Heaven.

'Mark out the limits of the mountain and declare it sacred.'
—Exodus 19:23

Once he's got it marked out, he doesn't want anybody messing with it. This is the 'sacred space' that he has created for the Gods to walk, or, if you prefer, for God to walk.

They all speak from Heaven.

Yahweh said to Moses,

"Tell the sons of Israel this, 'You have seen for yourselves that I have spoken to you from heaven.'"
—Exodus 20:22

Indeed.

The *baton of Heaven* is calling the tune for each of these cultures. They all try to pick it up, in their own way and to their own advantage, but they're still all listening to the same *celestial melody*.

Under Rome and *Aries*, it was the '*Battle Hymn of the Republic*.'

The *Celts* left no written record, but we can see from some of their artifacts the same *celestial themes* at work yet again.

Cernunnos is a *Celtic God* whose personal dossier we do not have. What we do have comes from the Romans, it is a name meaning '*Horned One*,' which is obvious from some of His depictions. In the image shown (Fig. 276), a buck matches antlers, while the God holds a torc in his right hand and a *Ram-headed Serpent* in his left. This is our theme of *Center and Circle*, or in this case, *Circle and Center*. The *Snake* is being grasped where we might artistically expect the Serpent was being lanced at their New Year festivals. We know the Dragon was associated with the May Day ritual (Fig. 21), one of two *Celtic New Years*.

Fig. 277

The Wheel of Time?

In the next image (Fig. 277) we see a central figure, Dagda, but He seems to be holding the *Wheel of Time* in His right hand, with a similar *Ram-headed Serpent* (Fig. 275) appearing to His lower right, below the *Wheel*.

Notice that only half of the *Wheel* is visible, just as we see only half of the Sky at any one time, whether as day and night, or summer and winter.

Center and Circle

Voyager 4

Time of the End

Note the circle whose center is the North Ecliptic Pole. Polaris sits on that circle

We've been following the path of both the *Vernal Equinox* and *North Celestial Pole* throughout this work, observing how its *constellational alignment* coincided with the prevailing *religious* and *artistic themes* of civilization across both Time and Space.

The *North Celestial Pole* had its stewardship monitored by a Dragon around the World, and it's the *common denominator* of our equation, a consistent measure against which the Ages would rise and fall. To some it would seem that the Dragon went on forever.

Of the *eight millennia* we have observed so far, the astronomical Dragon has accounted for approximately six as *celestial indicator*. One would hope he signed up for the seniority plan at the outset.

In 2788 BC there was a pole star. In 2100 AD there will be another. That will be the year Polaris 'precesses' to the point at which it will be closest to the *North Celestial Pole* and the *alpha* star of *Ursa Minor* officially becomes the new '*heart*' of the *Bear*, even if it is located on the *tip of the tail*.

All Hail the King!

For the first Time in almost five thousand years, Heaven will wear a New Crown (Fig. 278).

One (*Thuban*), two (*Polaris*). That's it. *Cepheus* will come to have his day as the *King of Heaven*, as *gamma*, *beta* and *alpha* approximate the pole position, but none of these come as close as *Thuban* or *Polaris*.

Vega will be the *brightest* Pole Star of our *precessional voyage*. It's already the second brightest star in the northern hemisphere, but it will not attain the polar accuracy of either of these other stars.

Center and Circle

Mythically speaking, *Creation's Lathe* was and is born of the fusion between Heaven and Earth, where the two stand in *cosmic balance* and *light and darkness* find perfect *celestial proportion*. Sky and planet, Zeus and Hera, together with a cast of thousands.

While the VE moved through the stars of *Gemini*, both talk and trade experienced a renaissance. While we're not naïve enough to believe that *mythic storytelling* began here, the expanded trade networks make it clear people were communicating with each other, with raw materials native to one area found in environments farafield. The *pathfinders* had their work cut out for them. Civilization's nervous systems, her trails, roads and waterways, were being overhauled. When trade involves artwork that embellishes the tools or crafts coming their way, questions are asked.

The origins of many of the mythic cultures we've looked at seem to have their roots entwined in the *Twin Tales* of the time,

Creation born of Siblings.

During *Taurus*, the vibration shifts and *agricultural concerns* become the mortar that bonds society together. The leaders of this time learn how to harness the population, their land and their rivers in a massive, collective effort.

During *Aries*, *Warlords* become the line of 'kings and priests' under which society is mortared, commencing with the changing climatic conditions of the late 22nd century and concluding with the Roman Republic's *professional standing army*.

The *Climatic Changes* didn't fully hit the *Indus River Valley* and **Harappan Civilization** until a little later, their 'mature period' running from 2600 to 1900 BC. Nevertheless, this civilization was to fall victim to the overall drying conditions that ended the *Agricultural Epoch*. Their *Indus Valley climate* grew significantly cooler and drier with a general diminishing of the power of the monsoons circa 1800 and 1700 BC. [137]

In 2600 BC all three-river civilizations, *Nile, Euphrates* and *Indus* had been in full swing, cultivating inroads in the building blocks of life, developing systems of mathematics, weights and measures, and extending areas of trade. Like *Egypt*, the *Indus Valley* was protected by mountains, desert and sea, insulating the culture, and allowing them the time and opportunity to grow.

As one Age gives way to another, the *currents of consciousness* are redirected and establish new channels. Faced with these changes, the foundations that hold society together degenerates. Chaos and uncertainty follow in their wake. This process has unfolded twice before in recorded history, and we currently stand at the gates once more, watching yet another '*End Time*' unfold.

During *Classical Civilization*, the *Ram's martial short temper* frightened the world and '*fear of the Lord,*' whether of a *Heavenly* or *Earthly Master*, becomes a highly coveted *Rod of Iron* for controlling the local populace, whether friend or foe.

Photo credit Mukul Banerjee

FIG. 279
Harappan Ox

As *Aries* draws to a close, a new vibration begins to emerge on the Horizon. Our *Eastern Inner Eye* begins an *extended meditation* while *Fishing*.

After *Aries*, the VE moved into *Pisces*, and there's another *New Spiritual Covenant* with *Heaven*, as the *Fish* take preeminence, bumping the *Ram* from the front of the line. How does a *Fish* outcompete a *Ram*?

Christ makes his first four disciples *Fishers of Men*, he feeds the multitudes with *Two Fish*, pulls the coin from the *Fish* to pay the tax, and walks on WATER. *Pisces* is a WATER sign, the new *Celestial 'Medium'* (element), helping to dampen the former FIRE.

FIG. 280
Indus, the End of Taurus?
Photo credit Bert Cozens

Athena's Web

Center and Circle

As each 'Age' comes to an end, there are periods of social disintegration, upheaval, and chaos. The pillars which have supported civilization begin to crumble and people suffer. These periods have sometimes been referred to as the *End of Time*, *End of the World*, or *End Times*. They do not mean, literally, the *End of the World*, but rather, the *end of a period of cultural continuity*, of *social consciousness*. The disciples are partially to blame for framing the question in this way, as we hear in Matthew 24:3. They ask Jesus,

"*Tell us, when is this going to happen, and what will be the sign of your coming and of the end of the world?*"

In the West, Christianity has represented the continuity of the last Epoch. The *Old Testament* (Hebrew Bible) was a child of the *Age of Aries*, just as the *New Testament* is a child of the *Age of Pisces*. Spirituality was regarded in different ways by Jews and Christians, yet both editions of the Bible contain references to the '*Time of the End*.'

In the *Old Testament*, they look ahead to the end of *their* era, just as we are curious about the end of *ours*. One prophet (among others) who looks ahead is *Daniel*. In his work, both his eyes and ours are raised to the stars.

The Book of Daniel was completed about 536 BC. It addresses issues still in their future and makes predictions about them. *Daniel* has a series of visions, many of which are explained in the same work. These predictions fill chapters 7 through 12. The style and images used here are later repeated and referenced in both *Matthew* and the *Book of Revelation*.

In Chapter 7, *Daniel* dreams he sees *four great beasts* emerge from the Sea. The first of these is said to be the Babylonian Empire (the prevailing power at the time of the writing of *Daniel*).

The second and third beasts represent the *Medes* and *Persians*, *Babylonian* successors. Finally, a fourth beast, described by *Daniel* as 'different' from the other three, crushes and tramples underfoot 'what remained.' This was said to be *Alexander*, who overthrew the *Persian Empire* in 333 BC.

These 'dynasties' represent what is left of the *Classical tradition*. The sands are running through the hourglass; the spiritual lessons left to be taught by *Aries* are winding down. There is both an urgency and a foreboding about the dark days ahead.

Chapters 10 through 12 go into considerable detail to describe what is believed to be the conflicts between the successors to the *Seleucid* and *Ptolemaic Empires*, two divisions of *Alexander's conquests*. Biblical researchers seem to agree that this takes us down to the end of *Antiochus Epiphanes* (175-165 BC), and the '*Time of the End*,' a period of '*great distress, unparalleled since nations first came into existence.*' This is *Aries* coming to a conclusion, preparing for the *advent of Pisces* and a new *spiritual covenant* for both Heaven and Civilization.

Let's look at *Daniel 7:1-7.*

In the first year of Belshazzar king of Babylon, Daniel had a dream and visions passed through his head as he lay in bed. He wrote the dream down, and this is how it began: Daniel said, "I have been seeing visions in the night. I saw that the four winds of heaven were

Photo credit Nafsadh

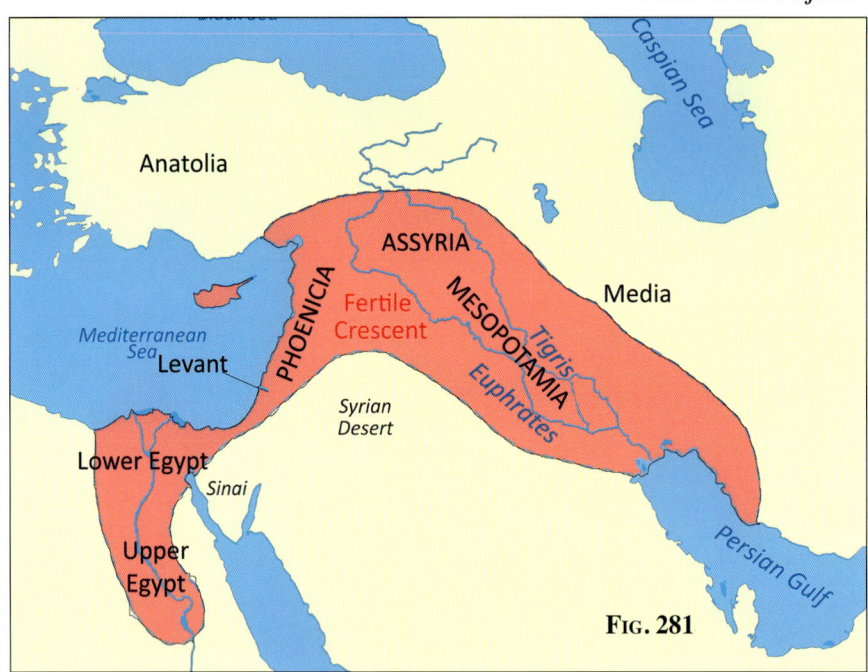

FIG. 281

The Fertile Crescent

Voyager 4

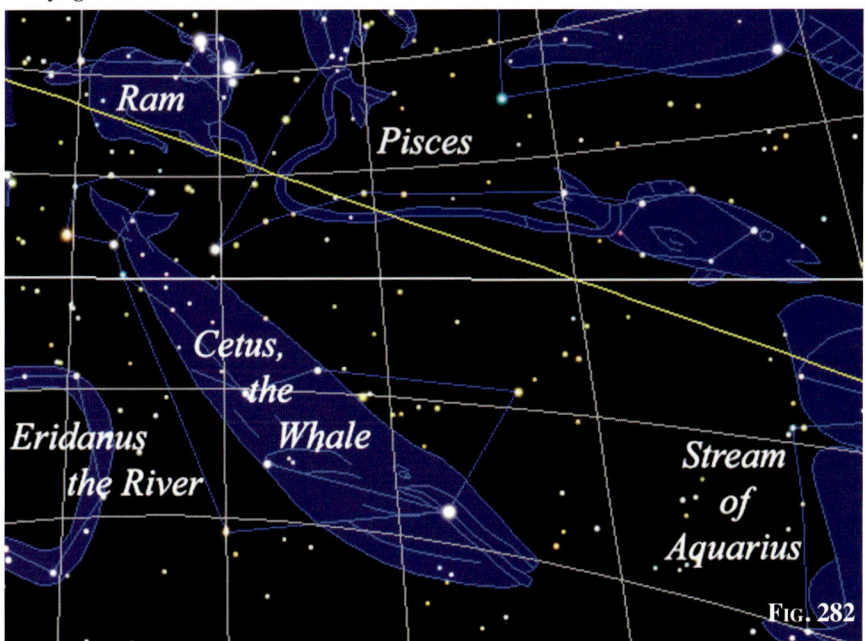

Ram by the River

Fig. 282

Center and Circle

He raised his eyes to gaze upon the starry realm and see a *Ram standing by a river*.

Before the breast of *Aries* lies a whole Seaworld of aquatic life, beginning with the *River Eridanus*, and loaded with *Fish*, *Dolphins* and *Whales*. Even the *Stream of Aquarius* is a tributary.

"*I saw the ram thrust westward, northward and southward. No animal could stand up to it, nothing could escape it. It did as it pleased and grew very powerful.*"

These are terrestrial events, the kingdom's martial expanse. We're working with '*the Man*' here.

"*This is what I observed: a he-goat came from the west, having covered the entire earth but without touching the ground,..*"

Like the *Golden Fleece* flying through the Sky, this is *Capricorn* covering the Earth but without ever setting hoof on it. Notice this is what Daniel '*observed*.' We have discussed this before and will again.

If *Capricorn* is in the *west*, *Cancer* must be *rising* (in the *east*) and the stars of the *Ram*, *Aries*, are therefore on the *Midheaven*, at the peak of their trajectory.

The *Ram* is at its 'high noon' (due South) position at *Celestial Center Stage*. Astrologically speaking, this is the position of the Midheaven, the current hierarchy or administrative power, king or pharaoh. With Aries here, the *Ram is* the establishment. The military is in charge.

As to the *Goat* being in *the west*, this is the position of the Descendant, or *adversary*. In astrology, it would be considered an '*open enemy*.'

stirring up the great sea; four great beasts emerged from the sea, each different from the other. The first was like a lion with eagle's wings; and as I looked its wings were torn off, and it was lifted from the ground and set standing on its feet like a man; and it was given a human heart. The second beast I saw was different, like a bear, raised up on one of its sides, with three ribs in its mouth, between its teeth.

'*Up!*' *came the command.* '*Eat quantities of flesh!*'

After this I looked, and saw another beast, like a leopard, and with four birds' wings on its flanks; it had four heads, and power was given to it. Next I saw another vision in the visions of the nights: I saw a fourth beast, fearful, terrifying, very strong; it had great iron teeth, and it ate, crushed and trampled underfoot what remained.

This first 'vision' of Daniel's is a dream. He tells us so.

But this next vision is different. Here he establishes his framework. The opening lines could be taken from a sky-watcher's note book.

In the third year of King Belshazzar a vision appeared to me, Daniel, after the one that originally appeared to me. I gazed at the vision, and as I gazed I found myself in Susa, the citadel in the province of Elam; gazing at the vision, I found myself at the Gate of the Ulai. I raised my eyes to look around me, and I saw a ram standing in front of the river. It had two horns; both were tall, but one taller than the other, and the one that rose the higher was the second.

Belshazzar was made co-regent by his father in 553 BC.[138] He would have been in the third year of his reign in 550 BC.

Thus far we have a classic astronomical setup. He is establishing the location and the year. He's standing by the Gate of Ulai.

Athena's Web   155

Center and Circle

Photo credit Hind, Arthur Mayger

The dream of Daniel

To the west of Babylon would lie Greece and the Mediterranean, which works equally well with our Biblical imagery.

"*...and between its eyes the goat had one majestic horn. It advanced toward the ram with the two horns, which I had seen standing in front of the river, and charged at it with all the fury of its might. I saw it reach the ram, and it was so enraged with the ram, it knocked it down, breaking both its horns, and the ram had not the strength to resist; it felled it to the ground and trampled it underfoot; no one was there to save the ram. Then the he-goat grew more powerful than ever, but at the height of its strength the great horn snapped, and in its place sprouted four majestic horns, pointing to the four winds of heaven.*

"*From one of these, the small one, sprang a horn which grew to great size toward south and east and toward the Land of Splendor* (Palestine). *It grew right up to the armies of heaven and flung armies and stars to the ground, and trampled them underfoot.*"

It grew up to the armies of heaven because that's the source of the imagery. Bringing stars to the ground is former dynasties.

"*It even challenged the power of that army's Prince; it abolished the perpetual sacrifice and overthrew the foundation of his sanctuary, and the army too; it put iniquity on the sacrifice and flung truth to the ground; the horn was active and successful.*" —Daniel 8:1-12

Whether as the *Horns of the Bull* or *Ram*, these are the symbols of military might, power, and most importantly, victory. They are metaphorical themes with which the Classical world had long been familiar. We'll examine the issue of the horns more in *The 8th Seal*.

The events *Daniel* predicts look forward to what was for *them* the future, but is for *us* the past.

This next dream outlines the same set of political events as the first, but uses a different set of metaphors. Let's look at the explanation the angel provides regarding the *Time of the End*.

In Daniel 8:19-22, *Gabriel interprets the second of Daniel's dreams* so we are made to understand. This makes it easy.

"Come," he said, "*I will tell you what is going to happen when the wrath comes to an end; this concerns the appointed End. As for the ram that you saw, its two horns are the kings of Media and of Persia. The hairy he-goat is the king of Javan* (Greece), *the large horn between its eyes is the first king. The horn that snapped and the four horns that sprouted in its place are four kingdoms rising from his nation but not having his power.*"

The stage is set. Here the 'End Times' are referred to as '*the appointed End.*' The vision is the future for the Middle East. *Daniel* is describing events during his service under Babylonian masters and the regional political turmoil which will follow. What Daniel outlines is what's left of the Age.

The two kingdoms that follow are the *Medes* and the *Persians* (the Ram with its two horns, they shared power). The Persians are over-powered by *Alexander* (the original horn of the goat), who died at the height of power. His territorial conquests were divided among his generals into *Macedonian*, *Seleucid*, and *Ptolemaic Kingdoms*, together with the original *Greek homelands*, making four smaller horns or kingdoms in all.

Yet the *New Testament* still speaks of *End Times*, long after these events are history. How can two different versions of the *End Times* both be correct? Because each refers to the end of a *different* astronomical epoch. The Old Testament refers to the end of the *Classical Age* (Aries), while the New Testament speaks of the end of the *current epoch* (Pisces).

We have seen what happened to the *agricultural empires* as the Age of Taurus came to a conclusion. Now we will examine what happens when a *martial age* draws to a close.

Perseus of Macedon
Photo credit British Museum
Fig. 284

There are a few ways we can frame its mythological demise. One would be to observe when the first star of the *Fish* is triggered by the VE (RAMC), and to say that the previous epoch ended at this time. That would be 407 BC when *alpha Piscium* became our *Eastern Star* (Fig. 285).

The second method is to simply observe the events that take place as the remaining star in the constellation of the *Ram* is triggered by the Vernal Equinox. This would be the *Ram's* mythological 'swan song' (so to speak) in 168 BC, the last time Aries appears 'on celestial stage.' This is when *Mesarthim, gamma Arietis,* the final horn of the *Ram* was our *Eastern Star.*

The third method would be to simply observe when *all* the stars of the *Ram* had been triggered, and then wait for the next star of the *Fish* to become the Eastern Star, marking an entry into the constellation 'unpolluted' by Arian (*Classical*) influence. That would be the third in the series, *nu Piscium,* aligning with the VE in 15 AD.

So which is it, you might ask: 407 BC, 168 BC or 15 AD? What makes you think there's only one right answer? To some degree, each is true. With each alignment, one vibration is winding down as another is starting up. In my humble opinion, the Hebrew prophets are looking ahead to the last star of the *Ram,* i.e. the end of the epoch of Yahweh.

The Wise Men who journey to visit the newborn Christ child are attuning themselves to the upcoming epoch of the *Fish*. They are anticipating a new vibration, one that both fulfilled prophecy, and brought together a new way of living and looking at life. With this celestial birth came a new feeling of excitement and expectation as they looked with wonder and curiosity toward the future.

This sort of mythological dilemma is one that skywatchers had to wrestle with, but one those working with spherical trigonometry did not. The stars' positions were being accurately tracked by the hours and minutes of Right Ascension, by degrees and minutes of longitude and latitude. The 'need' for images among the stars to help ascertain celestial location became caught in Creation's backwaters, endlessly spinning in circles as Time marched on.

Rather than the *Fish* becoming the new mythological marker of Time, the ancient system broke down, and the *Ram* maintained its hold on the star(t) point, the beginning of Spring. To this day, the image of the Ram continues to mark Spring, as the Ram follows (is defined as) the Vernal Equinox, or zero degrees of Aries.

It should be a *Fish*.

If I were Moses, I'd be breaking the tablets on the ground along about now and saying,

"NN*Nnnnnnnooooooo........*"

Culturally, this is one of those observational overlaps. The Classical World is coming to a close as this brave, new *Watery Realm* is being born.

Eastern mystery religions ranging from Buddhism to Isis, Magna Mater, and Mithraism begin to flower, vying with Christianity for spiritual and political control of the inner pathways.

The 'New Vibration' is being born, yet the old ways continue to linger. 407 BC, 168 BC and 15 AD are all 'waves' of the same incoming Oceanic tide.

407 BC is Rome's conquest of Veii, the conclusion to a 400-year-long struggle with their northern neighbor the Etruscans, an event Romans remember as an historical turning point, a 'second birth' from which Rome began its rise to martial ascendancy.

For the first time during this war the local landowners were paid for their *Martial efforts* from the public funds. This, in turn, meant that the governing officials could and did ask them to campaign right through the hard cold winter months. Farmers (the bulk of the army) were no longer allowed to return to their farms during the off-season.

This struggle, which marked the first definite step in Rome's career of world wide conquest, was remembered in Roman tradition as a turning point in the military history of the city, and the siege of Veii which ended it was magnified into a ten year investment, a Roman counterpart to the Greek leaguer round Troy. [139]

Immediately following these years, the Celts attacked and sacked Rome, with Camillus, the hero of Veii and legendary 'second founder of Rome' kicking their butts on the way out of town.

In the years following the turn of the century we have the sack of Veii by the Romans and the sack and occupation of Rome by the Celts.

What changed, in a conflict that had been on-again and off-again for over 400 years?

These tumultous times left Livy to somberly state,

"*Urged by the evil star which even then had risen over Rome...*"[140]

Voyager 4

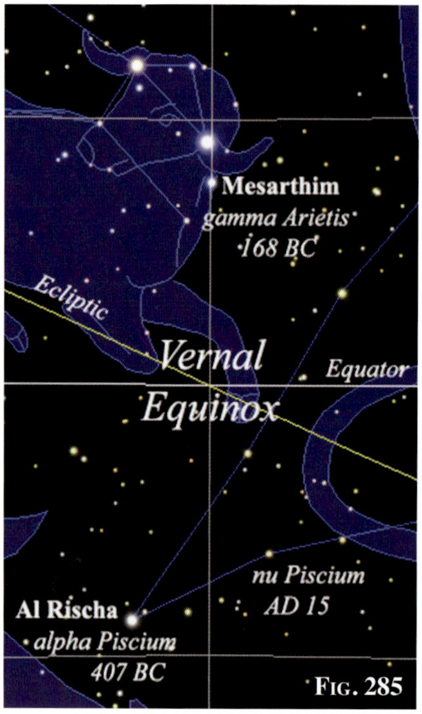

Fig. 285

The Ram starts to smell Fishy?

168 and **167 BC** were watershed years (Fig. 285). Twin spiritual flames of the classical period are permanetly dampened.

In the Hebrew tradition we have the rise and fall of **Antiochus IV Epiphanes**,[141] one of history's worst blights on the sanctity of the Jewish Temple. During his three-and-a-half year persecution, possession of the *Torah* (Judaic Law) was made a capital offense.

Shortly afterward, the king sent an old man from Athens to compel the Jews to abandon their ancestral customs and live no longer by the laws of their God; and to profane the Temple in Jerusalem and dedicate it to Olympian Zeus... [142]

Sacrifice, sabbaths and feasts were banned, circumcision forbidden. It was the law.

The Temple was filled with reveling and debauchery by the pagans, who took their pleasure with prostitutes and had intercourse with women in the sacred precincts, introducing other indecencies besides. [143]

Jewish traditions were specifically targeted, with Rabbis made under pain of death to sacrifice and consume animals prohibited by the Torah. A statue of Zeus erected on the altar oversaw the whole affair. Ultimately, this sacrilege would lead to the Maccabean Revolt, but the damage had already been done.

During these same years in the Balkans there was a battle that tipped the balance of power in the Mediterranean. In 168 BC, the Antigonid dynasty claimed direct descent from Alexander III of Macedon. This battle lost Macedonians control over their homeland, allowing them to be sold, partitioned and absorbed by the Roman Empire. This fateful battle decided the fate of the soul of ancient Greece.

Following the conquests of Alexander the Great in the 4th century BC, Greek culture had represented the military and political *inteligentsia* of the region, the cohesive culture of the East. This legacy was traced through

Center and Circle

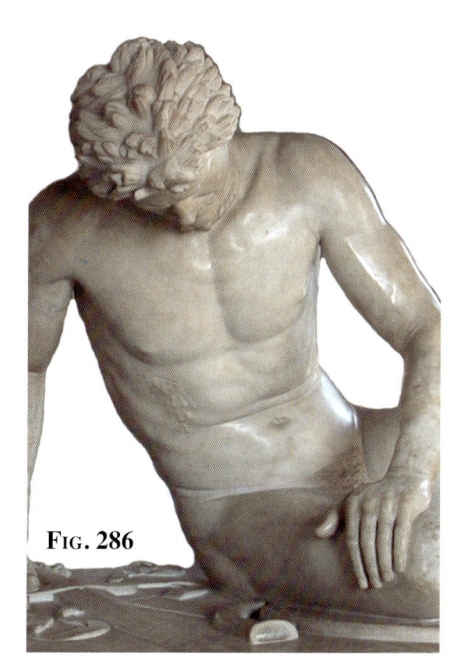

FIG. 286

Dying Gaul—End of an Age

the Macedonian kings. Following his father's death, Perseus made political and military decisions (including alliance by marriage) that promised to reinvigorate Greek culture, prestige and prosperity.

This worried the Romans who were concerned that the balance of power would be upset. The **Roman Republic** therefore declared war on Macedon. At the *Battle of Pydna* the mystique of the Greek phalanx utterly collapsed with the Greeks losing their freedom and their ultimate right to self-determination. The entire royal court of Perseus surrendered to the Romans and some 300,000 of his citizens were sold directly into slavery.

Macedon was divided into four sections that were heavily restricted from commercial intercourse with each other. There was a ruthless purge of nationalistic spirit. Although this was not the last Macedonian war, this was the one that broke its back. We'll detail some of the political events occurring in the century before Christ's birth so as to better demonstrate just how society began to collapse from within.

The *Age of Aries* had been ruled by *Mars*. This archetype is equated with the *God of War* and it was an *Age of Warfare*. Civilizations born of this time were filled with a passion for competitive life, hammered in the heat of battle, individual lives the ultimate wager.

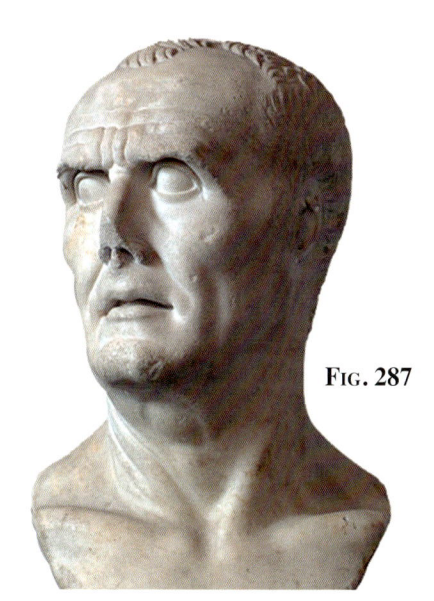

FIG. 287

Marius

It contained all the power, brute force, creativity and excitement of a life spent under the *Umbrella of War*, with its ephemeral euphoria, and eternal destruction.

To rise to these personal challenges, to brave the deed, to tap honor in the service of a higher moral code; such was the goal of most civilizations over the course of these two thousand years. There was a *Warlord* at the helm. Yet it's not war, *per se*, that causes the social deterioration of the *End Times*.

Classical Civilization's cornerstone had been founded on *successful Martial Dynasties*, many of which lasted for centuries.

Whether we consider the *Indo-Europeans*, *Pharaohs* of the New Kingdom, *Hittites*, *Mycenaeans*, *Celts*, *Chaldeans*, *Medes*, *Persians*, *Hindus*, *Carthaginians*, *Etruscans*, *Greeks* or *Romans*, we're looking at cultures that learned to live and die by the sword. Life was measured by success or loss in war.

By the first century BC, the *Roman Republic* had conquered most of the Mediterranean. They had absorbed the Carthaginian trade empire of Sicily, Sardinia, Corsica, Africa and Spain, as well as the Greek world and a portion of Asia Minor (Turkey). During this final century, under the leadership of *Marius*, *Sulla*, *Pompey*, *Julius* and *Anthony Caesar*, Gaul (France), and the East would also be brought under Roman control.

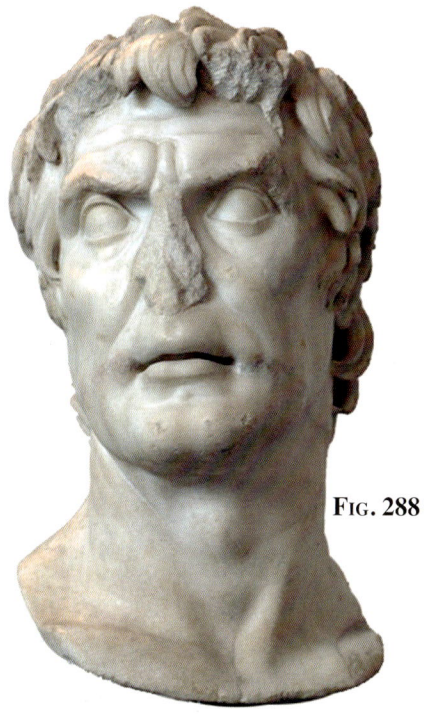

FIG. 288

Sulla

Athena's Web

Center and Circle

Fig. 289 — *Pompey*

The unrest begins as Slave Revolts erupt within the Republic (135-132 BC, 104-100 BC and in 73-71 BC with *Spartacus*). The murder of *Tiberius* (133 BC) and *Gaius Gracchus* (121 BC) are early waves of civil disorder. In 100 BC, *Marius*, the hero of the Jugurtha War, puts down rioting in Rome. In 91 BC, there's *Social War*, with *Marius* battling *Sulla* (Fig. 288), his former subordinate. First *Sulla* (88 BC), then *Marius* (87 BC), take Rome, with the latter systematically putting to death the leading members of the Roman aristocracy, displaying their heads in the Forum for a final roll call.

After the death of *Marius*, *Sulla* returns, and massacres the last of the *Marian armies* (82 BC), but not before they had raided the religious temples of Rome for revenues.

Following this, *Sulla* exercises reprisals against Romans that leave the *Marian atrocities* behind in the dust. From 57-53 BC, there's rioting in Rome again, and in 50 BC *Caesar crosses the Rubicon*, turning on Rome. Caesar defeats Pompey (48 BC). Caesar is assassinated (44 BC). *Brutus* and *Cassius* are defeated at Philippi (42 BC), *Anthony* at Actium (31 BC).

Finally in 27 BC, *Octavian*, now known as *Augustus*,[144] establishes himself as Emperor and begins *the Empire* and *Golden Age*, signaling a bloody *End to Aries*.

It left with it both the Republic and its heritage in ashes. Individual egos, together with their armies, had gone up in *flames*.

By the end of *Aries*, this FIRE Sign had burned itself out.

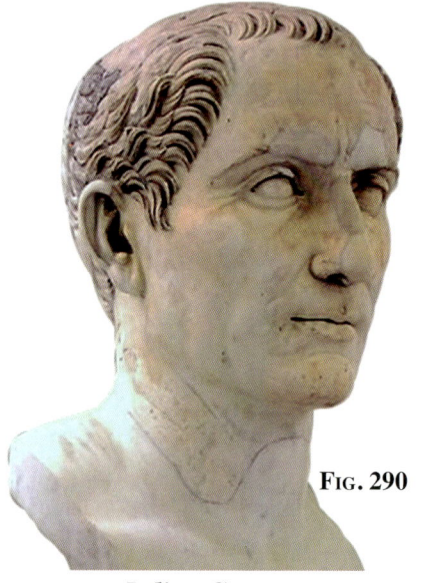

Fig. 290 — *Julius Caesar*

The aristocracy of the Classical tradition had drawn its membership from the patrician blue-bloods aligned under the law. Their Senatorial ranks had been decimated during the massacres of the final century, their vacancies filled by the growing mercantile class.

While both the *Roman Republic* (Age of Aries), and *Roman Empire* (Age of Pisces) had a Senate, they weren't the same. In these waning years, the Senate's power had been sapped during the political upheavals of the 1st century BC, becoming little more than a rubber stamp for Emperors, lacking political will, vitality or spirit. The *Pax Romana* had come at a price, although it opened safe trade routes from one end of the Empire to the other.

In the heated, accelerated environment of the *Classical Era*, life produced a surge of self-confidence, an immersion in drama, a flair for the arts and an ultimate need for the rule of *Justice* (*Libra*) in this barren *Martial landscape*.

The *Age of Pisces* would produce a merchant class (*Virgo*) spiritually subservient to the Church (*Pisces*). The new *Equinoctial Axis* coming on-line; the in-coming vibration.

The political thunderstorms were to clear the way for a new way of 'thinking and being.' Just as sympathies once shifted from *Taurus* to give way to *Aries*, so now *Aries* had to give way to *Pisces*.

While examining the passage of the Vernal Equinox though the constellations and asterisms along the Ecliptic, we've used two different information gathering techniques

Fig. 291 — *Mark Anthony*

by which heaven was organized. One of these was the *observational system*, which used a familiar Pole Star or other visual combinations as reference points. The other was the *mathematical systems* being developed by the Greeks.

With the advances in mathematics, the Ecliptic was increasingly used as the basis of measure, as equations overtook the more traditional observational techniques in popularity. In time, stellar observation would begin to lag far behind as mathematicians *calculated* their positions rather than *observed* them. There would be an unexpected result from this increase in accuracy. Many would come to lose touch with their visual connections to the stars. The need for Temples and visual aids began to fade until *Al Biruni* (943-1048 AD) complained at the end of the 1st millennia that none of those computing the positions of the stars on paper could identify them in the Sky. He said,

They're all pencil-pushers.

Or words to that effect.

In the foregoing passages, Daniel made a specific reference to 'observing' the stars. He 'raises his eyes' to look up to the skies.

The Vernal and Autumnal Equinoxes are the points of intersection between the Ecliptic and the Equator. In any given year, these points remain relatively fixed.

If we ponder our mathematical model, the concept is simplicity itself. It's one of the beauties of the system. One divides the circle's 360 degrees into twelve. From this we derive the familiar *Twelve Signs of the Zodiac*, each precisely 30 degrees in length.

But when we look to the Sky and the observational system, we find the process is not as easy. Apparently nobody told the stars that they had to fit into tidy divisions of 30 degrees each, boxed in by logic and the ease of mathematical elegance. Some constellations are larger or smaller than 30 degrees, while other star patterns overlap.

The first star of *Pisces* to be 'triggered' by the Vernal Equinox was *alpha Piscium* in 407 BC.

The *final star* in *Aries* would not be triggered until 168 BC.

FIG. 292

Astrolabe

Pisces begins before *Aries* ends. This means if we use the stars as our celestial benchmarks, there's an extended period of overlap.

Apparently, even Hipparchus was a ... 'convinced supporter of one of the leading doctrines of stellar religion,' astrology.

"Hipparchus," says Pliny, "*will never receive all the praise he deserves, since no one has better established the relationship between man and the stars, or shown more clearly that our souls are particles of heavenly fire.*" [145]

Remember, this quote comes from a Time when *Fire* was seen as the *supreme form* of *Spiritual Being*. On the pagan side, this view eventually evolves into the '*Divine Fire*' of the Stoics and the *Fire Ceremonies* of the Zoroastrians.

Here's an account of what *Hipparchus* (190-125? BC) would have heard in his day as he experienced the '*End Times*' personally while travelling.

Invaded by the Parthians about the year 140 BC, recaptured by Antiochus VII of Syria in 130, reconquered soon afterwards by King Phraates, Mesopotamia was terribly ravaged for more than a quarter of a century. Babylon, sacked and burned in 125, never recovered her former splendor; a progressive decay brought on her a death by slow consumption.

The new Iranian princes evinced no solicitude for the culture of Semitic priests. The vast brick-built temples, when the hand of the restorer was withdrawn, crumbled into dust, one by one were extinguished the lights of a civilization which extended backwards for forty centuries, and of the famous cities of Sumer and Accad there survived little but the name.[146]

Can an Age begin before another ends? Yes it can.

This is myth, not math.

In architecture, literature, religion and other forms of culture, an Age was drawing to a close. The *Classical Tradition*, like her temple ruins, witnessed the passing of an era, even as it gave birth to another — younger, stronger, and more vibrant.

Center and Circle

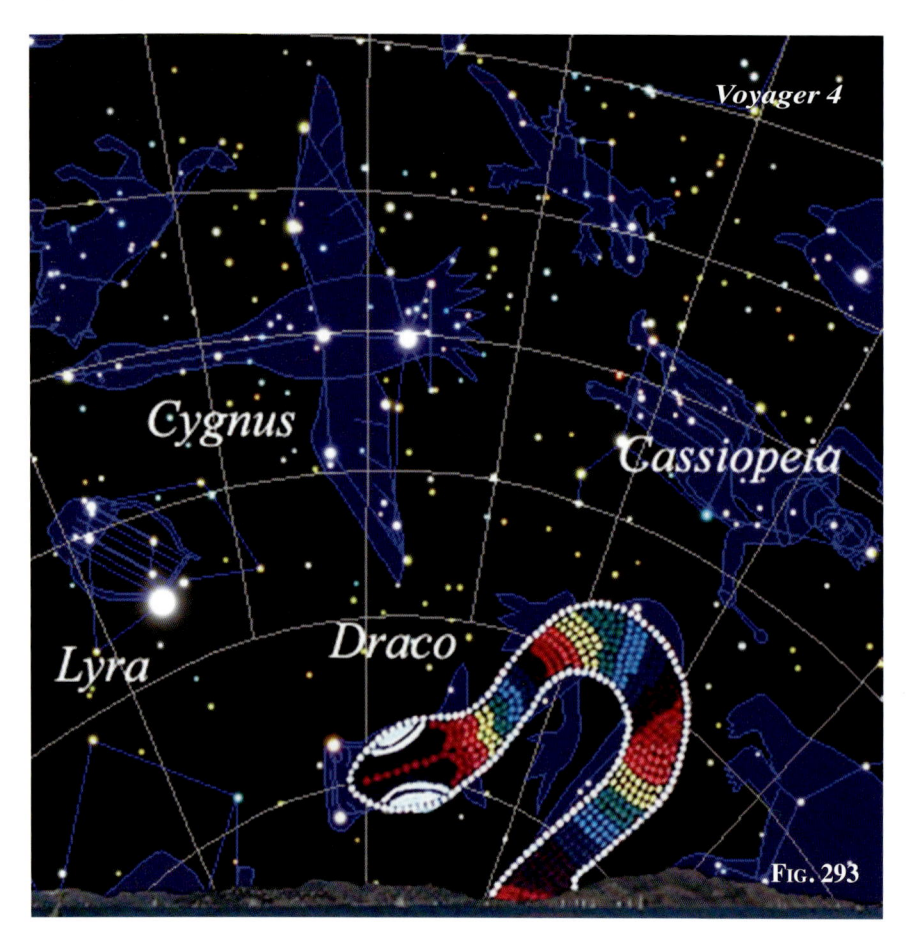

Rainbow Serpent Rising, Derby Australia 6000 BC

FIG. 293

Rainbow Dreams

Although many have heard about the *Rainbow Serpent*, not many people realize that it's not a *motif* confined to a single indigenous culture. We've already glimpsed this myth as African tradition, where the *Rainbow Serpent* is known as *Ayido-Hwedo* (Fig. 45). He's existed since the beginning of time. He was put there to serve *Nana-Buluku*, the One God.

As we recall, *Ayido-Hwedo* carried *Nana-Buluku* in his mouth, and the curves of rivers, mountains and valleys were created because that's the way *Ayido-Hwedo* moves. Rivers meander like serpent coils.

After All was created by *Nana-Buluku*, the load was so heavy He asked *Ayido-Hwedo* to coil beneath the Earth to help cushion it.

In return for his help, the Creator made *Ayido-Hwedo* an Ocean to live in because He couldn't stand the African heat. There *Ayido-Hwedo* has remained since the beginning of Time, cushioning the Earth with his *Tail in his Mouth*.

Oceans and the *Great Serpent* go back a long way together.

This memory comes from the *Dahomey* tribe of modern *Benin*, and as seen from that latitude (Fig. 295), a good portion of Draco does in fact lie below, or as the myth puts it, '*beneath the Earth*.' Although the Central portions of Africa are close to the equator, Draco is such a lengthy constellation that it is still a prominent stellar grouping even here.

Naturally, the image of the *Rainbow Serpent* biting himself is the same as *Ouroboros*, the Sea-Serpent (Fig. 296) that makes a great circle, nocturnally spinning around the pole like a cat chasing his own tail.

*Blessed are they
who travel in circles, for
they shall be called Big Wheels.*

FIG. 294

Compass Rose
Photo credit Anton Kokuiev

Center and Circle

In each of these myths, the *Rainbow Serpent* is tied to the *four cardinal points*, as every Circle is bound to its Center.

The **Native American** *tale* of the *Rainbow Serpent* [147] best illustrates this concept. According to their beliefs, there is a *Great Circle* which is so huge it holds everything (for it is the Universe). All that live inside the *Great Circle* are relatives (Hi there!).

When one stands at the Center, the four quarters are sacred. Native Americans offer prayers to each direction for each has a mystical power. As astrologers we agree, the East is the *Vernal Equinox*, while each of the other cardinal directions (North, South and West) also carry a corresponding significance.

According to Native Americans, where the *blue-black road* of the dead meets the *white road* of the living at the Center Place (what we would call the *North Celestial Pole*) is very holy. There coils the *Rainbow Serpent* (Draco). Her symbol is the sacred lightning spiral. For those who have awakened, She arises and arcs across the face of the world. It's possible to then climb on Her back and rise with Her into the *Sky-world* without dying.

For those who would dare, it's even possible to speak to the gods.

In **Australia**, the *Rainbow Serpent* is also associated with the *beginning of Creation*. These tales are most prevalent in northern Australia, but can be found throughout the continent. Northern Australia is where His (or Her) Stars would most easily be seen (Fig. 293).

Here, too, the mythic tradition speaks of a special relationship between the *Serpent* and the formation of the mountains and valleys *"in the beginning"* during the dreamtime. These forces helped to shape the Earth.

Yingarna was the *Mother of Creation* in one variation, and some *Rainbow Snake* portraits go back 8,000 years, [148] when Spring was moving through Gemini, making it one of the oldest known religious symbols on the planet.

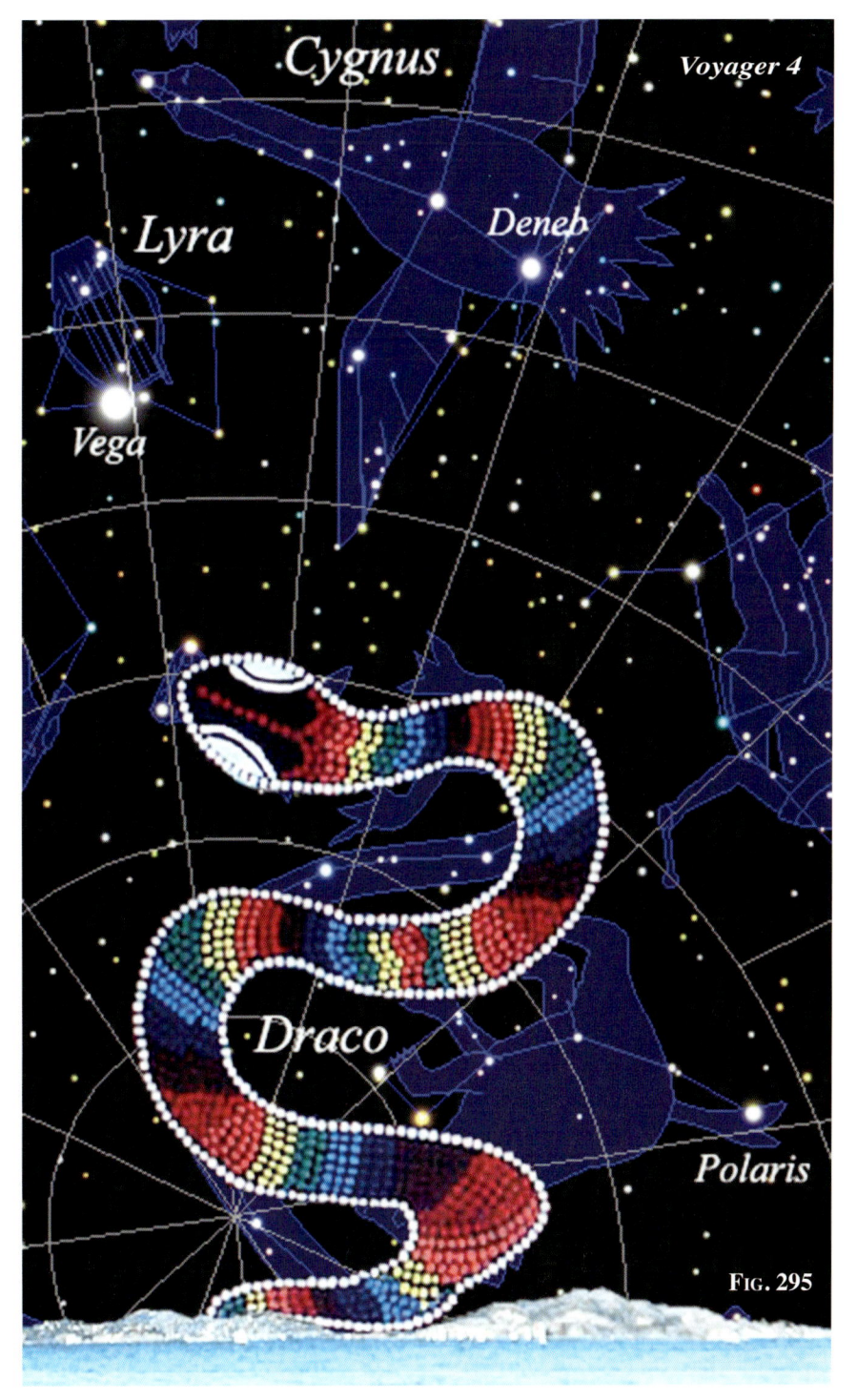

The Rainbow Serpent as seen from Benin Africa in 3000 BC

Athena's Web

Center and Circle

FIG. 296

Alchemical Serpent

Photo credit Carlos Adanero

One potential fly in the ointment though, is that some might argue our '*Rainbow Serpent*' is a circumpolar constellation and isn't visible from 'Down Under,' but that's not true. You can just see her head popping above the horizon in northern Australia. In Fig. 293, notice that the North Celestial Pole, where all the lines converge, lies below the horizon. Since this is due North, anything immediately on or next to the NCP will not be seen, as it never rises above the horizon. As is evident though, the head, and at other times the tail, would be clearly visible to all those interested in such things.

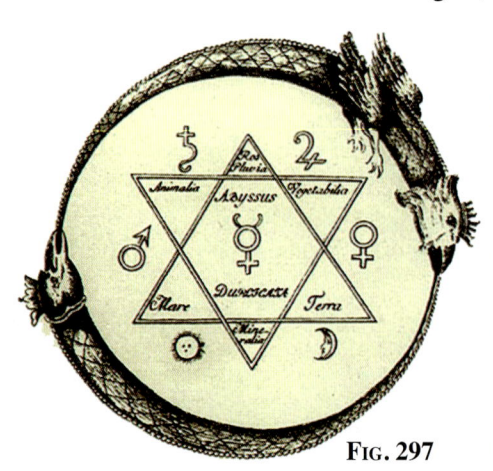

F**IG. 297**

Freemason's Twin Ouroboros

Photo credit Fieari

Even in Australia.

As the Norse stipulate, He is the *World-Wide Serpent*.

Indeed, the underlying theme of the archetype is obvious; some might call it Omnipresent. As the images on this page illustrate, the motif is simple and effective. It's the 'Hoop Snake' of Pecos Bill folklore, which could grab its own tail and roll after its prey, striking them with a poisonous bite. The only protection was to hide behind a tree and let the circling snake strike it, killing it. Trees and snakes even have mythic roots in the Southern US.

This folktale was an attempt to explain an image no longer understood, and indeed, no longer astronomically true. The constellational motif had moved on, divorced from the reality that had once provided it both truth and life. Precession shifted the picture, the Pole Star is now Polaris, and Thuban has been displaced. Long ago, Ursa Minor was considered to be a *Wing of the Dragon* (Fig. 309)—a depiction that goes back at least as far as Thales (c. 624-546 BC).

Most folks are familiar with the Big Dipper as a visual asterism, and of how big and bright it's stars are. Referring to both the Big and Little Bear, Manilius said,

"*Sprawling between them and embracing each the Dragon separates and surrounds them with its glowing stars lest they ever meet or leave their stations.*" [149]

—*Astronomica.*

But Draco's power, control and influence has waned since then. Even the Pyramid of Cheops was once aligned by sightings to the stars of Draco (Fig. 308).

F**IG. 298**

"O" is for Ouroboros

Mural by Leah B. Freeman

The year is 2788 BC. Thuban pivots on the Earth's axis. Throughout the centuries culminating in this observational climax, the VE had been very active courting various stars. In 3247 BC, *Aldebaran*, the *Bull's Eye*, passed the mark and signifies the 'center' of circles to this day. In 3056 BC, the other 'eye,' *Ain* took over, followed by *gamma Tauri* in 2963 BC (Fig. 306).

But the Autumnal Equinox responded with a light show of its own. *Antares*, the 'Heart' of the Scorpion, together with *beta Serpens*, caressed Autumn's mark and held the celestial baton in the 30th century BC.

F**IG. 299**

Celtic Torc with Dragon Heads

Photo credit Lisa Michel

164   Athena's Web

Center and Circle

Now, let's reassemble a few lost motifs of heaven. Thales spoke of Ursa Minor as part of the **Wing of the Dragon**. We also know that Libra, the Scales, was once seen as the extended *claws of the Scorpion* (Fig. 304). Gilgamesh speaks of the gate of the western hills as the entry way where

"*the Scorpions stand guard, half man and half dragon.*" [150]

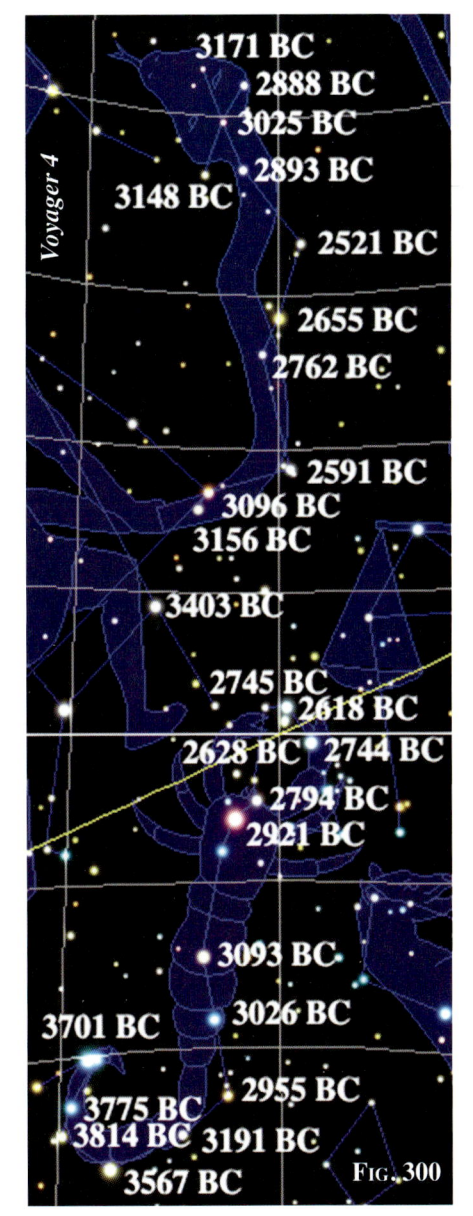

These images of Scorpio, Serpens and Draco interweave from below the Earth to ascend along the western stairway of heaven, like ivy climbing a tree.

One after the other, they pass the celestial baton as each becomes the stellar marker of the Fall, guardian of the doorway to Hades' realm.

In fact, this may solve another mystery, that of the '**Scorpion King**' of the Protodynastic Period of Egypt.[151] Whether he was one Pharaoh or two, he is usually relegated to the end of this epoch, dated circa 3200-3000 BC, known archaeologically as Naqada III. During the last centuries of this late predynastic period, there are no less than eight stars that cross the Autumnal Equinox (Fig. 303), a rich bounty for all those who look to heaven for their timely treasures.

The only pictorial evidence of the *Scorpion King's* existence comes from a mace head of a pharaoh holding a hand plough (Fig. 302), thought to be a ceremonial cutting of the first furrow in the fields, or of the opening of the dikes to flood them. While not negating either of these (what makes you think there's only one right answer?), I have a third proposal.

The plough is meant to represent a celestial furrow rather than a terrestrial one. Notice the scorpion seems to 'float' in front of pharaoh, an unusual place for ground-hugging scorpions to appear, but not so surprising if we consider our celestial Scorpion on the horizon. The hand plough is symbolic of the sacred cut of Sky and Earth, of a point in Time.

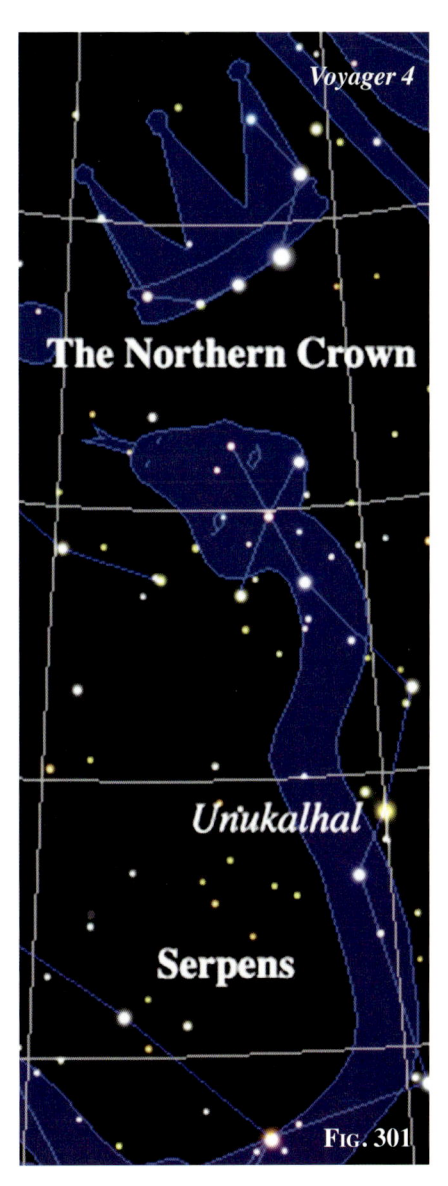

From the 4th and first half of the 3rd millenium BC, there was an incredible autumnal parade across the western cornerstone of heaven

Fig. 300

Throughout this period of stellar observation, the Serpent wore the Crown (Corona Borealis) *of Heaven. Figs. 300 & 301 are calculated for 2655 BC, when Unukalhal* (alpha Serpens) *marked Fall, the Autumnal Equinox.*

Fig. 301

The Scorpion King
Photo credit Udimu

Fig. 302

Athena's Web

165

Center and Circle
Voyager 4

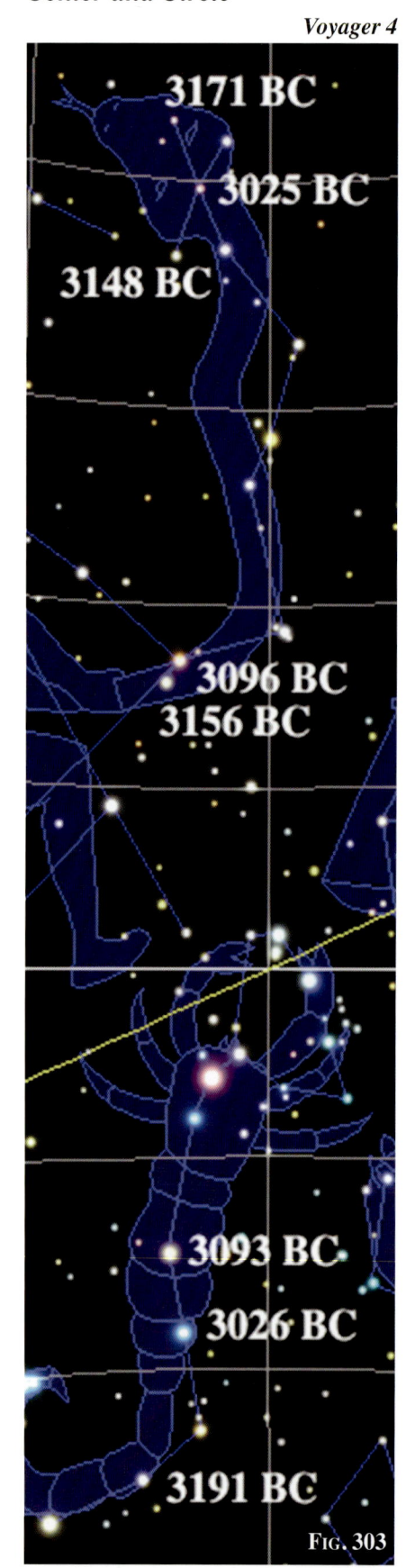

Eight stars in two centuries— Go West, young man, Go West!

FIG. 303 (labels on image: 3171 BC, 3025 BC, 3148 BC, 3096 BC, 3156 BC, 3093 BC, 3026 BC, 3191 BC)

And here we come back to a motif we've touched on before; that pharaoh, king or emperor is heaven's representative on Earth. The *Scorpion King*, whatever his 'real name,' reflected Heaven's stars on Earth at that time.

*If Heaven sings
slithering Serpent,
then slithering Serpent
the Earth replies.*

Note that two of the stars in the constellational image (Fig. 303), *mu Scorpii* (3026 BC) and *kappa Serpentis* (3025 BC) practically align, forming a perfect plumb line against which the correct astronomical position could be marked. This is the last quarter century of the predynastic phase, when the archaeologists feel the '*Scorpion King*' may have reigned.

And this brings us to yet another celestial serpent, the multi-headed *Hydra* of Twelve Labors fame. During this same epoch, the Hydra undulated along the celestial equator, our circle of *Center and Circle* fame (Fig. 304).

What we are seeing here is a *Serpentine Web* wrapping the Earth with many of its astronomically defining points. There's *Draco* with his extended wing guarding the NCP, *Serpens* climbing the Western Celestial highway to heaven, and the *Hydra* undulating along the celestial equator. *Lambda Hydrae* even marked due South at its transit in 3038 BC, defining three of our four cardinal directions.

As we have seen, another venomous creature guards Heaven's intersection. The author of Gilgamesh describes the *Scorpions* as being '*half man and half Dragon.*'

They are two and they are one.

Here's the upshot of all this. It is entirely conceivable, during the heart of this Age of Myth, that all these venomous creatures were seen as one great multi-headed entity, who, like the Hydra, undulates above and below the levels of the celestial 'Sea.' In the Native American tradition, it was *Grandmother Spider* and her one great Web.

This is *Ti'amat* at the center of her army, with the venomous constellations as her chief lieutenants. These stars were thought to be both addictive and destructive, but also very wise. Here we are witnessing at least one component of their stellar wisdom.

If we take a giant step backwards in time to when the stars of Gemini were crossing the Vernal Equinox, we discover a new contender for the highest throne of heaven.

While the Dragon is not displaced from his guardianship role, the club-wielding constellation of Heracles is contending with the summit of heaven at the beginning of this stellar passage (Fig. 305).

From circa 6300 BC to 4800 BC, a new theological outline was being put into place. The year was divided in two. Spring was important, but so was its right ascension 12 hour mark, the *Autumnal Equinox* or start of Fall, marked by Sagittarius and the constellation **Heracles** (Latin—*Hercules*). He was not yet called *Heracles* during the 7th, 6th and 5th millennia BC, as the Golden Era of Greek culture still lay many centuries in the future. In addition to being the 'half-way mark' of heaven, *Heracles* vied with Draco as the stellar pathfinder for the North Celestial Pole (NCP).

Center and Circle

Voyager 4

Because the *Kneeling One* was circumpolar during Gemini, he was visible throughout the year, and a handy asset for anyone seeking to understand seasonal motion and monitor its passage. In different ways, he defined two of the four cardinal points single-handedly.

What a man!

If we return to stories of *Heracles'* birth, (at the start of his alignment with 12 hours RA) we learn he was one of a set of twins, a clue for stellar myths born at this time. While yet an infant, Zeus placed Heracles at Hera's breast as she slept. *Because* his nursing was so rough she awoke and angrily pushed the child away, spilling her milk across heaven in what is now known as the Milky Way.

The Milky Way runs adjacent to the constellation Heracles.

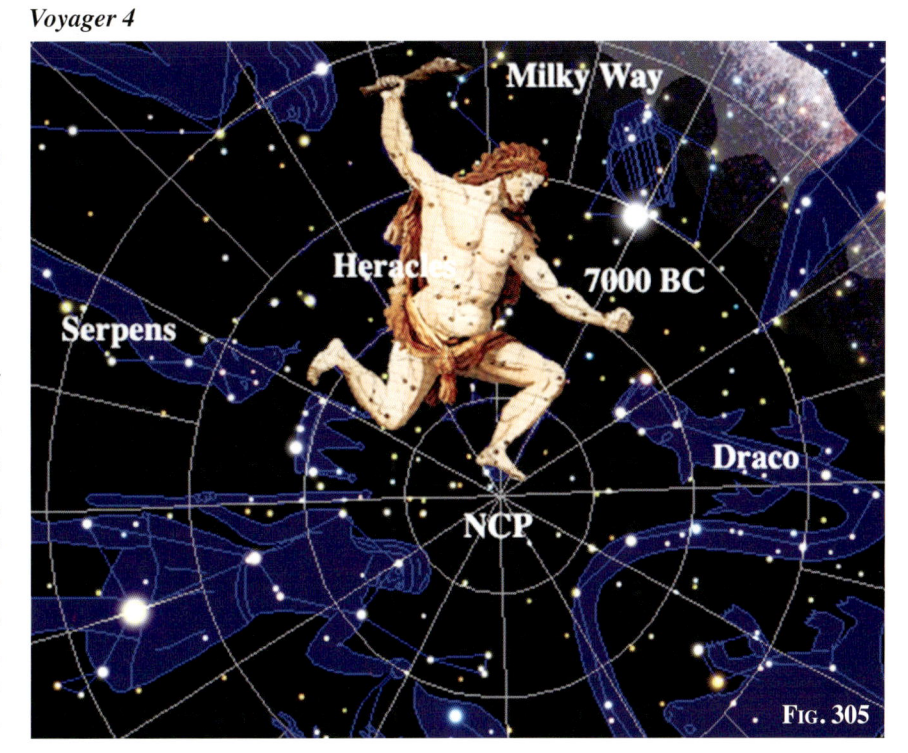

FIG. 305

A serpent in each hand—notice Heracles proximity to the NCP

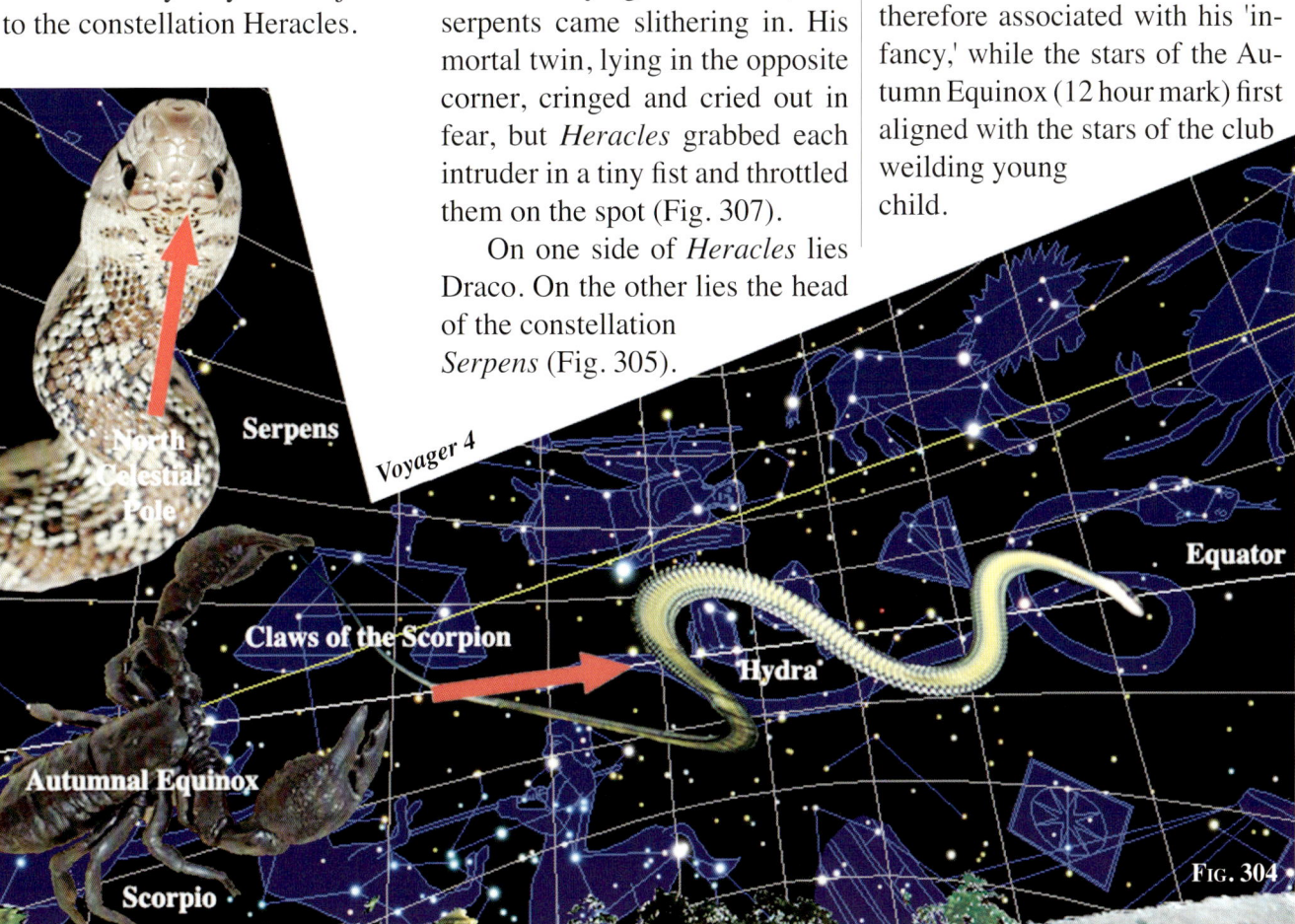

While lying in his crib, two serpents came slithering in. His mortal twin, lying in the opposite corner, cringed and cried out in fear, but *Heracles* grabbed each intruder in a tiny fist and throttled them on the spot (Fig. 307).

On one side of *Heracles* lies Draco. On the other lies the head of the constellation *Serpens* (Fig. 305).

An image of two snakes is therefore associated with his 'infancy,' while the stars of the Autumn Equinox (12 hour mark) first aligned with the stars of the club weilding young child.

Voyager 4

FIG. 304

During the 4th & 3rd millennia BC, venomous creatures guarded Heaven's highways

Athena's Web

Center and Circle

The Vernal Equinox is in the opposite corner where his Twin lay. Every year Spring danced and kept this Twin entertained. when Gemini was at zero hours Right Ascension, and *Heracles* at 12 hours Right Ascension.

Much later, both Judaic and Celtic tradition would split their year into two. Do we believe they were the first to do so?

Mythic theory suggests not.

But this was originally a Greek myth you say?

Not so.

When our oldest recorder of constellational data *Aratus* (315-240 BC) [152] first speaks of *Heracles*, he'd been all but forgotten. He's called *'the Kneeling One,'* whose name was known only to himself. By the waning centuries BC, the *Kneeling One* had passed his 'astronomical prime' and was no longer a contender for guardian of the Autumnal Equinox or NCP. His heavenly term had lapsed. As mathematics advanced the interests of astronomy, this popular motif later returned as martial hero, breathing new life into a constellational grouping for Greek and Roman stargazers. But the creative astronomical spark that had originally engendered these myths were born of much older cultures, under different divine aliases.

Myth says it is *Heracles* who must fight the dragon Ladon to obtain the Apples of the Hesperides. It's *Heracles* who battles the Hydra and must decapitate the mortal heads, burying the immortal one beneath a rock. It is *Heracles* who contests with the *Serpent Achelous* before he changes into a Bull and loses his horn.

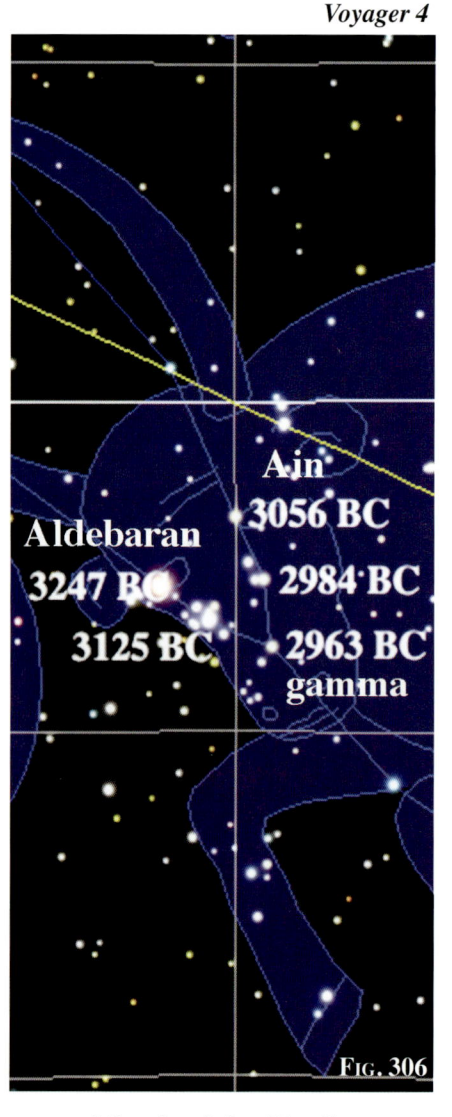

Voyager 4

Fig. 306

March of the Hyades

Widening our panorama, we observe that the *Rainbow Serpent* left a legendary trail across three different continents- Australia, Africa and North America. There were expanded mythological components in the firmament which augmented these images beyond what we know them to be today.

Venomous constellations once marked a time when their unified ranks commanded the astronomical corner stones of Creation.

Together with the Hyades on the Vernal Equinox (Fig. 306), this was a period of extended stellar superlatives for East and West.

These were the years prior to the construction of The Great Pyramid in the 26th century BC. We know Thuban was used to align the pyramid to the cardinal directions as it was being built. By standing inside the inner chambers, one could look out long, narrow air passageways that allowed these stars to be seen in the daytime, even in the heat of the desert (Fig. 308).

The name *Heracles* means 'the glory of **Hera'**. [153] It's because of Hera that Heracles was made to suffer the twelve labors, code for the twelve months of the year. It is only by living on Earth and grappling with each of the seasonal lessons that we learn to master life and move on. In his day, *Heracles* as the Son of Zeus (Child of the Sky) had lived up to his namesake, passed mortal muster, and was then allowed to take his 'rightful' place in the firmament. For those who knew how to read the constellation's secrets, the year could be observed, calibrated and mapped.

As with many of these mythic heroes, they are a reflection of the Heavens of their day. This is the pagan *'Face of God'* that keeps changing with the years. The history as well as the symbolism of what happened is sometimes included in the myth, mixing both celestial design together with more mundane terrestrial elements. Too often, folks are confused by the mythic elements. For instance, the myth of *Heracles* battling *Achelous* is not merely a serpent myth vying for the summit, it also tells history on another level, the geographic history of the river valley.

Center and Circle

To interact with *Heracles* is to wrestle with (integrate) time, stellar motion, and the calendar. A calendar enables people to coordinate their efforts on a societal level, providing the 'rhythm' of the year. It puts people in 'Seasonal (daily, weekly, monthly) Swing.'

FIG. 307

Heracles and two Serpents

Achelous [154] is the longest river in Greece, and runs through its western portions in a series of gorges that have long been considered wild and unruly. Floods were an annual event. Once it was finally tamed by a series of dikes and dams and using the calendar (Heracles taming the anthropomorphic forces), the early inhabitants of this river valley were able to transform it into a great agricultural Cornucopia.

Achelous had the ability to morph from river, to snake, to bull. He fared best in combat with Heracles as a river, since his broad bulk made him a very difficult opponent to overthrow.

When Achelous was in the guise of a serpent, the advantage passed to Heracles, who had experience subduing snakes as a infant child (so the story goes).

Unable to overcome our champion and using the last of his transformative abilities to change into a Bull (during the Age of Taurus), one of his horns was broken off by Heracles and transformed into a fruitful Cornucopia of delights. The horn is being dramatically woven into our myth as one of its fundamental images and so would suggest that these achievements were mastered sometime through the course of the 4th millennia BC, while the Vernal Equinox was passing through the horns of the Celestial Bull. It's even possible that it was at the end of this period, where the astronomical horn actually joins the head (Hyades) of the Bull, sometime between 3400 and 3100 BC.

FIG. 309

Wing of the Dragon before Ursa Minor?

If our translation of these archetypes is correct, it would seem that there was much happening through this epoch.

The Achelous River Valley was being transformed. Upper and Lower Egypt were being unified. Thor battled Jormungand, with the Bull's head being used as bait.

It was a happening time.

Were the twelve labors of Heracles an early zodiac lost to antiquity? If so, this may account for some of the curious images in the labors during his youth (*pi Herculis* — 7000 BC?). While some of these images are obviously astronomical (such as the Nemean Lion, Crab and Hydra), others have slipped the archetypal mold, with hind and boar, dung-filled stables and man-eating mares. Are we looking back on an older period, to a zodiac long-forgotten? Only Time will tell.

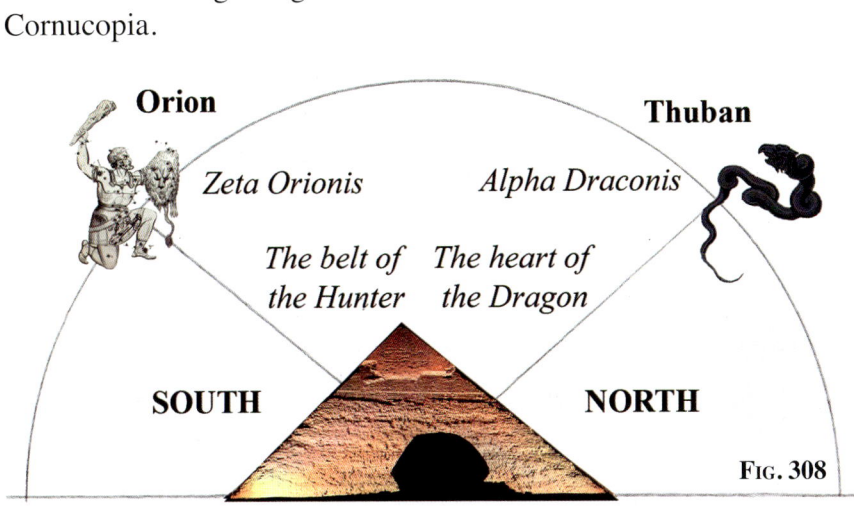

Orion — Zeta Orionis — The belt of the Hunter

Thuban — Alpha Draconis — The heart of the Dragon

SOUTH — NORTH

FIG. 308

Sighting on Thuban and Orion's belt — a pyramid aligned by stars

Athena's Web   169

Pisces

Photo credit Holly Hayes

Fig. 310

Chartes

Reflections

During the course of our work, we have observed not only the rise and fall of civilizations, but of constellations.

Across this extended period, there's been a significant shift in the way we view our relationship between Heaven and Earth. We have moved from an observational system of collecting stellar information (rising at different hours of the night to witness various celestial phenomena) to a mathematical construct of papyrus and reed born of mental disciplines, (before turning the light out when the hour gets late and we get tired).

The successful interpretion of the *Will of Heaven* has generally been held in high esteem. Ultimately, it's simple. What it all boils down to is getting the future right. The king and court want the correct combo. The stars provided a standardized means of calibrating and interpreting this celestial canvas. Many a priesthood invested generations in a concerted effort to read its script, the Egyptian and Semitic traditions chief among them. Sky wisdom can take many forms: forecasting the Seasons, harvest, tides, navigation (by land or sea), calendar, nocturnal watches, religious holidays and New Year's festivals, to name a few. All are born of the juxtaposition between Heaven and Earth.

The Skies were studied in earnest, with practical intent. Divinely deduced, it was the key to the future. The collection and sharing of this information was part of a vast network that re-tuned the celestial instrument every year by observationally checking their astronomical 'predictions.'

Had they been right?

Fig. 311
Photo credit Joachim Köhler

Piercing the Dragon in Christian garb

Center and Circle,
North Celestial Pole
and Vernal Equinox.

Heaven and Earth are unified, stand in balance, harmonize, align and reconstitute each other at the Equinoxes. The astronomical tickers reach zero, the old aspirations are discarded, the new embraced.

A new cycle is born.

The precise moment is important because that's where they believed the most powerful concentration of the magic was, at the inception. At each New Year's the astronomer/priests who studied celestial motion (and its mythic history) worked out what the Skies had to say about the upcoming piece of the puzzle, attempting to unravel the next knot in the string of this literary labyrinth. The mythic images for the future were not selected randomly or haphazardly, but were founded on an already established universal symbolic vocabulary and tradition. The Heavens were monitored, the times observed, the omens heeded. Ultimately, the divine spark behind the precision of the observations was that it was designed to accurately gauge Nature's rhythms and be in touch with her bountiful gyrations.

So far, we have observed as three constellations rose and fell along the path of the Vernal Equinox- Gemini, Taurus and Aries.

Twins, Bull and Ram.

Center and Circle

Through these three epochs, we have seen philosophies come and go. The Twins told oral traditions of Duality still echoed in distant lands. The Bull brought with it the bounty of a great agrian God, while the Ram offered a courageous path to personal mastery.

While these themes followed a set rhythm in the seasons, there was also fluidity within the ritual. Spring remembers the *hunt* for the Easter Egg, the *search* for the Apis Bull and the *quest* for the Golden Fleece. Each epoch had their reflective ritual for honoring the face of God looking down on them from above. (The question was, "How do you get Him to smile on us?")

It is the search for Spring, the eternal fountain of Youth, calibrated on a canopy of stars. It is Athena sitting at Her loom, weaving heroic stories into Life's tapestry. It measured Time for eras both past and future. It is caught in the nets of Marduk, Hephaestus and the first fishermen, as well as the webs of Grandmother Spider and Athena. The invisible strands of this weave snare us all.

Which brings us down to the beginning of a new epoch and a new era, with the end of the Ram.

According to what we have observed, it was Time for a new world view, but not one based on duality and the Tao, nor one founded on the passions of either Bull or Ram.

Heaven's Will now submerged its symbolism beneath the waters of the wine-dark sea, in the mysterious stars of the *Fish*.

While following the trail of our Dragon, we have observed this pattern repeated three times.

FIG. 312

Raise your eyes and look to the Skies for the answer

Art historians have tended to be the most excited about this thesis, because it begins to make sense of the artistic images found in their world view for thousands of years. Predictably, we know more about recent periods, and less about the Genesis, the first, and beginning (Gemini) of our story line. Nevertheless, it is easy to see how, in each period, indigenous civilizations turned to and embraced these celestial themes and their philosophies. The ancient world stood at a crossroads. The *Ram* was losing his steam as Aries wound down. There was a new excitement and a new concern about the up-coming vibration.

The classical world felt as though it was coming to the end of a period of 'being,' a way of life. At the end of Spring's passage through these woolly stars, there was indeed a social collapse as,

"...one by one were extinguished the lights of a civilization which (had) extended backwards for forty centuries." [156]

The end of an era.

Forty centuries; approximately two constellations. What we generally refer to as ancient history.

In our investigation, we've drawn the line at the beginning of Christianity because it's the start of the 'modern' era. It needs to be examined in its own right. According to the stargazers, each new constellation represents a new vibration, a new way of being. We can study the past objectively because we look at it from a more emotionally detached point of view.

Indigenous peoples used the Skies as a way of discerning the *Will of God*. It was their priest's job to interpret the *Divine dictum* and pass the information on to the community, whether king, council or clan. This was a world in tune with prophecy, prediction, and omens gathered in various ways. *Oedipus* is but one tale of a tragic hero who knows his fate (future) and does everything he can to avoid it, but to little avail. Classical theater is filled with heroes consciously wrestling with their fate.

Move over *Cassandra*.

Center and Circle

But prediction and prophecy did not cease with the passage of the Vernal Equinox through the stars of the Ram. They also made a series of predictions about the future.

Our future.

Living in a time at the 'End of an Age' tuned them into the Age following theirs, and its end. They stood at the threshold of a new beginning, together with all its implications. This is why the Wise Men are the first to seek, understand and find the spiritual guide for the then 'New' Age (*Pisces*).

It was the end of the classical tradition, and with it, the *Dawn of Christianity*.

Twins, Bull, Ram - *Fish*.

According to both Matthew and Mark the first four disciples, Simon, Andrew, James and John are all fishermen. The bait that the Messiah uses to hook them?

"*Follow me and I will make you fishers of men.*" Matt. 4:19

Say hello to the *Age of Pisces*.

Twice the Messiah 'feeds' the multitudes with the loaves and the *Fish*.

In the first instance, it's *Two Fish* and five loaves.

That's *two* Fish.

More Pisces.

The feeding of the multitudes is a metaphor (parable) for the 'teaching' of the multitudes, as confirmed by Matthew 16:9-12.

"*Do you not yet understand? Do you not remember the five loaves for the five thousand and the number of baskets you collected? Or the seven loaves for the four thousand and the number of baskets you collected? How could you fail to understand that I was not talking about bread? What I said was: Beware of the yeast of the Pharisees and Sadducees.*" Then they understood that he was telling them to be on their guard, not against yeast for making bread, but against the teaching of the Pharisees and Saducees.*"

The 'teaching' of *Fish* and the loaves Jesus offers is astrologically not difficult. The *Fish* are Pisces, the loaves of bread are the *sheaves of wheat* the *Virgin* holds. The sermon is a Pisces/Virgo presentation; the new astronomical polarity of heaven; a new contract (New Testament) with the Divine.

If you have *faith* (*Pisces*), you can *heal* yourself (*Virgo*). This is precisely what Jesus does as he walks from town to town. Where they have *faith*, there is *healing*. Where they do not have faith, he knocks the dust off his sandals (feet- Pisces) and keeps on walking. —(Matthew 10:14)

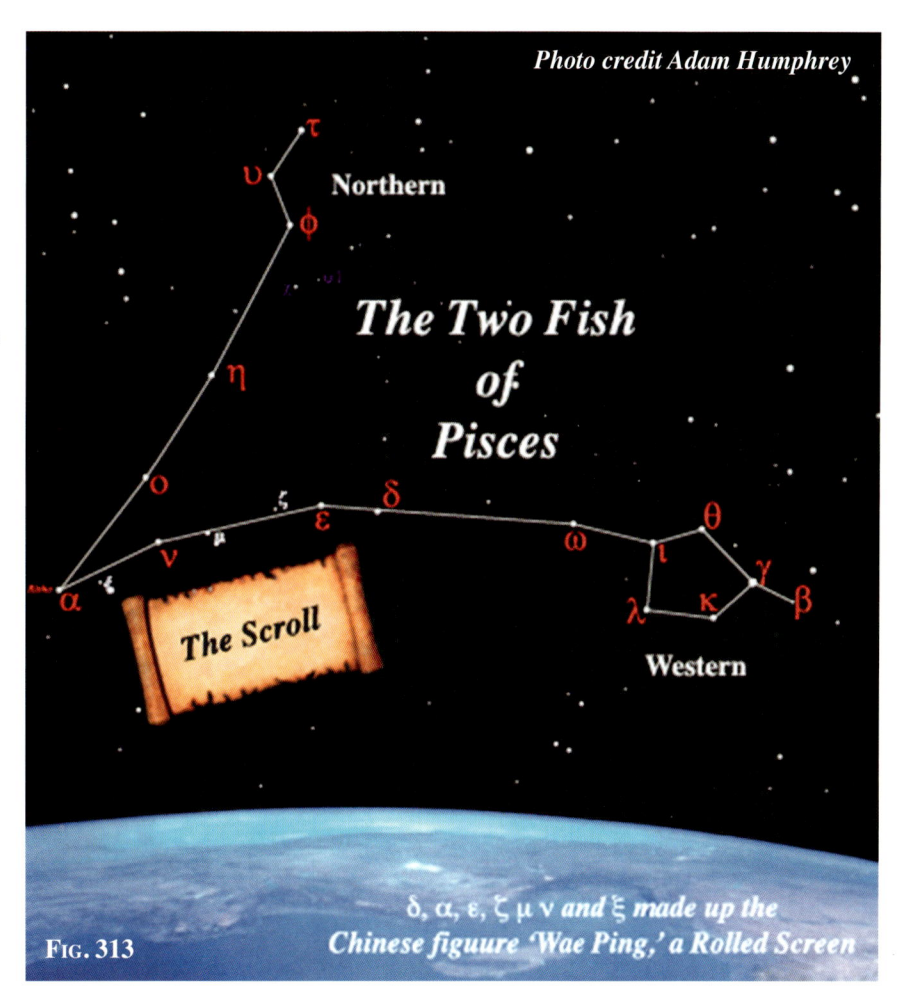

Photo credit Adam Humphrey

The Two Fish of Pisces

The Scroll

Northern

Western

δ, α, ε, ζ μ ν and ξ made up the Chinese figuure 'Wae Ping,' a Rolled Screen

FIG. 313

Alpha (α) through delta (δ). The Seven Seals of Revelation?

The *Book of Revelation* is the prophecy for the next step in the constellational path. It tells us so in its introductory verses. They are attempting to determine the future by studying its themes. We've examined the mythological *motifs* for the T*wins, Bull* and *Ram*- now it's time to do so with the *Fish*.

Athena's Web

Center and Circle

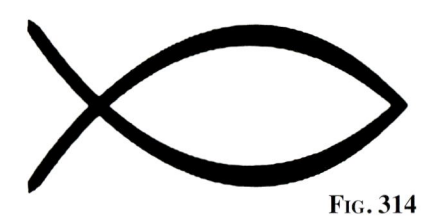

FIG. 314

Christian Fish

Because this last epoch is closer to us in time, there's a larger volume of historical material we can work with, and will.

Here's the skinny.

The *Seven Seals* of Revelation are the seven stars of the asterism, "*the Scroll*," (Fig. 313) a sub-set of the constellation *Pisces*. Essentially, it is the cord or 'fishing line' that extends from *Al Rischa*, the 'knot' (*alpha Piscium*), to the western *Fish* of the constellation.

The precessional dates when these stars align with the Vernal Equinox can be easily determined using contemporary software.

If, indeed, there's anything to these pagan mythologies, all we have to do is calculate the date each star yields and study its history. What was happening?

We stand at the end of Pisces, so most of the stars of this constellation are already behind us. The stars of the *Seven Seals* (aka *The Scroll*) certainly are.

To study Revelation in this context one needs a fair understanding of the social currents, history, mythology, astronomy, geography and religion of their day.

Having digested the material in this work, you now have most everything you need to know, with the soul exception of the Christian images, themes and motifs.

The pattern is the same. You're three-quarters of the way there.

You're about to learn of a period of apocalyptic writing that was popular for several centuries. The term apocalypse means '*describing or resembling the complete destruction of the world*.' Prophets were looking ahead to the end, not of the world, but of the VE's passage through the stars of *Aries*, and in Revelation, *Pisces*. The Hebrew *prophets* are looking ahead *to the end of their own era* (Aries), while *Jesus* and *Revelation* have that Age behind them and are looking ahead *to the end of Pisces*.

In Matthew 15:24, Jesus tells us that he was sent for only "the *lost sheep* of the house of Israel." The '*lost sheep*' are the *Age of Aries* slipping away from the Hebrews as Time runs out.

Establishing *the Dragon's Path* over the course of 8,000 years has taken a while and yet we have only just skimmed the surface. We began with China, introducing seasonal indicators and discovering a new wisdom. Other than that we have largely confined ourselves to Mediterranean mythologies,

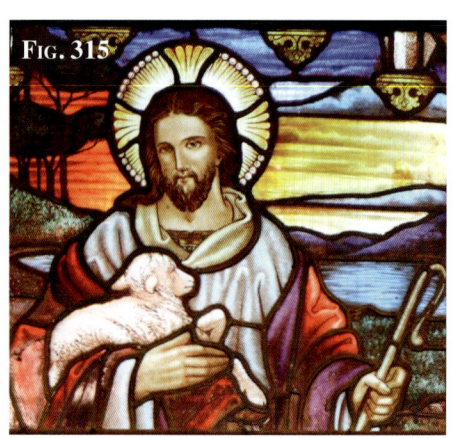

FIG. 315

With the lost sheep

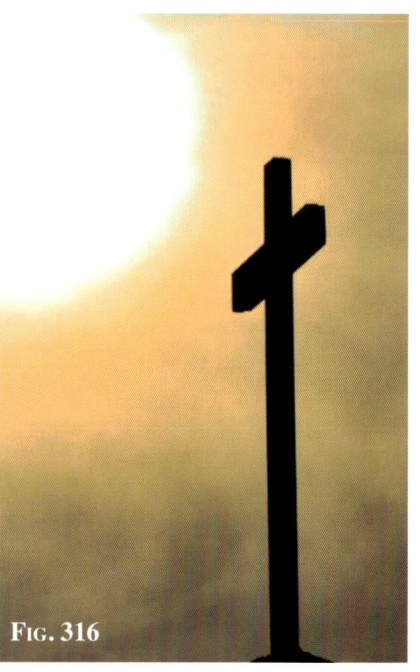

FIG. 316

The dawn of a new era
Photo credit Dean Martin

while glimpsing African, Indian, Norse, Native and Mesoamerican beliefs. The highway was once clearly marked. For those who worked with a calendar (and how many civilizations haven't?), star wisdom was essential, practical and productive.

Yet with the onset of *Pisces*, the mythological motif for Time was withdrawn, the Moon abandoned, the storyline confused.

We will examine this *Epoch of Pisces*, but as a separate work. It is the door through which we must pass if we want to better understand the dawning of *Aquarius*.

There is one small caveat, however. While we have managed to by-pass the *Seven Seals*, there is one *Seal* still being opened, and its day is nigh. Do you dare break *The 8th Seal*?

Center and Circle

Photo credit David Cooley

F~IG.~ 317

Guardian, Axis and Stars

Center and Circle

The Imperial Pearl

Photo credit Durham University Museums, England

We opened with a Chinese myth illustrating seasonal rotations with regard to the Dragon (Figs. 25 to 29). He rose with the rain clouds in the Spring, sank back into the pools in the Fall, and during the Winter rested his head on the Earth along the northern horizon. It was silently implied, not stated, that this was the Imperial Dragon, our Draco. Over the centuries and across the continents, we have witnessed variations on this theme, incorporating a number of perspectives. It's only fitting that we should open and close with the Chinese, as they are the Dragon's children, the inheritors of His Wisdom. No other people have claimed such a long history or devotion to this singular entity.

Naturally, as we've looked back across the years, the emphasis throughout has been on observational astronomy. It is what those of us living here on Mother Earth see when we gaze upwards at night. The pole star circling high overhead, the Sun rising in the East, the Moon setting in the West; these are all observational phenomena of what we see with our own two eyes.

But there's another way of canvassing the heavens, and it has come to overshadow even what we see. This new focus is, of course, *spherical trigonometry*, and the mathematics needed to illuminate the cycles of Heaven.

The Imperial, five-toed Dragon and Pearl

But the dragon has one more ace up his sleeve to play.

We know Thuban drew to its closest approach to the North Celestial Pole in 2788 BC. In 2100 AD, Polaris will surmount the supreme summit, coming as close as it's ever going to get. *Kochab* (*beta Ursae Minoris*) was once claimed by the Greeks as the pole star. They called it *Polos*. It reached its peak in 1166 BC, but only rose to 83 and a half degrees latitude, some six and a half degrees from the actual pole. Polaris will climb to 89 degrees 35 minutes and 37 seconds in 2100 AD before reaching its summit, less than half a degree from an exact match, while in 2788 BC Thuban came even closer, reaching 89 degrees 57 minutes and 25.9 seconds, or about two and a half *minutes* of arc from the summit.

While *Polos* was the closest star of its day, it does not really compare at six degrees with these two at less than half a degree (Fig. 278). The width of the face of the Sun or Full Moon is about 30 minutes of arc or half of a degree. Polos was about twelve Full Moons distant from the NCP.

176 Athena's Web

Center and Circle

While Draco claimed the prize for longevity in maintaining his role as polar guardian, he has seen contenders come and go. We just noted *beta Ursae Minoris* as pole star while Heracles in a earlier incarnation contested the summit.

Polaris has now taken over the position as present pole star, pushing the Dragon aside as primary caretaker.

So what's this ace?

It turns out that the Dragon, in addition to marking our north pole for so long, also marks the mathematical center upon which astronomy builds its temple.

Among the Dragon's coils nestles a gem of incalculable worth, a gem that only the Imperial Dragon could command.

Are there mythical hints, stories of a deeper knowledge derived via mathematical calculation?

As children of the Dragon, the Chinese have studied the King Snake for so long they have defined specific attributes they use to describe them. To the Chinese, nine is a lucky number. The Dragon has nine distinguishing qualities and they are:

*the Dragon's
head is like a camel's,
its horns were like a deer's,
its eyes like a hare's,
its ears like a bull's,
its neck like an iguana's,
its belly like a frog's,
its scales like those of a carp,
its paws like a tiger's and
its claws like an eagle's.*[155]

To make the Dragon even more auspicious, he has nine times nine scales.

The Dragon has whiskers on either side of his mouth, and floating under its chin is a bright Pearl. On top of its head is the '*poh shan*' or foot rule, without which it cannot ascend to heaven.

The Dragon is fond of eating sparrows and swallows. He is afraid of swords made of iron. He shuns contact with the centipede and silk dyed of five colors. The Dragon can only be killed by the careful placement of a single iron needle.

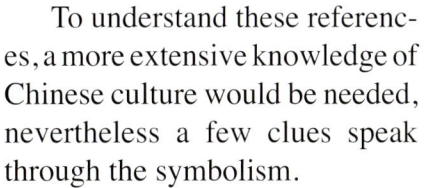

To understand these references, a more extensive knowledge of Chinese culture would be needed, nevertheless a few clues speak through the symbolism.

Like the Western Dragon, Draco can be captured or 'killed' by a single needle as it pinpoints the center of Heaven around which the constellation turns.

Because the Chinese cosmology is based on the number five, the five-toed Dragon controls each of these elements. It is also why he avoids silks dyed of five colors. They can capture his secret. While holding the Earth in its claw, it commands all of Creation.

Voyager 4

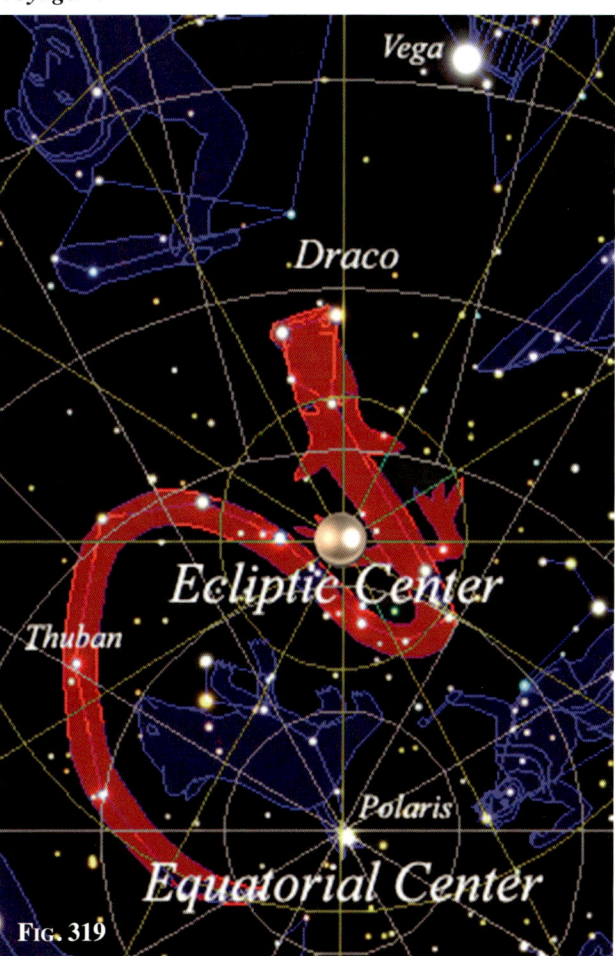

Fig. 319

Is the Ecliptic Center the Pearl below the chin?

*He has the whole world
in his claw...*

But it is the Pearl that gives the final secret away. The Ecliptic Pole lies within the first great coil of Draco, or, as the myth states, *floating below the chin* (Figs. 318, 319). The Pearl's treasure is its wisdom, in the calendar, the agricultural year and navigation. Having determined the center of the ecliptic circle, all heaven may be mapped, calibrated and determined. This is astronomy and all it relates to. For those who believe in Heaven's influence over Earth, this 'Pearl' cannot be overestimated.

Athena's Web

Center and Circle

Photo credit Jay Reinfeld FIG. 320

Close and bolt the door behind you when you're finished

Center and Circle

In the Twilight

We have covered over seven thousand years in a little under 200 pages as part of an on-going investigation lasting almost forty years. As the dust starts to settle, let's look back and reflect for a few moments.

Our mythological examination has detected a pattern in the stars, marked by the cadence of Time, and honored as a reflection of Heaven's Will by those in antiquity. Careful examination of a few of these myths has suggested a new method of interpretation, a new way to date them in Time.

While meditating on these various weaves over the years, I heard a faint voice whispering in my ear. The clues along the trail led to Gemini, a Time when oral traditions and storytelling would have received an infusion of interest on the world stage.

The symbolism of stellar script was being practised, refined, and disseminated around the globe. From the couplets of Sumerian ritual to Odin's splitting the Great Egg with his sword, or Vishnu's sleeping within the Great Egg. Pan-ku was born of the Great Egg, while the celestial twins, Castor and Pollux wear caps from pieces of the Egg they were born from.

To be born of the Egg is to be born, or live, during this period of time.

F<small>IG</small>. 321

According to astrologers, Gemini represents the archetype of communication. During Gemini, a greater celestial focus would have been expressed through language, communication, divine speech, magical incantations and the oral tradition. Is there any record, any memory of a Time when there was a focus on speech, and the development of a single, world wide language?

From the first chapter of the Bible, Genesis 11: 1, 5-9.

"Throughout the Earth men spoke the same language, with the same vocabulary..."

"Now Yahweh came down to see the town and the tower that the sons of man had built." "So, they are all a single people, with a single language!" said Yahweh.

"This is but the start of their undertakings! There will be nothing too hard for them to do. Come, let us go down and confuse their language on the spot so that they can no longer understand one another."

Yahweh scattered them thence over the whole face of the Earth, and they stopped building the town. It was named Babel, therefore, because there Yahweh confused the language of the whole Earth. It was from there that Yahweh scattered them over the whole face of the Earth."

between Times...

Astrologers believe that during each Age, a new vibration emerges, molding civilization in its wake. Our view over Atlantis suggests that the Earth's energies were being cultivated during Gemini, with a world wide community coming together, all inhabitants of One Planet. This is the 'town' they speak of. It is the new city mentioned in Revelation. But with each Age, the vibration comes in strong after birth, builds to a mature culmination, and then begins to fade away like the leaves. If these lines are being intrepreted correctly, they suggest a Time during which people learned to use Ziggurats, read the stars, and fix problems leaving them with *"nothing too hard...to do."* Geminis are clever. However, as the Age came to a conclusion, our capacity to comunicate crumbled, as social initiatives began to disintegrate.

That was the last time we had an AIR element defining our Age, until now. As we enter the constellation Aquarius, it's time to pick up the social banner and re-build a world city, a collective community, a new Jerusalem, in tune with Heaven's Will. Not a script for times gone by, but a new world where connection, communication and cooperation can all again be our guiding medium, for now and well into the new vibration.

There's a new day dawning.

Photo credit Peter Juerges FIG. 322

With special thanks to Professor Ed Phinney

Closing the Circle

Know Thyself *

* Inscription over the Temple of Apollo at Delphi. See page xi.

Center and Circle

Star Crossings

Star Name	Greek	constellation	R.A.	Year	Date	bi	var	Conjunct
	theta	Herculii	12 hrs	BC 8596	6-Nov			
	omicron	Herculii	12 hrs	BC 7440	22-Aug	2	var	
	rho	Herculii	12 hrs	BC 7402	23-Mar	2		
	nu	Herculii	12 hrs	BC 7368	1-Dec		var	
	xi	Herculii	12 hrs	BC 7230	24-Jan		var	
	pi	Herculii	12 hrs	BC 7006	6-Apr			
	mu	Herculii	12 hrs	BC 6746	14-Jul	2		
	omega	Sagittarii	12 hrs	BC 6284	2-Jan			
	rho 1	Sagittarii	12 hrs	BC 6135	22-Feb	2	var	
	eta	Herculii	12 hrs	BC 6127	6-Mar	2		
	phi	Herculii	12 hrs	BC 5887	26-Oct		var	
Albaldah	pi	Sagittarii	12 hrs	BC 5806	13-Apr	2		
	epsilon	Herculii	12 hrs	BC 5695	13-Jul	2		
Sarin	delta	Herculii	12 hrs	BC 5619	28-Jul	2		
	tau	Sagittarii	12 hrs	BC 5484	17 Agu			
Rasalhague	alpha	Ophiuci	12 hrs	BC 5396	25-Jan	2		
Nunki	sigma	Sagittarii	12 hrs	BC 5358	1-Feb	2		
Ascella	zeta	Sagittarii	12 hrs	BC 5326	4-Mar	2		
	phi	Sagittarii	12 hrs	BC 5179	3-Jul	2		
	xi	Hydrae	6 hrs	BC 5160	7-Jul	2		
	zeta	Herculii	12 hrs	BC 5152	2-Feb	2		
Ras Algethi	alpha 1	Herculii	12 hrs	BC 5004	5-Aug	2	var	
Kaus Borealis	lambda	Herculii	12 hrs	BC 4954	16-Feb			
Kaus Meridionslis	delta	Sagittarii	12 hrs	BC 4682	24-Jan	2		
	1	Geminorum	0 hrs	BC 4634	9-Mar	2		
Kaus Australis	epsilon	Sagittarii	12 hrs	BC 4565	29-Jan	2		
Nash	gamma 2	Sagittarii	12 hrs	BC 4421	13-Aug			
	eta	Sagittarii	12 hrs	BC 4379	11-Jul	2	var	
	zeta	Tauri	0 hrs	BC 4246	7-Jul	2	var	
Kornephoros	beta	Herculii	12 hrs	BC 4208	18-Oct	2		
Sabik	eta	Ophiuci	12 hrs	BC 3908	14-Aug	2		

Athena's Web

	gamma	Herculii	12 hrs	BC 3895	26-Aug	2	var
	nu	Hydrae	6 hrs	BC 3846	17-Oct		
Al Nath	beta	Tauri	0 hrs	BC 3836	25-Dec	2	
	iota 1	Scorpii	12 hrs	BC 3814	11-Mar		
Kajam	omega	Herculii	12 hrs	BC 3800	21-Mar	2	var
	kappa	Scorpii	12 hrs	BC 3775	27-Sep		var
Lesuth	lambda	Scorpii	12 hrs	BC 3701	8-Feb	2	var
	upsilon	Scorpii	12 hrs	BC 3652	30-Oct		
Sargas	theta	Scorpii	12 hrs	BC 3567	21-Oct		
	kappa	Herculii	12hrs	BC 3472	16-Jan	2	
	mu	Hydrae	6 hrs	BC 3457	9-Sep		
	zeta	Ophiuci	12 hrs	BC 3403	6-Mar		
Aldebaran	alpha	Tauri	0 hrs	BC 3247	21-Apr	2	var
	eta	Scorpii	12 hrs	BC 3191	9-Mar		
	rho	Serpentis	12 hrs	BC 3171	11-Jun		
Yed Posterior	epsilon	Ophiuci	12 hrs	BC 3156	11-Jul	2	
	gamma	Serpentis	12 hrs	BC 3148	13-Sep	2	
	theta 2	Tauri	0 hrs	BC 3125	8-Feb	2	var
Yed Prior	delta	Ophiuci	12 hrs	BC 3096	30-Aug	2	
	epsilon	Scorpii	12 hrs	BC 3093	17-Jul		
Ain	epsilon	Tauri	0 hrs	BC 3056	16-Jul	2	
	lambda	Hydrae	6 hrs	BC 3038	3-Oct	2	
	mu 1	Scorpii	12 hrs	BC 3026	2-Mar		var
	kappa	Serpentis	12 hrs	BC 3025	16-May		
	tau	Scorpii	12 hrs	BC 2995	23-Nov		
	delta 1	Tauri	0 hrs	BC 2984	27-Jul	2	
	gamma	Tauri	0 hrs	BC 2963	16-Apr	2	
	zeta 2	Scorpii	12 hrs	BC 2955	6-Jun		
	zeta 1	Scorpii	12 hrs	BC 2946	30-May		var
Antares	alpha	Scorpii	12 hrs	BC 2921	22-Sep	2	var
	psi	Scorpii	12 hrs	BC 2921	13-Mar		
	rho	Ophiuci	12 hrs	BC 2914	8-Sep	2	
	beta	Serpentis	12 hrs	BC 2893	21-May	2	
	iota	Serpentis	12 hrs	BC 2888	17-Aug	2	
	omicron	Scorpii	12 hrs	BC 2812	14-May		
	sigma	Scorpii	12 hrs	BC 2794	30-Jun	2	var
	upsilon 1	Hydrae	6 hrs	BC 2771	20-Jan		
	epsilon	Serpentis	12 hrs	BC 2762	19-Nov		

Center and Circle

	xi	Scorpii	12 hrs	BC 2747	16-Aug	2	
Lesath	nu	Scorpii	12 hrs	BC 2745	27-Dec	2	
Dschubba	delta	Scorpii	12 hrs	BC 2744	25-Sep	2	var
	chi	Serpentis	12 hrs	BC 2730	30-Dec		var
	lambda	Serpentis	12 hrs	BC 2722	29-Apr		
	omega	Serpentis	12 hrs	BC 2706	7-Feb		
Unukalhai	alpha	Serpentis	12 hrs	BC 2655	18-Jun	2	
	omega 2	Scorpii	12 hrs	BC 2635	9-Aug		
	omega 1	Scorpii	12 hrs	BC 2628	28-Jun		
Graffias	beta 1	Scorpii	12 hrs	BC 2618	29-Oct	2	
	mu	Serpentis	12 hrs	BC 2591	24-Jan	2	
	kappa	Hydrae	5 hrs	BC 2551	19-Nov		
	delta	Serpentis	12 hrs	BC 2521	10-May	2	var
	pi	Scorpii	12 hrs	BC 2391	27-Oct	2	var
	rho	Scorpii	12 hrs	BC 2304	4-Jan	2	
Taygeta	19	Tauri	0 hrs	BC 2175	20-Jul	2	
Electra	17	Tauri	0 hrs	BC 2174	1-Apr	2	
	iota	Hydrae	6 hrs	BC 2173	21-Aug		
Celaeno	16	Tauri	0 hrs	BC 2170	6-Aug	2	
Alphard	alpha	Hydrae	6 hrs	BC 2151	13-May	2	
	tau2	Hydrae	6 hrs	BC 2034	6-Apr	2	
	nu	Persei	0 hrs	BC 1882	7-Jan		
	tau 1	Arietis	0 hrs	BC 1785	28-Jun	2	var
	delta	Persei	0 hrs	BC 1761	16-May		
	zeta	Arietis	0 hrs	BC 1669	15-Feb		
	theta	Hydrae	6 hrs	BC 1636	17-Aug	2	
Botein	delta	Arietis	0 hrs	BC 1624	9-Apr		
Menkar	alpha	Ceti	0 hrs	BC 1613	30-Dec		var
Mirphak	alpha	Persei	0 hrs	BC 1433	10-Mar		
	epsilon	Arietis	0 hrs	BC 1377	5-Jan	2	
	rho 2	Arietis	0 hrs	BC 1343	8-Feb		var
Algol	beta	Persei	0 hrs	BC 1341	8-Jul		
	kappa	Draconis	6 hrs	BC 1296	19-May		var
	sigma	Arietis	0 hrs	BC 1292	30-Nov		
Kaffaaljidhma	gamma	Ceti	0 hrs	BC 1236	22-Nov		
Dubhe	alpha	Ursae Majoris	6 hrs	BC 1235	26-Jun	2	
	pi	Arietis	0 hrs	BC 1229	6-Jul	2	
	zeta	Hydrae	6 hrs	BC 1218	4-Jan		

	omicron	Arietis	0 hrs	BC 1158	23-Dec		
	rho	Hydrae	6 hrs	BC 1098	15-Apr	2	
	gamma	Persei	0 hrs	BC 1080	20-Jan		
	mu	Arietis	0 hrs	BC 1077	31-Mar	2	
	nu	Ceti	0 hrs	BC 1067	27-Nov	2	
Kocab	beta	Ursae Minoris	18 hrs	BC 1062	4-Nov	2	
	epsilon	Hydrae	6 hrs	BC 1057	28-Nov	2	var
	eta	Hydrae	6 hrs	BC 1055	22-Dec		var
	rho	Puppis	6 hrs	BC 1001	26-Jul	2	var
	nu	Arietis	0 hrs	BC 995	19-May		
	sigma	Hydrae	6 hrs	BC 975	23-May		
	36	Draconis	18 hrs	BC 930	21-Aug	2	
	delta	Hydrae	6 hrs	BC 909	22-Jul	2	
	xi 2	Ceti	0 hrs	BC 893	6-Dec		
	xi	Arietis	0 hrs	BC 812	31-Jul		
Mira	omicron	Ceti	0 hrs	BC 792	9-Jul		
Asellus Australis	delta	Cancri	6 hrs	BC 790	29-Jun		
	omega	Draconis	18 hrs	BC 754	11-Feb	2	
Asellus Borealis	gamma	Cancri	6 hrs	BC 704	5-Sep	2	
	theta	Ursae Majoris	6 hrs	BC 674	15-Feb	2	
	theta	Arietis	0 hrs	BC 625	20-Jan		804 BC
	phi	Cygni	18 hrs	BC 532	22-Jan		
	theta	Aquilae	18 hrs	BC 519	1-Feb	2	
	eta	Arietis	0 hrs	BC 519	22-Jun		741 BC
Dabin	beta	Capricorni	18 hrs	BC 453	24-Aug		
Algedi Secunda	alpha 2	Capricorni	18 hrs	BC 439	7-Jan		
Al Rischa	alpha	Piscium	0 hrs	BC 407	22-Jun	2	110 BC
Hamal	alpha	Arietis	0 hrs	BC 402	27-Jul		710 BC
	kappa	Arietis	0 hrs	BC 395	20-Feb		687 BC
Achernar	alpha	Eridani	0 hrs	BC 345	26-May		
Altair	alpha	Aquilae	18 hrs	BC 294	30-May	2	
Albiero	beta 1	Cygni	18 hrs	BC 263	19-Aug	2	
Baten Kaitos	zeta	Ceti	0 hrs	BC 251	6-May	2	
	iota	Arietis	0 hrs	BC 246	20-Feb		414 BC
	xi	Piscium	0 hrs	BC 235	27-Aug		24 AD
	lambda	Arietis	0 hrs	BC 231	28-Oct	2	580 BC
	chi	Ceti	0 hrs	BC 213	29-Jan	2	
	46	Draconis	18 hrs	BC 199	28-May		198 BC
	eta	Lyrae	18 hrs	BC 182	7-Jun	2	

Center and Circle

Sharatan	beta	Arietis	0 hrs	BC 182	21-Feb	2	447 BC
Mesarthim	gamma	Arietis	0 hrs	BC 168	13-Apr	2	390 BC
	omicron	Piscium	0 hrs	BC 51	1-Jul		9 AD
	nu	Piscium	0 hrs	AD 15	25-May		167 AD
	pi	Piscium	0 hrs	AD 121	28-May		62 AD
	eta	Piscium	0 hrs	AD 238	1-Jan	2	71 AD
	mu	Piscium	0 hrs	AD 238	6-Aug	2	347 AD
Deneb Okab	delta	Aquilae	18 hrs	AD 298	20-Jan	2	
	theta	Ceti	0 hrs	AD 326	31-Jan	2	
	rho	Piscium	0 hrs	AD 347	16-Jul		49 AD
	delta 2	Lyrae	18 hrs	AD 433	8-Nov	2 var	
	upsilon	Piscium	0 hrs	AD 493	12-Sep		76 BC
	zeta	Piscium	0 hrs	AD 563	4-Sep	2	576 AD
	phi	Piscium	0 hrs	AD 593	10-Dec	2	93 AD
	chi	Piscium	0 hrs	AD 630	8-Aug		234 AD
Sheliak	beta	Lyrae	18 hrs	AD 637	13-Mar	2	
	tau	Piscium	0 hrs	AD 643	16-Nov		42 BC
	epsilon	Lyrae	18 hrs	AD 653	26-Nov	2	
	psi 3	Piscium	0 hrs	AD 659	16-Jul		299 AD
	psi 2	Piscium	0 hrs	AD 696	18-Aug		299 AD
	psi 1	Piscium	0 hrs	AD 740	12-Jul	2	313 AD
Nodus Secundus	57	Draconis	18 hrs	BC 6459	21-Dec	2	679 AD
	epsilon	Piscium	0 hrs	AD 774	30-Sep	2	740 AD
Vega	alpha	Lyrae	18 hrs	AD 849	22-Jul	2 var	
	delta	Piscium	0 hrs	AD 1052	4-Feb	2	986 AD
	61 or 62	Piscium	0 hrs	AD 1058	31-Dec		1001 AD
Diphda	beta	Ceti	0 hrs	AD 1134	24-Jan		
	omega	Piscium	0 hrs	AD 2013	12-Feb	2	1815 AD
AD		Ursae Minoris	max	AD 2100	13-Mar	2 var	
	TX	Piscium	0 hrs	AD 2265	14-Sep	var	2124 AD
	lambda	Piscium	0 hrs	AD 2350	23-Jan		
	iota	Piscium	0 hrs	AD 2391	1-Dec	2	2168 AD
	theta	Piscium	0 hrs	AD 2626	1-Jan		
	kappa	Piscium	0 hrs	AD 2644	27-Jun	2 var	
	gamma	Piscium	0 hrs	AD 2836	7-Oct		
Fum al Samakah	beta	Piscium	0 hrs	AD 3097	5-Aug		2821* AD

Center and Circle

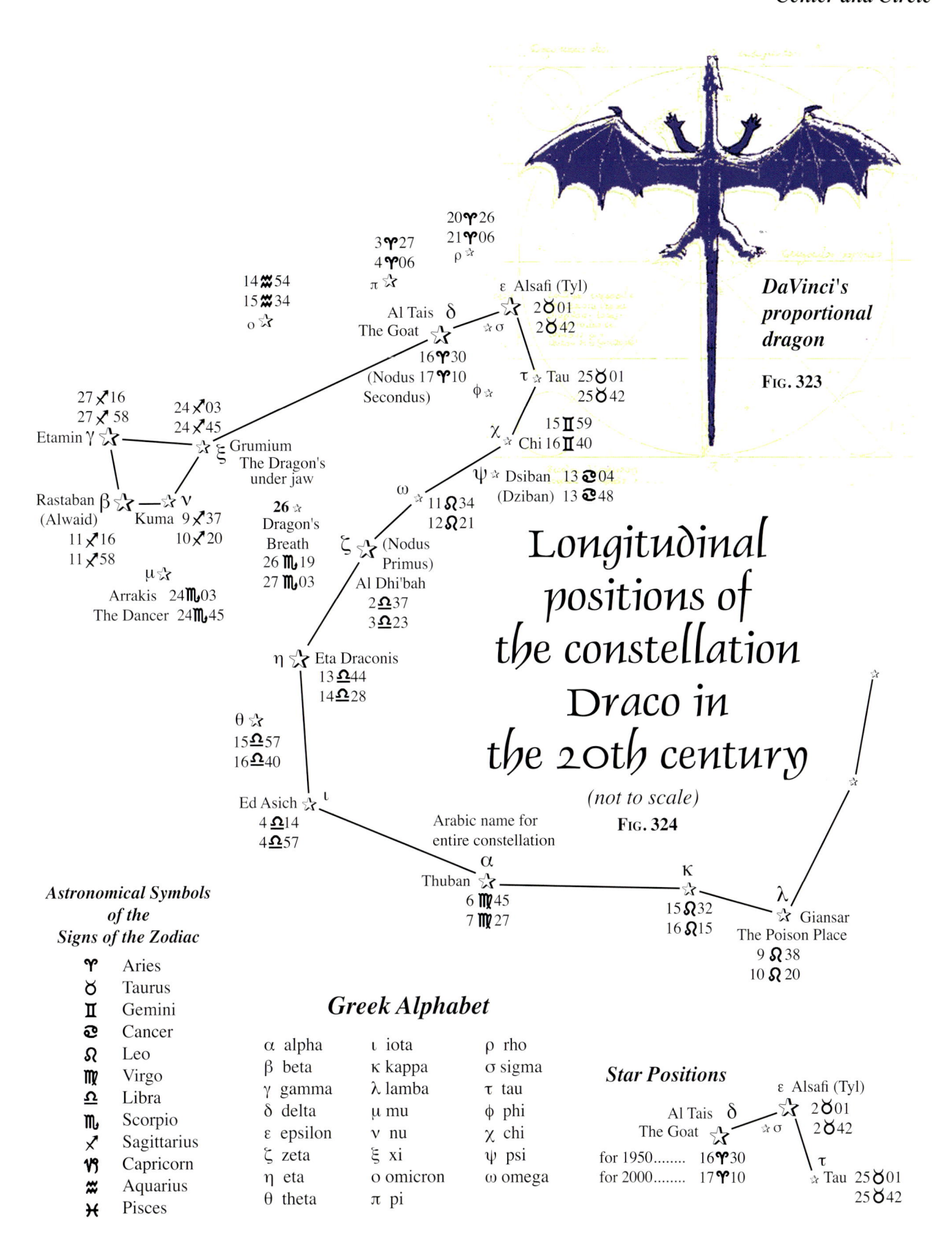

DaVinci's proportional dragon

Fig. 323

Longitudinal positions of the constellation Draco in the 20th century
(not to scale)

Fig. 324

Star labels and positions:

- 14 ♒ 54 / 15 ♒ 34 o ☆
- 3 ♈ 27 / 4 ♈ 06 π ☆
- 20 ♈ 26 / 21 ♈ 06 ρ ☆
- ε Alsafi (Tyl) 2 ♉ 01 / 2 ♉ 42
- Al Tais δ The Goat 16 ♈ 30 (Nodus Secondus) 17 ♈ 10
- σ ☆
- τ ☆ Tau 25 ♉ 01 / 25 ♉ 42
- φ ☆
- χ ☆ Chi 15 ♊ 59 / 16 ♊ 40
- 27 ♐ 16 / 27 ♐ 58 Etamin γ ☆
- 24 ♐ 03 / 24 ♐ 45 ξ Grumium The Dragon's under jaw
- Rastaban β ☆ (Alwaid) 11 ♐ 16 / 11 ♐ 58
- ν ☆ Kuma 9 ♐ 37 / 10 ♐ 20
- μ ☆ Arrakis 24 ♏ 03 The Dancer 24 ♏ 45
- **26** ☆ Dragon's Breath 26 ♏ 19 / 27 ♏ 03
- ζ ☆ (Nodus Primus) Al Dhi'bah 2 ♎ 37 / 3 ♎ 23
- ω ☆ 11 ♌ 34 / 12 ♌ 21
- ψ ☆ Dsiban (Dziban) 13 ♋ 04 / 13 ♋ 48
- η ☆ Eta Draconis 13 ♎ 44 / 14 ♎ 28
- θ ☆ 15 ♎ 57 / 16 ♎ 40
- Ed Asich ι ☆ 4 ♎ 14 / 4 ♎ 57
- Arabic name for entire constellation α Thuban ☆ 6 ♍ 45 / 7 ♍ 27
- κ ☆ 15 ♌ 32 / 16 ♌ 15
- λ ☆ Giansar The Poison Place 9 ♌ 38 / 10 ♌ 20

Astronomical Symbols of the Signs of the Zodiac

- ♈ Aries
- ♉ Taurus
- ♊ Gemini
- ♋ Cancer
- ♌ Leo
- ♍ Virgo
- ♎ Libra
- ♏ Scorpio
- ♐ Sagittarius
- ♑ Capricorn
- ♒ Aquarius
- ♓ Pisces

Greek Alphabet

α alpha	ι iota	ρ rho
β beta	κ kappa	σ sigma
γ gamma	λ lamba	τ tau
δ delta	μ mu	φ phi
ε epsilon	ν nu	χ chi
ζ zeta	ξ xi	ψ psi
η eta	o omicron	ω omega
θ theta	π pi	

Star Positions

Al Tais δ The Goat
for 1950........ 16 ♈ 30
for 2000........ 17 ♈ 10

ε Alsafi (Tyl) 2 ♉ 01 / 2 ♉ 42
σ ☆
τ ☆ Tau 25 ♉ 01 / 25 ♉ 42

Athena's Web 187

Center and Circle

This is the myth as it comes down to us, right out of history and off the WEB. The artifact which records this story is from the 19th Egyptian Dynasty, our 13th century BC (*VE in Aries*), but it seems to speak of an earlier star period, one that began thousands of years before this record was left, echoing a *Time of the Twins*. It describes a pastoral way of life after an Age of sustainable livihoods, corruption and decay enter the picture, upsetting social balance and domestic harmony. A great rift occurs, a departure ensues, and a crocodile-filled river comes to separate them. The younger brother later dies, but is reborn as a Bull, who in turn dies, is born again as Bovine, who materializes as amazing furniture, etc.

What pieces can *you* glean in this,

The Tale of the Two Brothers?

Anpu and Bata

Once, there were two brothers. Anpu was the name of the elder and Bata was was the name of the younger. When their parents died, Anpu was already married and had a house of his own, but his little brother was to him, as it were, a son; so he took his little brother to live with him. When the little brother grew into a young man, he was an excellent worker. He it was who made for him his clothes; he it was who followed behind his oxen to the fields; he it was who did the plowing; he it was who harvested the corn; he it was who did for him all the matters which were in the field. There was not an equal in the land. Behold the spirit of a god was with him.

Every morning, the younger brother followed his oxen and worked all day in the fields and every evening he returned to the house with vegetables, milk, and wood. And he put them down before his elder brother who was sitting with his wife; and he drank and ate, and after he lay down in his stable with the cattle. And at the dawn of the next day he took bread which he had baked, and laid it before his elder brother; and he took with him his bread to the field, and he drove his cattle to pasture in the fields.

And as he walked behind his cattle, they said to him, "Good is the herbage which is in that place"; and he listened to all that they said, and he took them to the good place which they desired. And the cattle which were before him became exceedingly excellent, and they greatly multiplied in number.

Now at the time of plowing his elder brother said to him, "Let ourselves make a good yoke of oxen ready for plowing, for the land has come out from the water; it is ready for plowing. Furthermore, come to the field with corn, and we will begin the plowing tomorrow morning." this the elder brother said; and his younger brother did all things as his elder brother had told him to do.

And when the morning came, they went to the fields with all of their things; and their hearts were greatly pleased with their tasks they had to do for the beginning of their daily work.

After this, as they were in the field, they stopped for corn, and the elder brother sent his younger brother, saying, "Hurry, bring us the corn from the town for planting."

And the younger brother returned home to find the wife of his elder brother, as she was sitting brushing her hair. He said to her, "Get up, and give me corn, so that I may run back to the field, for my elder brother is in a hurry, to not delay. She said to him, "Go open the bin, and take as much as you wish, so that I may not let my braids of hair fall while I am brushing them."

The youth went into the stable; carrying a large measure, for he wished to take much corn; he loaded the measure with wheat and barley; and he left carrying it on his shoulders. She said to him, "How much of the corn that is wanted, is that which is on thy shoulder ?" He replied to her, "Three bushels of barley, and two of wheat, in all five; these are what

I carry upon my shoulders". And she seductively spoke with him, saying, "There is great strength in you, for I see your strength every day." And her heart knew him with the knowledge of the passion of youth. And she arose and came close to him, and spoke with him, saying, "Come, stay and play with me, and it shall be well for you, and I will make beautiful clothes for you."

Then the youth became like a panther of the south with fury at the seductive evil of her words to him; and she greatly feared for the consequences. And then he spoke angrily to her, saying, "Look, you are like a mother to me, your husband is like a father to me, for he who is older than I has raised me. What is this wickedness that you have said to me? Never say it to me again. But, meanwhile, I won't tell anyone of it, for I will not let it be said by the mouth of any man." He lifted up his burden, and he went to the field and came to his elder brother; and they took up their work, to labor together at their task.

Rebuked, and because Anpu loved his brother very much, his wife became jealous and wanted to destroy Bata.

Now afterward, at evening, his elder brother was returning to his house; and the younger brother was following behind with his oxen, loading himself with all the things of the field. Driving the oxen before him, he took them to lie down in their stable which was in the farm.

Meanwhile the wife of the elder brother was afraid of what she had said. So she took a parcel of fat, and used it to make it look as though she was one who is evilly beaten, intending to say to her husband, "It is your younger brother who has done this wrong." Her husband returned in the evening as was his normally did each day: and as he came into his house, he found his wife ill from the violence: she did not give him water to wash his hands as she normally did, she did not make a light for him, and his house was in darkness, as she was lying down seemingly very sick.

Her husband said to her, "Who has done this to you?" She said, "No one has spoken with me today except your younger brother. When he came to take the corn for you he found me sitting alone; he said to me, 'Come, let us stay and play together, tie up your hair': This he said to me. I did not listen to him, but I said to him: 'Look, am I not your mother, is not your elder brother like a father to you?' And he was afraid, and he beat me to stop me from telling you, and if you should let him live I shall die. Now look, he is coming in the evening; and I complain of these wicked words, for he did this even in the daylight."

And now the elder brother became like a panther in the south; he sharpened his knife; he took it in his hand; he stood behind the door of the stable to kill his younger brother as he came in the evening to return his cattle to the stable.

Now the sun went down, and he loaded himself with vegetables in his usual manner. He came in, and the first cow entered the stable, and she said to her keeper, "Look your elder brother is standing in the dark before you with his knife to kill you; run from him." He heard what his first cow had said and didn't enter. The next entering, the cow said it again. He looked beneath the door of the stable; he saw the feet of his elder brother as he was standing behind the door, with his knife in his hand. He threw down his load to the ground, and fled swiftly as his elder brother chased after him with his knife.

The the younger brother cried out to Ra Harakhti (the Sun-god), saying, "My good lord! you are the one who divides the evil from the good." And Ra, the sun, about to rise heard his cry; and so Ra made a wide canal of water between him and his elder brother, and it was full of crocodiles; with the one brother on one bank, whilst the other was on the other bank; and the elder brother hit his hands together at being unable to kill him. And the younger brother called to his elder brother on the other bank, saying, "Stand still until sun rises for the day; and when Ra rises, I shall swear my innocence to you before him, and as he can distinguish between good and evil. And has not your wife been as a mother to me? And I shall leave you forever; Now, since you want to kill me I shall avoid every place where you are; I shall go to the valley of the Acacia.

Now when the land was lightened, and the next day appeared, Ra Harakhti rose, and one brother looked at the other. And the youth spoke with his elder brother, saying, "Why have you come after me to kill me secretly, when you have not heard the words of innocence from my mouth? For I am truly your brother, and you are to me as a father, and your wife even as a mother: is this not true? Anpu answered, "Why did you beat up my wife and almost kill her?" Bata answered, "I did no such thing. Have I not told you that I have always looked upon her as my mother?"

"Truly, when I was sent to bring corn for us, your wife said to me, 'Come, stay and play with me;' for see this truth has been turned over for you into its opposite." And he made him understand of all that happened with him and his wife. And he swore an oath by Ra Harakhti, saying, "Your coming to kill me secretly with your knife was an abomination." Then the youth took a knife, and cut off of his flesh, and cast it into the water, and the fish swallowed it. He fell and fainted; and his elder brother cursed his own heart greatly; he stood weeping for him from far off; as he knew he could not pass over to where his younger brother lay, because of the crocodiles. And the younger brother called unto him, saying, "Whilst you have dreamed an evil thing, wilt you not also dream a good thing, just like that which I would do for you? When you go to your house you must look after your cattle properly. And now as to what you shall do for me; I know you shall come to seek after me, if you see it enough. And this is what shall happen; I shall draw out my soul, and I shall put it upon the top of the flowers of the acacia, and when the acacia is cut down, and it falls to the ground, and you come to look for it, if you search for it even for seven years do not let your heart grow wearied. For thou will find it, and then you must put it in a cup of cold water, and know then that I shall live again, that I will make better that which has been done wrong. And you shall know of this, that is to say, that good things are happening to me, for when one person shall give a cup of beer to you in your hand, and it shall tremble; do not stop then, for truly it shall come to pass with you."

So Anpu went home. He found his wife near the river washing off the black and blue dye with which she had painted herself. Filled with great anger, Anpu killed his wife and cast her to the dogs. Then he sat down, poured ashes on his head, and mourned for his younger brother. Bata reached the Valley of the Acacia. There was no one with him; he slew wild animals of the desert for his food and built himself a house under the sacred acacia tree, the tree sacred to the gods, which bore his soul upon the topmost flower. And after this he built himself a tower with his own hands, in the valley of the acacia; it was full of all good things, that he might provide for himself a home.

One day as he walked out of his house, he met the Nine Gods who knew of his innocence and goodness. Ra said to the god Khunumu, "Look, make a woman for Bata that he may not remain alone. And Khunumu made for Bata a wife to dwell with him. She was indeed more beautiful than any other woman in the whole land. She was like a goddess as the essence of every god was in her and Bata loved her very much. The seven Hathors came to see her: they said with one mouth, " She will die a sharp death."

And Bata loved her very exceedingly, and she dwelt in his house; he passed his time in hunting the beasts of the desert, and brought and laid them before her. He said, "Go not outside, lest the sea seize you; for I cannot rescue you from it, for I am a person like you; my soul is placed on the head of the flower of the acacia; and if another find it, I must fight with him." And he opened unto her his heart in all its nature.

Now after saying these things Bata went to hunt in his daily manner. And the young girl went to walk under the acacia which was by the side of her house. Then the sea saw her, and cast its waves up after her. She ran from before it. She entered her house. And the sea called unto the acacia, saying, "Oh, would that I could seize her!" And the acacia brought a lock from her hair, and the sea carried it to Egypt, and dropped it in the place of the fullers, the makers of Pharaoh's linen. The smell of the lock of hair entered into the clothes of Pharaoh; and they were angry with the fullers of Pharaoh, saying, "The smell of ointment is in the clothes of Pharaoh." And the people were rebuked every day, they knew not what they should do. And the chief fuller of Pharaoh walked by the bank, and his heart was very evil within him after the daily quarrel with him. He stood still, he stood upon the sand opposite to the lock of hair, which was in the water, and he made a servant go into the water and bring it to him; and there was found in it a smell, exceedingly sweet. He took it to Pharaoh; and they brought the scribes and the wise men, and they said unto Pharaoh, "This lock of hair belongs to a daughter of Ra Harakhti: the essence of every god is in her, and it is a tribute to thee from another land. Let messengers go to every strange land to seek her: and as for the messenger who shall go to the valley of the acacia, let many men go with him to bring her." Then said his majesty, "Excellent, this is exceeding is what has been said to us;" and they sent them. And many days after these things the people who were sent to strange lands came to report to the king: but those that went to the valley of the Acacia did not return, for Bata had killed them, but he let one of them return to give a report to the king. His majesty sent many men and soldiers, as well as horsemen to hold Bata, and to bring her back. And there was a woman amongst them, and to her had been given in her hand beautiful ornaments of a woman. And this time the girl came back with her, and they rejoiced over her in the whole land.

And his majesty loved her exceedingly, and raised her to high estate; and he spoke to her saying that she should tell him concerning her husband. And she said, "Let the acacia be cut down, and let one chop it up."

And they sent men and soldiers with their weapons to cut down the acacia; and they came to the acacia, and they cut the flower upon which was the soul of Bata, and he fell dead suddenly.

And when the next day came, and the earth was lightened, the acacia was cut down. And Anpu, the elder brother of Bata, entered his house, and washed his hands; and a person gave him a cup of beer, and it became troubled; and another one gave him another of wine, and the smell of it was evil. Then he took his staff, and his sandals, and likewise his clothes, with his weapons of war; and he went forth to the valley of the acacia. He entered the tower of his younger brother, and he found him lying upon his mat; he was dead. And he wept when he saw his younger brother truly was lying dead. And he went out to seek the soul of his younger brother under the acacia tree, under which his younger brother lay in the evening.

He spent three years in seeking for it, but found it not. And when he began looking in the fourth year, he desired in his heart to return into Egypt; he said in his heart, "I will go tomorrow morning". Now when the land lightened, and the next day appeared, he was walking under the acacia; he was spending his time in seeking the flower. And he returned in the evening, and labored at seeking it again. Then he at last found a seed. He returned with it. Look, this was the soul of his younger brother. He brought a cup of cold water, and he threw the seed into it: and he sat down, as he usually did. Now when the night came his soul of his brother sucked up the water; Bata then shuddered in all his limbs, and he looked on his elder brother; his soul was in the cup. Then Anpu took the cup of cold water, in which the soul of his younger brother was; Bata drank it, his soul stood again in its proper place, and he became as he had been. They embraced each other, and they spoke together.

And Bata said to his elder brother, "Behold I am to become as a great bull, which bears every good mark; no one knows its history, and you must sit upon my back. When the sun arises I shall be in the

* * * Hyades *	Pleaides * * *	* Ram substitution	* Back	Ram Horns * *
3000 BC	2000 BC		1000 BC	0 BC

place where my wife is, that I may return answer to her; and you must take me to the place where the king is. For all good things shall be done for you; for one shall load you with silver and gold, because you bring me to Pharaoh, for I become a great marvel, and they shall rejoice for me in all the land. And you shall go to your village."

And when the land was lightened, and the next day appeared, Bata took the form of the bull, like he had told his elder brother. And Anpu sat upon his back until the dawn. Together they came to the place where the king was, and they made his majesty to know of him; he saw him, and he was exceeding joyful with him. The king made for him great offerings, saying, "This is a great wonder which has come to pass." There were rejoicings over him in the whole land. They presented unto him silver and gold for his elder brother, who went and stayed in his village. They gave to the bull many men and many things, and Pharaoh loved him exceedingly above all that is in this land.

And after many days after these things, the bull entered the purified place; he stood in the place where the princess was; he began to speak with her, saying, "Behold, I am alive again." And she said to him, "And, pray, who are you?" He said to her, "I am Bata. I perceived when you told them that they should destroy the acacia of Pharaoh, which was my abode, that I would die. Look! I am alive again, I am as an ox." Then the princess feared exceedingly for the words that her husband had spoken to her. And he went out from the purified place.

And his majesty was sitting, making a good day with her: she was at the table of his majesty, and the king was exceeding pleased with her. And she said to his majesty, " Swear to me by God, saying, 'What ever you shalt say, I will obey it for your sake.'" He listened carefully to all that she said, even this. "Let me eat of the liver of the ox, because he is fit for nothing:" said the woman to him. And the king was exceeding sad at her words, the heart of Pharaoh grieved greatly. And after the land was lightened, and the next day appeared, they proclaimed a great feast with offerings to the ox. And the king sent one of the chief butchers of his majesty, to cause the ox to be sacrificed. And when he was sacrificed, as he was upon the shoulders of the people, he shook his neck, and he threw two drops of blood over against the two doors of his majesty. The one fell upon the one side, on the great door of Pharaoh, and the other upon the other door. They grew as two great Persea trees, and each of them was excellent.

And one went to tell unto his majesty, "Two great Persea trees have grown, as a great marvel of his majesty, in the night by the side of the great gate of his majesty." And there was rejoicing for them in all the land, and there were offerings made to them.

And when the days were multiplied after these things, his majesty was adorned with the blue crown, with garlands of flowers on his neck, and he was upon the chariot of pale gold, and he went out from the palace to behold the Persea trees: the princess also was going out with horses behind his majesty. And as his majesty sat beneath one of the Persea trees, it spoke to his wife: "Oh thou deceitful one, I am Bata, I am alive, though I have been evilly entreated. I knew who caused the acacia to be cut down by Pharaoh at my dwelling. I then became an ox, and you caused me to be killed."

And many days after these things the princess stood at the table of Pharaoh, and the king was pleased with her. And she said to his majesty, "Swear to me by God, saying, 'That which the princess shall say to me I will obey it for her.'" And he listened carefully to all she said. And she commanded, " Let these two Persea trees be cut down, and let them be made into goodly planks." And after this his majesty sent skillful craftsmen, and they cut down the Persea trees of Pharaoh; and the princess, the royal wife, was standing looking on, and they did all that was in her heart unto the trees. But a chip flew up, and it entered into the mouth of the princess; she swal-

Photo courtesy Tour Egypt

A page from the 'Tale of the Two Brothers'   FIG. 325

lowed it, and after many days she bore a son. And one went to tell his majesty, "There is born to you a son." And they brought him, and gave to him a nurse and servants; and there were rejoicings in the whole land. And the king sat making a merry day, as they were about the naming of him, and his majesty loved him exceedingly at that moment, and the king raised him to be the royal son of Kush.

Now after the days had multiplied after these things, his majesty made him heir of all the land. And many days after that, when he had fulfilled many years as heir, his majesty flew up to heaven. And the heir said, "Let my great nobles of his majesty be brought before me, that I may make them to know all that has happened to me." And they brought also before him his wife, and he judged with her before him, and they agreed with him. They brought to him his elder brother; he made him hereditary prince in all his land. He was thirty years king of Egypt, and he died, and his elder brother stood in his place on the day of burial.

Excellently finished in peace, for the ka of the scribe of the treasury Kagalu, of the treasury of Pharaoh, and for the scribe Hora, and the scribe Meremapt. Written by the scribe Nena, the owner of this roll.

He who speaks against this roll, may the god Djehuti smite him.

http://en.wikisource.org/wiki/Tale_of_Two_Brothers

Center and Circle

Quiz

on the

Tale of the Two Brothers

(see p. 188)

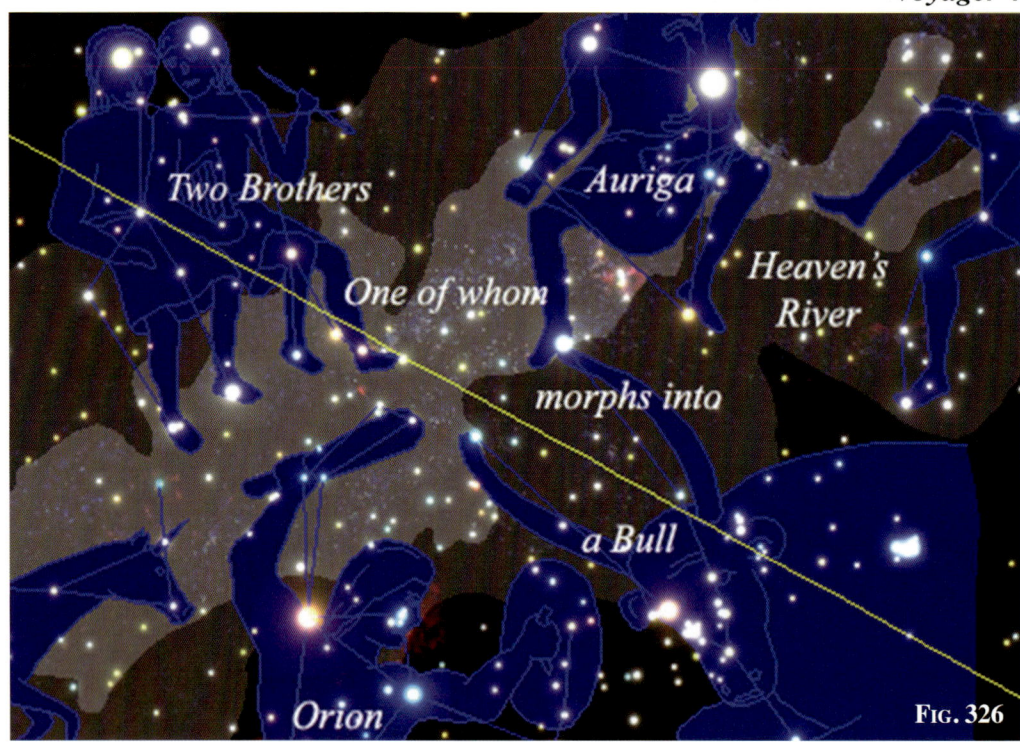

The Milky Way was known to many cultures as the River of Heaven

The narrative is preserved on the D'Orbiney Papyrus that belonged to Seti II (1209-1205 BC) of the 19th Egyptian dynasty while he was crown prince. But it speaks of an older time, when two epochs came together, and then forever parted.

From Brothers, united in a common enterprise for mutual benefit, to breaking of trust and communication across a great gulf, to Bata's metamorphosis into various reincarnations of Great Bull. As should be obvious to all by now, Time regularly changes the mythic picture.

The *Ennead* is the *Circle of Twelve Constellations* who govern the land. They now recognize Bata as the *Bull of the Ennead* (*VE in Taurus*), the new leader of the pack. Why?

What's the 'great gulf of water' that lies between the two brothers?

"*Thereupon Pre heard all his petitions, and Pre caused a great* (gulf of) *water to come between him and his elder* (brother), *infested with crocodiles*, so that one of them came to be on one side and the other on the other (side)."

Why does Anubis strike the back of his hand twice? What would striking the front of your hands be?

Counting the coils (days)

Why does Bata throw his flesh into the river? What are the crocodiles?

Bata is warned by '*his lead cow*'. What is the significance of this image to the story line?

After Bata is warned by the lead cow, he looks under the door and sees only the *feet* of his brother. Bata turns and runs, leaving the feet behind. How does this reflect astronomical motion (VE)?

What is the significance of his 'heart' being located at the top of the Valley of the Pine? How does this complete our recurring theme of *Center and Circle*?

What is the significance of including the phrase '*so the story goes*' at the very beginning of our '*Tale of Two Brothers*?'

Photo credit Tour Egypt   **Center and Circle**

FIG. 328

To say or to speak. How does this 'verb' take on meaning in the following lines?

"What means / this great offense which you have said to me? Don't say it to me again, but I shall tell it to no one."

"*Said, say, shall tell...*"

How many times do 'hands' make appearances in the tale? Why do I ask?

What does it mean through the *stars of Gemini*' to have "your brother associated with you after the manner of a father?" How do we see this reflected in the both the Greek and Egyptian pantheons?

Anubis stands behind the door of his stable to kill his younger brother upon his return in the evening. If we are focused on duality themes, why would Gemini setting (twilight) be significant?

How does this compare with 'standing before God' (Pre-Harakhti), "as soon as the Sun rises" in order that he might be 'judged'. How would this differ from Gemini rising?

Why does Anubis take his staff?

Extra credit: Why is the Bull repeatedly killed by the Chief Lady of the land? What happens each time?

FIG. 329

Cultivating the land was the focus of many ancient cultures

How does the 'Valley of the Pine' compare with Humbaba's 'Cedars of Lebanon' (*Epic of Gilgamesh*)? Is the Great Tree to be compared with all the other trees of the forest?

Leave your handwritten answers on my desk together with an apple bright and early first thing Monday morning.

FIG. 330

Scenes from daily Egyptian life

Athena's Web   195

Center and Circle

Pages	Range	Perspective
13-14	6000 BC to 0 BC/AD	Early Avian Goddess to Julius Caesar

* Early Bird Goddess	* Peak of culture	*Early solar calendar	Egyptian Unification *
6000 BC	5000 BC	4000 BC	3000 BC

| 31-32 | 6000 BC to 0 BC/AD | Archaeological and mythic overlays |

* Birds, Eggs and Twins-----------------	------* Nabta Playa ---	* Bovine -----	--- worship --* Knowth, Newgrange --
6000 BC	5000 BC	4000 BC	3000 BC

| 38-39 | 5000 BC to 2000 BC | Bovine and Egyptian evolution |

4800 * BC- Nabta Playa	Yoke/first Solar calendar * 4236 BC	...agricultural development...
5000 BC	4500 BC 4000 BC	3500 BC

| 45 & 100 | 3500 BC to 0 BC/AD | Tigris-Euphrates focus |

---------------- Sumerians --------- Akkad -- *	---* Babylonian culture --------	---- Assyrian*Chaldean Empire
3000 BC	2000 BC 1000 BC	0 BC

| 74-75 | 1500 BC to 1500 AD | Mesoamerican traditional groupings |

*1500 BC Pre-Classic--------------------	--	--- to 300 BC. *Classic -----------------
1500 BC	1000 BC 500 BC	0 BC/AD

| 88-89 | 26000 BC to 2012 AD | Great Year divided by Five |

Sun of Jaguar / Gold	Sun of Wind / Silver	Sun of Fire Rain /
23615 BC	18490 BC	13365 BC

| 92-93 | 5000 BC to 2000 BC | Anatomical stars of Taurus |

Last star of Gemini- *4634 BC	Tip of the Horns * 4246 BC	...Pre-Dynastic Period...
5000 BC	4500 BC 4000 BC	3500 BC

| 100 | 3500 BC to 0 BC/AD | Tigris-Euphrates focus |

---------------- Sumerians --------- Akkad -- *	---* Babylonian culture --------	Assyrian*Chaldean Empire
3000 BC	2000 BC 1000 BC	0 BC

| 103-104 | 2400 BC to 0 BC/AD | Greek myth and culture |

Last Pleiad- *2170 BC	--Phrixus*Cadmus*Jason---	--Mycenaeans ---------------- ---Troy *
2400 BC	2000 BC 1600 BC	1200 BC

| 119-120 | 3000 BC to 0 BC/AD | Phoenician culture |

* Trade with Egypt and Cyprus.........	....	...Phoenician merchants......................
3000 BC	2500 BC 2000 BC	1500 BC

| 141-142 | 2400 BC to 0 BC/AD | The Land of Egypt |

Old King. to 2152 BC ----- 1st Inter. --	-- Middle --------- 1720 BC - Hyksos ----	---- to 1550 BC - New Kingdom ---------
2400 BC	2000 BC 1600 BC	1200 BC

196   *Athena's Web*

Center and Circle

Time Lines

* Pyramids	* Abraham	Troy *	* Homer	Julius Caesar *
3000 BC — 2000 BC — 1000 BC — 0 BC

| * Stonehenge, Phase One -------------- | ---------------- * 1600 Stonehenge, final | Caesar visits Britain- 54 BC* |
3000 BC — 2000 BC — 1000 BC — 0 BC

| Unification * 3200 BC | Pyramid of Cheops * 2750 BC | ...demise of the * Old Kingdom |
3500 BC — 3000 BC — 2500 BC — 2000 BC

| -------------------------------------- | ---------------------------- to 950 AD* | Post-Classic ------------ to 1521 AD --- |
0 BC/AD — 500 AD — 1000 AD — 1500 AD

| Bronze Age | Sun of Water / Heroic Age | Tonatiuh / Race of Iron |
8239 BC — 3114 BC — 2012 AD

| Aldebaran * 3247 BC | * 2963 BC Last star of the Hyades | Last star of the Pleiades * 2170 BC |
3500 BC — 3000 BC — 2500 BC — 2000 BC

| ---Dark Ages--- | ---AIrchaic--- Greek Golden Age-- | -------- Battle of *Corinth 146 BC |
1200 BC — 800 BC — 400 BC — 0 BC/AD

| .... 1150 BC .... Golden Age to 850 BC .... | .... 333 BC ...... Carthage 146 BC |
1500 BC — 1000 BC — 500 BC — 0 BC

| ----------- to 1070 BC -- 3rd Inter ------ | ---------- to 657 BC -- Late ---------------- | -- to 332 BC -- Alex to Cleo ----- 30 BC |
1200 BC — 800 BC — 400 BC — 0 BC

Athena's Web   197

Center and Circle

Figures & Illustrations

Key:
AC = Artwork or Artist Credit
CC-BY-SA = Creative Commons Attribution-Share Alike 3.0 Unported license
GNU = Free Documentation License
FL = Flickr
PC = Photo Courtesy or Credit
PD = Public Domain
PP = Personal Photo, taken by Don
PS = Created in PhotoShop
WC = Wikimedia Commons

Cover   Fiske Dragon artwork, compliments Fiske Planetarium, U. of Colorado, Boulder
Fig. 0   Dragon Plaque. PC Tony Jones of Horsham, UK
Fig. 1   Theseus and Minotaur outside the Louvre in Paris. PC Luis Benkard
Fig. 2   Greece. NASA. PD
Fig. 3   Crete. NASA. PD
Fig. 4   Theseus and the Minotaur. PC with kind permission of FL: petrus.agricola
Fig. 5   Seven-headed Dragon. Leonardo da Vinci. PD
Fig. 6   Red Dragon. AC Craig B. Musselman
Fig. 7   White Dragon rising after birth from the Egg. PC FL: LILFr38
Fig. 8   Teacher in classroom. PC Claire Whittenbury. FL: whitt
Fig. 9   Ottoman astronomers. WC. PD
Fig. 10   Chinese Water Dragon. PS
Fig. 11   Looking North in 2788 BC. PS
Fig. 12   Earth. AC Andrew Zachary Shiff
Fig. 13   Angle of the Sun's rays striking the Earth. PS
Fig. 14   Etruscan Dragon (*sans* Earth) . PC FL: Sebastia Giralt
Fig. 15   Precessional Path of NCP. WC. PC Tauʻolunga.
Fig. 16   Spinning like a top. PS
Fig. 17   Polar Axis. PS
Fig. 18   Precession along the Vernal Equinox. Voyager v. 1
Fig. 19   Precession along the North Celestial Pole. Voyager v. 1
Fig. 20   Marduk chasing Ti'amat. PC Georgelazenby. WC. CC-BY-SA
Fig. 21   May Day Dragon, Judy Allen and Jeanne Griffiths, *Book of the Dragon*, p. 108. PD
Fig. 22   'Blueprint' Temple of Luxor. PC Tour Egypt, www.touregypt.net
Fig. 23   Stained Glass Compass Rose. PC TR Corwin
Fig. 24   Chinese Dragon, *Symbols, Signs and Signets,* Lehner, p. 140. Used by permission
Fig. 25   Dragon Spring. Starry Night Education, a Simulation Curriculum Corporation
Fig. 26   Dragon Autumn. Starry Night Education, a Simulation Curriculum Corporation
Fig. 27   Dragon Winter. Starry Night Education, a Simulation Curriculum Corporation
Fig. 28   Dragon, Dec. 31st. Starry Night Education, a Simulation Curriculum Corporation
Fig. 29   Sleepy Dragon. Starry Night Education, a Simulation Curriculum Corporation
Fig. 30   Constellation Gemini from "*Atlas Coelestis*" Sir James Thornhill, Plate 13- PD
Fig. 31   Constellational map of the VE 6300-4800 BC. Starry Night Education. PS
Fig. 32   The Dragon and the NCP in 5000 BC. Voyager v. 1. PS
Fig. 33   Two-headed Goddess. Dr Marija Gibutas. AC Jeanne Fallot

Fig. 34   Bird, Double-headed Deity. Dr Marija Gibutas. AC Jeanne Fallot
Fig. 35   Spiral Serpent. Dr Marija Gibutas. AC Jeanne Fallot
Fig. 36   Bird Goddess of 6000 BC. Dr Marija Gibutas. AC Jeanne Fallot
Fig. 37   Bird as Creation. Dr Marija Gibutas. AC Jeanne Fallot
Fig. 38   Bird with Egg. Dr Marija Gibutas. AC Jeanne Fallot
Fig. 39   Serpent Bowl. Dr Marija Gibutas. AC Jeanne Fallot
Fig. 40   Vase with Bird and Egg. Dr Marija Gibutas. AC Jeanne Fallot
Fig. 41   Serpent Bowl - top. Dr Marija Gibutas. AC Jeanne Fallot
Fig. 42   Nommo, Dogon celestial Twins. *Mythologies*, Bonnefoy. AC Jeanne Fallot
Fig. 43   Caduceus, *Symbols, Signs and Signets*, Lehner, p. 199. Used by permission
Fig. 44   Serpent & Egg, Vishnu resting on Ananta-Sesha, 18th century. PD
Fig. 45   Dan Ayido Hwedo. AC Jeanne Fallot
Fig. 46   Nabta Playa, Egypt. WC. GNU. PC FL: Raymbetz.
Fig. 47   Nabta Playa. PC University of Texas, Austin. WC
Fig. 48   Outline Nabta Playa. PC University of Texas. WC
Fig. 49   Ophion. *Symbols, Signs and Signets*, Lehner, p. 84. Used by permission
Fig. 50   Temple of Ba' al. PC Daniel Fernandez. FL: twiga-swala
Fig. 51   Chalk Mt. Observatory. PC High Altitude Observatory of the
          National Center for Atmospheric Research, funding NSA.
Fig. 52   Tulsa's Gladys-Leo Observatory. PC Leo Cundiff
Fig. 53   Tulsa's Gladys-Leo Observatory. PC Leo Cundiff
Fig. 54   Univ. of Saskatchewan Observatory Skylight. PC Sevenstar Studio
Fig. 55   Isle of Man Observatory. PC Isle of Man Astronomical Society
Fig. 56   Welcome Ramadhan. PC Ahmed Rabea. WC. CC-BY-SA
Fig. 57   Nut, Shu and Geb. PC Tour Egypt, www.touregypt.net
Fig. 58   Thoth, God of the Moon. PC Tour Egypt, www.touregypt.net
Fig. 59   Twin-headed Serpent, California King Snake. PC FL: Scott Hanko
Fig. 60   Sirius or Isis. PC Tour Egypt, www.touregypt.net
Fig. 61   Pyramid of Khufu. PC Nina Aldin Thune. FL: mrsnina-no
Fig. 62   Egg of P'an Ku. PC Van Camp. WC. CC-BY-SA
Fig. 63   The incredible, edible egg. PP
Fig. 64   Newgrange. PC Boyne Valley Tours, www.boynevalleytours.com
Fig. 65   Newgrange light. PC Boyne Valley Tours, www.boynevalleytours.com
Fig. 66   The Ring towards the Loch of Harray. PC David Barlow. FL: Dave_Barlow
Fig. 67   Satellite photo of Earth at Sunrise. NASA. PD
Fig. 68   Standing stone, silent sentinel. PC FL: ColinsCamera
Fig. 69   Celtic Solar Cross. PS
Figs. 70, 71 & 72   Eight-spoked Solar System, Shadow Wheel (71) and Tunnel (72).
          AC Inner Traditions from *The Stones of Time* by Martin Brennan,
          Inner Traditions, Rochester, VT 05767. www.InnerTraditions.com
Fig. 73   Equinox Light at Loughcrew. PC Sean Rowe. FL: sjrowe53
Fig. 74   Newgrange. PC Shira. WC. CC-BY-SA
Fig. 75   A gnome. PS. PC BoyneValleyTours.com, www.boynevalleytours.com
Fig. 76   Minoan fleet frieze from Akrotiri. PC Luxo. WC. PD
Fig. 77   Satellite photo British Isles. NASA. PC Luxo. PD
Fig. 78   The Siren. AC Edward Armitage. PD

Center and Circle

Fig. 79   Cyclades map of Greece, Asia Minor and Crete. NASA. PD
Fig. 80   Cyclades figurines looking up. PS
Fig. 81   Hagar Qim, Malta. PC Heritage Malta
Fig. 82   Malta model. PC Heritage Malta
Fig. 83   Greek Drakospito roadsign. PP
Fig. 84   Roof of Drakospito, Evia, Greece. PP
Fig. 85   Sunlight on Equinoxes. PC Heritage Malta
Fig. 86   Drakospito on Evia. PP
Fig. 87   Dionysus legend. PC MatthiasKabel. WC. GNU
Fig. 88   Newgrange spiral. PC Boyne Valley Tours, www.boynevalleytours.com
Fig. 89   Equinox entrance. PP
Fig. 90   Spirals of Malta. PP
Fig. 91   Lascaux Bull. PC Johan Wilbrink. FL: YIP2
Fig. 92   Constellation Taurus. Starry Night Education, a Simulation Curriculum Corporation
Fig. 93   Hyades. Starry Night Education, a Simulation Curriculum Corporation
Fig. 94   Moon in Taurus. Starry Night Education, a Simulation Curriculum Corporation
Fig. 95   Apis Bull. Museo Gregoriano Egizio. PC Heidemarie Niemann. FL: HEN-Magonza
Fig. 96   Apis Bull. The power behind the throne. PC Néfermaât. WC
Fig. 97   Apis Bull. PC Tour Egypt, www.touregypt.net
Fig. 98   Badarian woman, 4500 BC. PC Bobak Ha'Eri. WC. GNU
Fig. 99   Precession & Horns. Starry Night Education, a Simulation Curriculum Corporation. PS
Fig. 100   Father Nile. PC Nathan Petersen. FL: petersnm
Fig. 101   Palette of Narmer. PC Jeff Dahl. WC. PD
Fig. 102   Nubian procession. PC Tour Egypt, www.touregypt.net
Fig. 103   Map of Egypt and Nubia. PC Mark Dingemanse. WC
Fig. 104   Cylinder Seal Showing A God in A Boat on River. An early steatite seal showing a god in a river scene of reeds and plants, birds or chevrons, star of divinity, three dots and an early inscription. It could be late Dynastic I or early Dynastic II period. period. Although the sun god Utu-Shamash, and others as well, were said to travel the heavens in a sacred boat, this naturalistic scene is more likely a legend of southern Sumeria which had more waterways. Myths place the seal in southern Sumeria and hint that it is very early in date. Sumerian Dynastic Period. Leroy Golf Sumerian Seals. www.antiquesatoz.com/golf/golfsumeriaseal.htm
Fig. 105   Sumerian sky observer. PC Stewart Cherlin. FL: Stew says
Fig. 106   A small black cylinder from Old Babylon showing a kneeling woman making an offering to a seated god with crescent-topped staff, standard behind him, star of divinity in the sky and a three column, boxed inscription.
               Black steatite, c. 1900-1700 BC   Leroy Golf Sumerian Seals
               www.antiquesatoz.com/golf/golfsumeriaseal.htm
Fig. 107   Sumer map. PC Ciudades_de_Sumeria.svg. Modifications by Phirosiberia. CC-BY-SA
Fig. 108   Gilgamesh. PP
Fig. 109   Gilgamesh, Enkidu & Bull. © the Schøyen Collection, MS 1689. Used by permission
Fig. 110   Heavenly Bull Fight. Starry Night Education, a Simulation Curriculum Corporation. PS
Fig. 111   Gilgamesh myth flood tablets. British Museum. PD
Fig. 112   Marduk on the back of Ti'amat, clay cylinder seal. PD
Fig. 113   Babylonian Ziggurat. PC © 2005 Joshua McFall. FL: jmcfall

Fig. 114	Stars of Draco 2788 BC. *Voyager v. 1*. PS
Fig. 115	Uraeus. PC Tour Egypt, www.touregypt.net
Fig. 116	Pharaoh's Uraeus. PC Tour Egypt, www.touregypt.net
Fig. 117	Desert Sphinx. PC Tour Egypt, www.touregypt.net
Fig. 118	*Vitruvian Man*. Leonardo da Vinci. WC. PD
Fig. 119	Nature's reflection. PC Dean Martin, Val Caron, Ontario. FL: Dean Martin (simplyred4x4)
Fig. 120	Viking prow. PC Harold Moses. http://www.flickr.com/photos/57729348@N02/
Fig. 121	Thor battles Midgard Serpent. Fuseli. PC Troy/New Visions2010 PD. FL: NewVisions2010
Fig. 122	Vasuki & Mt. Mandara. PC Vishnu Avatara. WC. PD
Fig. 123	Sunrise from Mt. Olympus. PP taken while on our honeymoon
Fig. 124	Takshaka, aka Taxaka, King of the Serpents. PC LR Burdak. WC
Fig. 125	Draco, plate 50-51. PC James Thornhill. PD
Fig. 126	*Liber Astronomiae*, Guido Bonatti. Illustration intentionally reversed. PD
Fig. 127	Vishnu standing on the intertwined bodies of seven snakes. PC Arthur Millner
Fig. 128	Socketed Iranian Dragon, 19th century BC. AC Jeanne Fallot
Fig. 129	Inanna or Ishtar. British Museum. PC AmberinSea Photography FL: Amberinsea
Fig. 130	Gate of Ishtar. Berlin Museum. PP
Fig. 131	Yggdrasil World Tree. AC Jen Delyth ©1990/2000 www.celticartstudio.com
Fig. 132	Yggdrasil, the World Ash. PS
Fig. 133	Heimdallr, guardian of the Rainbow Bridge. PD
Fig. 134	Heracles and his Lion-skin. PC Jim Naureckas. FL: edenpictures
Fig. 135	Michaelanglo's God. PD
Fig. 136	Yule Fire. PC Marinell Turnage. FL: mturnage
Fig. 137	Mayan World Tree. PC Madman2001. WC. CC-BY-SA
Fig. 138	Feathered Serpent. PC Matt Logan. FL: Mattron
Fig. 139	Pyramid. PC FL: Hanneorla. www.flickr.com/photos/hanneorla/collections. Spiral Serpent. Museum of Anthro, Mexico City. PC Travis Shinabarger. FL: Travis S.
Fig. 140	Serpent Quetzacoatl. PC James A Glazier and James A Ferguson. FL: jaglazier
Fig. 141	Feathered Serpent. PC James A Glazier and James A Ferguson. FL: jaglazier
Fig. 142	Hun-Hunahpu shoots Vucub-Caquix in the Ceiba. PC Glazier and Ferguson. FL: jaglazier
Fig. 143	Mayan World Tree relief. PC Madman2001. WC. PD
Fig. 144	Gourds of the Calabash. PC Polyparadigm. WC. GNU
Fig. 145	Skulls in the Calabash Tree. PC Benjamin Jakabek. FL: BRJ INC
Fig. 146	Adam, Eve and a Serpent. PC Museo Diocesano de Solsona, Lerida, Spain. PD
Fig. 147	Expulsion from the Garden in Worcester cathedral. PC FL: Granpic
Fig. 148	Taurus setting 4000 BC, tips of the Horns. *Voyager 4*. PS
Fig. 149	Draco rising 4000 BC. *Voyager 4*. PS
Fig. 150	Taurus setting 3000 BC. Head of the Bull, *Voyager 4*. PS
Fig. 151	Draco rising 3000 BC. *Voyager 4*. PS
Fig. 152	Apophis and a Cat-like God (*Re* the Sun-God). PC Tour Egypt, www.touregypt.net
Fig. 153	Taurus setting 2000 BC. Back of the Bull. *Voyager 4*. PS
Fig. 154	Draco rising 2000 BC. *Voyager 4*. PS
Fig. 155	Taurus setting 1000 BC, back of the Ram. *Voyager 4*. PS
Fig. 156	Draco rising 1000 BC. *Voyager 4*. PS

Center and Circle

Fig. 157   Time exposed photograph of the night sky. PC Brother Childs
Fig. 158   Hopi Horizon. PC Columbia University. PD
Fig. 159   Athena's bird. PC NaySay. WC. GNU. PD
Fig. 160   Apophis, the cut on the neck and the Tree. PC Tour Egypt, www.touregypt.net
Fig. 161   Apophis, closer. PC Tour Egypt, www.touregypt.net
Fig. 162   Expulsion. St Edmundsbury Cath., PC G.Licence. FL: A&G Creative
Fig. 163   Tree of Knowledge. PC Rodrigo Alvarez-Icaza. FL: r.Al(-)
Fig. 164   Ceiba Tree. A World Tree worthy of its name. PS
Fig. 165   Mayan Tree of Life with horoscope wheel overlay. PS
Fig. 166   Tree added to Mayan Tree of Life. PS
Fig. 167   Vucub-Caquix. PC J. Elizabeth Clark. FL:Urban Sea Star
Fig. 168   The Aztec Calendar Stone. PC Rafael Saldana. FL: ikarusmedia
Fig. 169   Eclipse Moon. PC Jonathan Kaiser. FL: jkaiser's
Fig. 170   Venus, a Heavenly Star. PS
Fig. 171   The golden orb of the Sun, *Numero Uno*. PS
Fig. 172   Venus displaying her symmetry. PS
Fig. 173   Tonatiuh, the Current Epoch of the Aztec. PC FL: Joseph A Ferris III
Fig. 174   Atonatium, Sun of Water. PC FL: Joseph A Ferris III
Fig. 175   Quiauhtonatiuh, Sun of Fire Rain. PC FL: Joseph A Ferris III
Fig. 176   Ehecatonatium, Sun of Wind. PC FL: Joseph A Ferris III
Fig. 177   Ocelotonatiuh, Sun of Jaguar. PC FL: Joseph A Ferris III
Fig. 178   Hesiod. PC Helen Sanders. FL: f1jherbert
Fig. 179   Gold Crown. PC Oscar Anton. FL: oshkar
Fig. 180   Silver Crown. PC Martin Tremblay. FL: martin t.
Fig. 181   Bronze Crown. AC and design by Miriam Boy. FL: silverandmoor
Fig. 182   Our Lady of Victory. PC Mike Fitzpatrick, Annapolis, MD. FL: Piedmont fossil
Fig. 183   Iron Crown, Absalon. PC Marco Franchino, 2011. FL: little_frank
Fig. 184   Father Zeus. PC Sandro Menzel. FL: smenzel
Fig. 185   Detail from 1632 edition of Ovid's *Metamorphoses*. WC. PD
Fig. 186   Publius Ovidius Naso, *aka* Ovid. WC. GNU
Fig. 187   Bull Horns, Head (Hyades) and Shoulder (Pleiades). *Voyager 4*. PS
Fig. 188   Celestial gulf between the backs of Bull and Ram. *Voyager 4*. PS
Fig. 189   *Europa*. Rembrandt. PC Google Art Project. PD
Fig. 190   Europa and the Bull. Author unknown. PD
Fig. 191   The Celestial Back of the Bull. *Voyager 4*. PS
Fig. 192   Minoan fresco of the Bull-leaping 'rodeo'. PC FL: Lapplaender
Fig. 193   Minoan Snake Goddess. PC FL: Chris73. WC. CC-BY-SA
Fig. 194   Zeus carries Europa across the Sea. AC Soa Lee, http://soanala.com
Fig. 195   Wadjet in Her Lion skin. PC FL: dominotic
Fig. 196   Menes detail from the Narmer Palette. PC FL: Udimu
Fig. 197   Deshret and Hedjet Crowns. PC Paul Hessell. FL: fessell810
Fig. 198   Wadjet as a Serpent. PC FL: OnceAndFutureLaura
Fig. 199   Deshret Crown of Upper (Southern) Egypt. PC Immanuel Giel. WC
Fig. 200   Minoan dolphin fresco. PC Armagnac. WC
Fig. 201   Gold Minoan signet ring. PD
Fig. 202   Solar glyph overlaying the Sun. PC Inner Traditions. PS

Fig. 203 King of the Beasts- Jason Morgan (this is a painting, not a photo!)
http://www.onlineartdemos.co.uk/pages/print_pages/Lion-print.html
Fig. 204 Solar glyph superimposed upon Newgrange markings. PC Inner Traditions. PS
Fig. 205 Bast. PC Richard Klug. FL: Klugrd
Fig. 206 Minoa's Bull of Heaven. PC FL: Nick Kaye. www.flickr.com/photos/nickkaye/
Fig. 207 Ipuwer's parchment. PC Tour Egypt, www.touregypt.net
Fig. 208 Akkadian Victory Stele, 2300 BC. PC Thomas Schuman. FL: Schumata
Fig. 209 Achill Ram. PC http://www.maguiregallery.com/barrie/achill%20ram.htm
Fig. 210 Phrixus attempting to hang on. PC Peter Roan. FL: peterjr1961
Fig. 211 Ram Rhyton. PC Luke Penrod. FL: nordic47
Fig. 212 Phrixus and Helle. PC Immanuel Giel. WC. PD
Fig. 213 Desert landscape. PC FL: Moyan_Brenn. flickr.com/aigle_dore
Fig. 214 Jason and the Bulls. PC Petra Schemmel. FL: Wolkenrose
Fig. 215 Jason sand sculpture. PC Stewart. FL: StewieD
Fig. 216 Jason reaching out for the Fleece. PC Peter Roan. FL: peterjr1961
Fig. 217 Etruscan Dragon. PC FL: Sebastia Giralt
Fig. 218 Dragon's Head at Zenith circa 1400 BC. *Voyager v. 1*. PS
Fig. 219 Six hours later- how the Dragon turns. *Voyager v. 1*. PS
Fig. 220 Twelve hours later- Head 'against the Tree'. *Voyager v. 1*. PS
Fig. 221 Cadmus and Dragon. Black-figured amphora from Euboea, ca. 560–550 BC. PD
Fig. 222 *Sacrifice of Isaac*. Rembrandt. PD
Fig. 223 The Brazen Serpent. *Dore Illustrations*, Dover 1974, p. 44. Used by permission
Fig. 224 Moses before Pharaoh. *Dore Illustrations*, Dover 1974, p. 32. Used by permission
Fig. 225 Ecstacy, impersonating Spring. PC Maxfield Parrish. PD
Fig. 226 PC St. Paul Lutheran Church, Sterling, Illinois, 61081
Fig. 227 Golden Calf taken in Yoron-jima, Japan. PC Stefanie & Justin McFall. FL: jmcfall
Fig. 228 Maypole celebration. PC hampel-auctions.com. PD
Fig. 229 Bull silhouette. PS
Fig. 230 Ram bust. PS
Fig. 231 Semitic Ba'al. PC Jastrow. WC. PD
Fig. 232 El, Father of Ba'al. PC Michael Martin. FL: magika42000
Fig. 233 Temple of Baal at night. PC Rafael Gomez www.micamara.es/siria
Fig. 234 Ba'al in his Earthly temple. PC FL: petrus.agricola
Fig. 235 Astarte, Assyrian goddess. PC Marie-Lan Nguyen. WC. PD
Fig. 236 The Scroll, Torah or Law of Moses. PC FL: Lawrie Cate
Fig. 237 Phoenician writing. PC Johan Anglemark. FL: jophan
Fig. 238 Phoenician ship, 2nd century AD. PC Elle plus. WC. PD
Fig. 239 Phoenician settlements. PC FL: Benowar
Fig. 240 Phoenician Tree of Life. PC Tina Negus. FL: tina negus
Fig. 241 Beltaine Fire Ceremony. PC Jason Fowler. FL: jasonfowler
Fig. 242 Leaping bonfire. PC FL: Graham Green
Fig. 243 Loch Ness. PC Ayaz Asif. www.ayaz.com FL: wwwayazdotcom
Fig. 244 Megalithic sunset. PC David Barlow. FL: Dave_Barlow
Fig. 245 Aphrodite of Milos, Louvre. PC Shawn Lipowski. PD
Fig. 246 Venus. AC William-Adolphe Bouguereau. PD
Fig. 247 Severing the Head. *Starry Night Education*, a Simulation Curriculum Corporation

Fig. 248   Lion of Gate of Ishtar. PC Josep Renalias. WC. CC-BY-SA
Fig. 249   Gate of Ishtar, Berlin. PP
Fig. 250   Nebuchadnezzar on New Year's Day. PC Ingrid Ramsay. FL: Ingrid 1959
Fig. 251   Nebuchadnezzar and Daniel. PC W. A. Spicer. WC. PD
Fig. 252   The Gate of Ishtar Dragon. PC Prof. Richard T. Mortel, Riyadh. FL: Prof. Mortel
Fig. 253   Nebuchadnezzar II by William Blake. PC Web Gallery of Art. WC
Fig. 254   Community carrying the ridge pole. Author unknown
Fig. 255   Zipacna & the 400 boys, from *Zipacna's Travels*. PC Bruce Rimell. www.biroz.net
Fig. 256   Classic Chicha. PC FL: Raquel Gil Fotografía (kela)
Fig. 257   M-45, the Pleiades. PC Rob Gendler. www.robgendlerastropics.com
Fig. 258   Pigeons. PC Allison Lucas. hearingvoices.com/news/2008/10/brooklyn-pigeons
Fig. 259   The Parthenon. PC FL: Tiago C Lima. www.flickr.com/photos/tclima/
Fig. 260   A Bacchanal. Titian. WC. PD
Fig. 261   Amenophis III, peak of the New Kingdom. PC FL: Miltiade
Fig. 262   Amun-Re. PC Tour Egypt, www.touregypt.net
Fig. 263   Amun-Re. PC Tour Egypt, www.touregypt.net
Fig. 264   Criosphinx (Ram-Sphinx). PC Frederic Chanal. http://www.nach.biz/photos
Fig. 265   Uraeus-Ram Deity from Kush. PC Joan Ann Lansberry
Fig. 266   Time of the Hyksos. *Voyager 4*. PS
Fig. 267   The New Kingdom's Epoch. *Voyager 4*. PS. Ram's head PC Joan Ann Lansberry
Fig. 268   Celestial map of the constellation Aries. *Voyager 4*. PS
Fig. 269   The covenant of Abraham. PC Rex Harris. FL: Sheepdog Rex
Fig. 270   The opening of the Book of Judges. PP
Fig. 271   Agni - the Persian and Hindu God of Fire. WC. PD
Fig. 272   Agni's Red Ram. WC. PD
Fig. 273   Meet Mr Mars. PS
Fig. 274   Alexander with horns. PC Wayne Sayles
Fig. 275   Detail Gundestrup Cauldron. PC Dbachmann. WC. PD
Fig. 276   Gundestrup Cauldron. PC Bloodofox. WC. PD
Fig. 277   Dagda, Gundestrup Cauldron. PC Dbachmann. WC. PD
Fig. 278   Ecliptic Pole Path. *Voyager 4*. Insert NASA, Mysid. PS
Fig. 279   Harappan Ox. PC Mukul Banerjee. www.mukulbanerjee.com
Fig. 280   Oxen. PC Bert Cozens. FL: Hear and Their
Fig. 281   Fertile Crescent map. PC FL: NafSadh
Fig. 282   The Ram by the River (note upside-down Cetus). *Voyager 4*. PS
Fig. 283   The dream of Daniel. PC Hind, Arthur Mayger. WC. PD
Fig. 284   Perseus of Macedon. British Museum. WC. PHGCOM. PD
Fig. 285   Vernal Equinox in 407 BC, 186 BC and AD 15. *Voyager. 4*. PS
Fig. 286   Dying Gaul. PP
Fig. 287   Marius. PC Glyptothek Munich. PD
Fig. 288   Sulla. PC Glyptothek Munich. PD
Fig. 289   Pompey Magnus. PC Faculty of Classics, University of Cambridge.
Fig. 290   Julius Caesar. PC Andreas Wahra. PD
Fig. 291   Mark Anthony. PC Amadscientist. WC. PD
Fig. 292   Astrolabe. PP
Fig. 293   Rainbow Serpent, Australia. www.aboriginaltourism.com.au. *Voyager 4*. PS

Fig. 294	Compass Rose. PC Anton Kokuiev. FL: Kassandra_gg
Fig. 295	Rainbow Serpent, Africa. www.aboriginaltourism.com.au. *Voyager 4*. PS
Fig. 296	*Serpiente alquimica*, alchemical snake. PC Carlos adanero. WC. PD
Fig. 297	Freemason's Twin Oroborus. PC Fieari. WC. PD
Fig. 298	AC Ouroborus mural by Leah B. Freeman. FL: lbfreeman
Fig. 299	Torc. PC Lisa Michel
Fig. 300	Autumnal parade. *Voyager 4*. PS
Fig. 301	When the Serpent Wore the Crown. *Voyager 4*. PS
Fig. 302	The Scorpion King. PC FL: Udimu
Fig. 303	Eight stars in two centuries. *Voyager 4*. PS
Fig. 304	Venom rising. *Voyager 4*. PS
Fig. 305	Heracles circa 7000 BC. *Voyager 4*. PS
Fig. 306	March of the Hyades. *Voyager 4*. PS
Fig. 307	Heracles in the crib with two serpents. PP
Fig. 308	PC Pyramid, Donna Corless; PC Draco, chaouki; PC Orion, Sidney Hall. PS
Fig. 309	Wings of the Dragon. *Voyager 4*. PS
Fig. 310	Pisces stained glass window. Chartres. PC FL: Art History Images (Holly Hayes)
Fig. 311	"Ulm Münster," St George by Hans Acker. PC Joachim Köhler. WC. GNU. PD
Fig. 312	Cyclades Star*gazer. PP
Fig. 313	Scroll stars of Pisces. PC Adam Humphrey
Fig. 314	Christian symbol of the Fish
Fig. 315	Jesus and Lamb stained glass. Alfred Handel. PC Toby Hudson. FL: toby hudson
Fig. 316	Sun and Cross. PC Dean Martin. FL: Dean Martin (simplyred4x4)
Fig. 317	St. George and the Dragon. PC David Cooley. FL: DaveFlker
Fig. 318	The Imperial, five-toed Dragon. Reproduced by kind permission of Durham University Museums. Ref DUROM.1991.217
Fig. 319	Pearl. Ecliptic and Equatorial centers. *Voyager 4*. PS
Fig. 320	European Dolmen. PC Jay Reinfeld
Fig. 321	Confusion of Tongues (Tower of Babel). AC Gustave Dore. WC. PD
Fig. 322	The Bridestones. PC FL: Peter Juerges. www.peterjuerges.co.uk
Fig. 323	Proportional dragon. Leonardo DaVinci. PD
Fig. 324	Longitude positions of Draco's stars in the 20th century. PP
Fig. 325	Page from *A Tale of Two Brothers*. PC Tour Egypt, www.touregypt.net
Fig. 326	Constellations Gemini and Taurus, with Milky Way. *Voyager 4*. PS
Fig. 327	Counting the days. PC Tour Egypt, www.touregypt.net
Fig. 328	Egyptian life. PC Tour Egypt, www.touregypt.net
Fig. 329	Bringing in the hay. PC Tour Egypt, www.touregypt.net
Fig. 330	Egyptian harvest. PC Tour Egypt, www.touregypt.net
Fig. 331	Settling down with a Mythos in Athens. PC Lisa Michel
Fig. 332	Sky Dragon. Cloud-rider captured in Tzaneen, South Africa. PC FL: Piet Grobler
Fig. 333	Sea Dragon. PC Matteo Aventaggiato. FL: TeoMat
Fig. 334	Fallen Angel. *Symbols, Signs and Signets*, Lehner, p. 216. Used by permission.
Back cover	Parthenon Don. PC Lisa Michel

Center and Circle

Roots of the Tree

Apollodorus
    Loeb Classical Library, Harvard University Press, Cambridge, MA
        ISBN 0-674-99135-4

Astrology and Religion among the Greeks and Romans
    Franz Cumont, Publisher: G.P. Putnam's sons, 1912
        ISBN 0-48620581-9

The Jerusalem Bible, Reader's Edition
    Alexander Jones, Ed., Doubleday & Company, Inc. Garden City, NY
        ISBN 0-38501156-3

The Bible As History,
    Werner Keller, Bantam Books, 1983
        ISBN 0-55327943-2

The Book of Calendars
    Frank Parise, Ed., Facts on File, New York, NY
        ISBN 0-98196-467-8

The Book of Days
    R. Chambers, Ed, London & Edinburgh
        ISBN 0-76618-356-4

The Book of the Dragon Judy Allen and Jeanne Griffiths,
    Chartwell Books, Inc. 110 Enterprise Ave Secaucus, NJ 07094
        ISBN 0-89009241-9

Crowell's Handbook of Classical Mythology
    Edward Tripp, Crowell Company, NY
        ISBN 0-69022608-x

The Dragon in China and Japan
    Marinus Willem de Visser,
    Cosimo, Inc., Amsterdam, 1858
        ISBN 978-1-60520410-9

Echoes of the Ancient Skies,
    The Astronomy of Lost Civilizations
    Dr. E.C. Krupp,
    Harper & Row Publishers, NY, 1983
        ISBN 0-06015101-3

The Epic of Gilgamesh, trans. by Nancy K. Sandars
    Penguin Putnam, 1972,
        ISBN 0-14044249-9

The First Eden, The Mediterranean World and Man
    David Attenborough,
    Little, Brown and Co. Boston, 1987
        ISBN 5-55093767-0

Egyptian Tales: translated from the Papyri
    William M. F. Petrie, Hard Press
        ISBN 1-40693929-3

Photo credit Lisa Michel

FIG. 331

Don on his honeymoon in Athens with a Mythos

Egypt's Sunken Treasures
    Martin-Gropius-Bau, Berlin
                                  ISBN: 3-79133544-8

The Homeric Hymns and Homerica, Works and Days
    Hesoid, Loeb Classical Library, Harvard Univ. Press, Cambridge, MA, 1977
                                  ISBN 0-67499063-3

A History of Rome : Down to the Age of Constantine
    Cary, M.; Scullard, H.H., MacMillan Press Ltd, 1980
                                  ISBN 0333278305

In Search of Ancient Astronomies
    edited by Dr. E.C. Krupp, McGraw-Hill Book Company, 1978
                                  ISBN:0-07-035556-8

New Larousse Encyclopedia of Mythology
    Intro: Robert Graves, Prometheus Press, 1973
                                  ISBN 0-60002420-2

The Mathesis of Firmicus Maternus
    translated by Jean Rhys Bram, Noyes Classical Studies, 1975
                                  ISBN 0-81555037-5

Maya Cosmos, Three Thousand Years on the Shaman's Path
    Freidel, Schele and Parker. New York: William Morrow & Co.
                                  ISBN 0-68810081-3.

Middle Eastern Mythology
    Hooke, S.H., Penguin, Harmondsworth, Myth and Ritual, 1963
                                  ISBN 0486435512

Mythologies
    compiled by Yves Bonnefoy, University of Chicago Press
                                  ISBN 0-22606454-9

People of the Silence (Native Americans)
    Kathleen O'Neal and W. Michael Gear, Tor Books
                                  ISBN-0-81251559-5

The Phoenicians
    Gerhard Herm, William Morrow and Company, Inc., 1975
                                  ISBN 0-68802908-6

Star Names, Their Lore and Meaning
    Richard Hinckley Allen, Dover Press, 1963
                                  ISBN 0-48621079-0

Symbols, Signs and Signets
    Ernst Lehner, Dover Press, 1969
                                  ISBN 0-48622241-1

Tetrabiblos
    Ptolemy, Loeb Classical Library, Harvard Univ. Press, Cambridge, MA, 2001
                                  ISBN 0-67499479-5

The Treasures of Darkness, A History of Mesopotamian Religion
    Thorkild Jacobsen, Yale University Press, 1976
                                  ISBN: 0-30001844-4

Unger's Bible Dictionary
    Merrill F. Unger, Moody Press, Chicago, 1957
                                  ISBN: 0-80249066-2

foot notes

page	footnote	source
vi	00	*The Homeric Hymns and Homerica, Works and Days,* Hesiod, Loeb Classical Library, 1977, pp. 453-455.
8	0	*Middle Eastern Mythology,* Hooke, S.H., (Penguin, Harmondsworth 1963) Myth and Ritual (OUP, London 1933).
13	1	*The Book of the Dragon*, Allen & Griffiths, Chartwell Books, Inc, p. 36.
	2	keystone
17	3	*Tetrabiblos*, Ptolemy, F.E. Robbins, Loeb Classical Library, Harvard University Press, Cambridge, MA.
19	4	-*Mythologies*, compiled by Yves Bonnefoy, p. 33.
	5	ibid., p. 50.
	6	*Pre-history Africa & the Badarian Culture: Evidence of the Badarians into Pre-historic Egypt (4500-3800 BC)*, p. 42. http://www.homestead.com/wysinger/badarians.html
20	7	*The Dictionary of Mythology.* Coleman, J. A., Arcturus, London, 2007
21	8	http://en.wikipedia.org/wiki/Nabta_Playa
22	9	*New Larousse Encyclodepia of Mythology,* by Robert Graves Section on Phoenician Mythology by L. Delaporte, p. 82.
	10	ibid.
	11	ibid.
	12	ibid.
	13	ibid.
	14	*Robert Graves, The Greek Myths: Vol. 1*, Penguin Books, London, 1960, p. 27.
23	15	http://www.spacetoday.org/SolSys/Earth/OldStarCharts.html
	16	*New Larousse Encyclodepia of Mythology,* by Robert Graves Section on Phoenician Mythology by L. Delaporte, pp. 78-79.
24	17	*Egyptian Tales*: William Matthew Flinders Petrie, translated from the Papyri, 1895, p. 66.
27	18	*The Treasures of Darkness, A History of Mesopotamian Religion* Thorkild Jacobsen, Yale University Press, 1976, pp. 98-99.
28	19	Sumerian Lexicon: Version 3.0, Halloran, John A., http://sumerian.org/sumerlex.htm
	20	*New Larousse Encyclodepia of Mythology,* by Robert Graves Section on Egyptian Mythology by J. Viaud, p. 13.
	21	ibid.
	22	*Star Names, Their Lore and Meaning*, by Richard Hinckley Allen, Dover Pulications, Inc. New York, 1963, p. 229.
29	23	*Sexagesimal*, Wiki- http://en.wikipedia.org/wiki/Sexagesimal

Keystone to the arch

page	footnote	source
	24	*New Larousse Encyclodepia of Mythology,* by Robert Graves Section on Egyptian Mythology by J. Viaud, p. 10.
30	25	*The Stones of Time*, Calendars, Sundials, and Stone Chambers of Ancient Ireland, Martin Brennan, Inner Traditions, 1994, p. 7.
	26	ibid., p. 127.
31	27	ibid., p. 15.
	28	ibid., pp. 16-17.
33	29	ibid., p. 40.
	30	*Dictionary,* Version 1.0.2., Apple Computer, Inc.
34	31	*The Stones of Time*, Calendars, Sundials, and Stone Chambers of Ancient Ireland, Martin Brennan, 1994, p. 20.
	32	ibid., p. 23.
	33	ibid., p. 25.
	34	ibid., p. 23.
35	35	ibid., p. 24.
36	36	*Crowell's Handbook of Classical Mythology* by Edward Tripp, Thomas Y. Crowell Company, New York, 1970, p. 192.
	37	*The Homeric Hymns and Homerica, Works and Days,* Hesiod, Loeb Classical Library, 1977, p. 31.
39	38	http://www.goldenageproject.org.uk/51redpaint.php.
40	39	*The Homeric Hymns and Homerica, Works and Days,* Hesiod, Loeb Classical Library, 1977, p. 3.
42	40	*The First Eden, The Mediterranean World and Man* David Attenborough, Little, Brown and Co. Boston, p. 72.
43	41	ibid., p. 74.
45	42	ibid.
	43	ibid.
	44	ibid.
46	45	*Egypt's Sunken Treasures*, Martin-Gropius-Bau, Berlin, Prestel, p. 35.
49	46	*The Epic of Gilgamesh*, Nancy K. Sandars, Penguin Putnam, 1972, p. 70.
50	47	*The Babylonian Genesis*, Alexander Heidel, Univ. of Chicago Press, 1963, p. v.
52	48	*The Epic of Gilgamesh*, Nancy K. Sandars, Penguin Putnam, 1972, p. 71.
	49	ibid., p. 80.
	50	ibid., p. 71.
	51	ibid., p. 88.
53	52	ibid., p. 98.
	53	ibid.
54	54	*The Treasures of Darkness, A History of Mesopotamian Religion* Thorkild Jacobsen, Yale University Press, 1976, p. 177.

Center and Circle

page	footnote	source
55	55	ibid., p. 182.
	56	ibid.
56	57	*New Larousse Encyclodepia of Mythology,* by Robert Graves, Section on Egyptian Mythology by J. Viaud, p. 14-15.
61	58	*New Larousse Encyclodepia of Mythology,* by Robert Graves, Section on Indian Mythology by Masson-Oursel and Morin, p. 340.
62	59	ibid.
64	60	*The Treasures of Darkness, A History of Mesopotamian Religion* Thorkild Jacobsen, Yale University Press, 1976, rear dust cover
66	61	Snorri Sturluson's Prose Edda, "The Norse Myths" by Kevin Crossley-Holland.
67	62	https://en.wikipedia.org/wiki/Germanic_peoples
	63	http://odinsvolk.ca/O.V.A.%20-%20COSMOLOGY.htm
68	64	http://en.wikipedia.org/wiki/World_tree.
	65	ibid.
	66	ibid.
	67	http://en.wikipedia.org/wiki/Tree_of_life.
	68	ibid.
69	69	http://en.wikipedia.org/wiki/Tree_of_life#China.
70	70	http://en.wikipedia.org/wiki/Tree_of_life.
	71	http://en.wikipedia.org/wiki/Yule_log.
	72	ibid.
	73	ibid.
71	74	http://meta-religion.com/World_Religions/Ancient_religions/Central_america/popol_vuh.htm#top.
72	75	ibid.
	76	ibid.
	77	*Crowell's Handbook of Classical Mythology,* by Edward Tripp Thomas Y. Crowell Company, 1970, pp. 401-402.
74	78	http://meta-religion.com/World_Religions/Ancient_religions.Central_america/popol_vuh.htm#top.
	79	ibid.
75	80	*Star Names, Their Lore and Meaning,* by Richard Hinckley Allen. Dover Pulications, Inc. New York, 1963, p. 212.
	81	http://meta-religion.com/World_Religions/Ancient_religions/Central_america/popol_vuh.htm#top.
	82	ibid.
	83	ibid.
76	84	ibid.
	85	ibid.
78	86	ibid.
79	87	*The Jerusalem Bible, Reader's Edition,* Alexander Jones, Ed., Doubleday & Company, Inc, NY, 1968, p. 5.
83	88	*Middle Eastern Mythology,* Hooke, S.H., (Penguin, Harmondsworth 1963)

page	footnote	source
		Myth and Ritual (OUP, London 1933).
85	89	*The Jerusalem Bible, Reader's Edition*, Alexander Jones, Ed., Doubleday & Company, Inc, NY, 1968, p. 6.
86	90	http://en.wikipedia.org/wiki/World_tree.
88	91	http://meta-religion.com/World_Religions/Ancient_religions/Central_america/popol_vuh.htm#top.
	92	http://meta-religion.com/World_Religions/Ancient_religions/Central_america/popol_vuh.htm#top.
91	93	http://en.wikipedia.org/wiki/Great_Year.
92	94	Julian day is used in the Julian date (JD) system of time measurement for scientific use by the astronomy community. Julian date is recommended for astronomical use by the International Astronomical Union (IAU).
95	95	*The Homeric Hymns and Homerica, Works and Days,* Hesiod, Loeb Classical Library, pp. 11-17.
96	96	*Metamorphoses*, Ovid, Penguin, 1980, pp. 31-32.
97	97	ibid., pp. 72-73.
98	98	http://en.wikipedia.org/wiki/Minoan_chronology
100	99	http://en.wikipedia.org/wiki/Wadjet.
	100	ibid.
103	101	*The Epic of Gilgamesh*, Nancy K. Sandars, Penguin Putnam, 1972, p. 80.
104	102	*The Dialogue of Ipuwer and the Lord of All*, by R. Enmarch, The Griffith Institute, Oxford 2005.
105	103	http://etcsl.orinst.ox.ac.uk/cgi-bin/etcsl.cgi?text=t.2.1.5#.
106	104	*The Bible as History*, Werner Keller, Bantam Books, 1983, pp. 28-29.
108	105	*The Jerusalem Bible, Reader's Edition*, Alexander Jones, Ed., Doubleday & Company, Inc, NY, 1968, p. 8.
	106	ibid., pp. 7-8.
	107	*The Epic of Gilgamesh*, Nancy K. Sandars, Penguin Putnam, 1972, p. 84.
109	108	Apollodorus, Loeb Classical Library, Harvard University Press, pp. 75-77.
110	109	ibid., p. 75.
111	110	*The Golden Fleece*, by Padraic Colum, Macmillan, 1949.
112	111	*The Agronautica*, http://www.gutenberg.org/files/830/830-h/830-h.htm.
114	112	*Age of Fable: Vols. I & II: Stories of Gods and Heroes.* Thomas Bulfinch, 1913.
	113	ibid.
115	114	ibid.
123	115	*Unger's Bible Dictionary*, Moody Press, Chicago. 1957, *Baal et al.*, p. 413.
125	116	*New Larousse Encyclodepia of Mythology,* by Robert Graves, Section on Phoenician Mythology by L. Delaporte, p. 75.
	117	*Unger's Bible Dictionary*, Moody Press, Chicago. 1957, p. 293.

page	footnote	source
126	118	ibid.
127	119	*The View Over Atlantis,* by John Michell. Sago Press, 1969, p. 40.
or	119	2 Samuel 5:11
128	120	*The Stones of Time*, Calendars, Sundials, and Stone Chambers of Ancient Ireland, Martin Brennan, Inner Traditions, 1994.
129	121	*The New View Over Atlantis.* John Michell, Thames and Hudson, 1995, p. 52.
	122	*The Stones of Time*, Calendars, Sundials, and Stone Chambers of Ancient Ireland, Martin Brennan, Inner Traditions, 1994.
130	123	*Ancient Legends, Msytic Charms, and Superstitions of Ireland* Lady "Speranza" Wilde, London, 1888.
133	124	invokes the constellation of the Ram
134	125	Head of the Ram
	126	Chain-of-command
135	127	shivering light
	128	http://www.mindspring.com/~mysticgryphon/bitakitu.htm
	129	*The First Great Civilizations*: Life in Mesopotamia, the Indus Valley and Egypt by Jacquetta Hawkes, Hutchinson, ISBN 0091165806.
140	130	http://meta-religion.com/World_Religions/Ancient_religions/Central_america/popol_vuh.htm#top.
143	131	*Astronomica*, Manilius, 1st century AD, Bk V, pp. 311-313.
	132	*Chronicle of the Pharaohs* by Peter A. Clayton, Thames and Hudson, London, 1997, pp. 90-97.
144	133	ibid.
145	134	*New Larousse Enclyclodepia of Mythology,* by Robert Graves Section on Egyptian Mythology by J. Viaud, p. 29.
146	135	*Ancient Egyptian Literature*, Miriam Lichtheim, University of California Press, 1976, ISBN 0-050-03615-8, pp. 105-106.
	136	*The Qur'an with Annotated Interpretation in Modern English, by Ali Unal, The Light, ISBN 09 08 07 06 1234, p. 483.
153	137	http://www.harappa.com/har/indus-saraswati.html
155	138	http://tdwotd.blogspot.com/2010/10/belshazzar.html.
158	139	*A History of Rome*, H.H. Scullard, MacMillan Press Ltd, 1980, p. 72.
	140	*The Early History of Rome, Books I-V*, Livy, Penguin Classics, 1971, p. 381.
	141	1 Macc 1:20-24; 2 Macc 5:11-21.
	142	2 Maccabees 6: 1
	143	2 Maccabees 6: 4
160	144	*A History of Rome*, H.H. Scullard, MacMillan Press Ltd, 1980.
161	145	*Astrology and Religion among the Greeks and Romans*, Franz Cumont, G.P. Putnam's Sons, 1912, pp. 40-41.
	146	ibid., p. 71.
163	147	*People of the Silence* (Native Americans), pp. 19-20

page	footnote	source
	148	http://www.blackdrago.com/rainbowserpent.htm.
164	149	*Astronomica*, Manilius, Bk. I, p. 29.
165	150	*The Epic of Gilgamesh*, Nancy K. Sandars, Penguin Putnam, 1972, p. 98.
	151	http://www.ancientegypt101.net/2009/05/protodynastic-egypt.html.
168	152	*Sky Signs: Aratus' Phaenomena*, translated by Stanley Lombardo, North Atlantic Books, Berkeley, p. 3.
	153	*Crowell's Handbook of Classical Mythology*, by Edward Tripp Thomas Y. Crowell Company, 1970, p. 272.
169	154	ibid., p. 5.
170	155	*Legend in Japanese Art*, by L. Joly, Kegan Paul, London 1967, p. 34.
174	156	*Astrology and Religion among the Greeks and Romans*, Franz Cumont, G.P. Putnam's Sons, 1912, p. 41.

Photo credit Piet Grobler   FIG. 332

Blessing to all who enter here.

Welcome to the Web.

Athena's Web.

Photo credit Matteo Aventaggiato FIG. 333

Sea Dragon

*When that day comes—
it is God who speaks—
the heavens will have
their answer from me,
the earth its answer from them...*

Hosea 2:23

FIG. 334

Carpe Noctem!   April 19, 2015

Athena's Web

Bonny Doon, CA 95060. www.AthenasWeb.com, AthenasWeb@gmail.com, (413) 219-4214

Don Cerow

has been pursuing both mythic and stellar paths for 40 years. Between these pages lie a few of the time-honored secrets he has uncovered hidden between the columns over the decades.

Photo credit Lisa Michel

"It's Harry Potter, the DaVinci Code and Indiana Jones all in one - except this one is real!"